NATURAL ENVIRONMENT RESEARCH COUNCIL

INSTITUTE OF GEOLOGICAL SCIENCES

MEMOIRS OF THE GEOLOGICAL SURVEY OF GREAT BRITAIN
ENGLAND AND WALES

Geology of the Country around Canterbury and Folkestone

(*Combined memoir in explanation of One-inch Geological sheets* 289, 305 *and* 306, *New Series*)

By J. G. O. Smart, B.Sc., G. Bisson, A.R.S.M., B.Sc.,
and B. C. Worssam, B.Sc.,

with Chapters on Palaeontology of the Gault *by*
R. Casey, D.Sc., *and on* Palaeontology of the Chalk *by*
R. V. Melville, M.Sc., *and other contributions by*
P. A. Sabine, A.R.C.S., Ph.D., M. Mitchell, M.A., G. P. Jones, M.Sc., *and*
H. A. Hope Macdonald, M.Sc.

LONDON
HER MAJESTY'S STATIONERY OFFICE
1966

ISBN 0 11 880065 5

PREFACE

THIS MEMOIR describes the geology of the district which is depicted on the Canterbury (289), Folkestone (305) and Dover A (306) sheets of the New Series One-inch Geological Map of England and Wales. These maps are at present in the press, the last two being combined to form one sheet due to the small area represented by the Dover A (306) Sheet. As well as providing an explanation of the geological maps, the memoir contains information concerning geological aspects of the history of Romney Marsh, the exploration of part of the Kent Coalfield, and the investigation of landslips near Folkestone.

The district was originally surveyed on Old Series One-inch Sheets 3 and 4 by W. B. Dawkins, F. Drew, C. E. Hawkins, T. McK. Hughes, W. Topley and W. Whitaker. Sheet 3 was published in 1868, and a new edition showing the post-Tertiary deposits followed in 1875; it was revised in 1889. Sheet 4 was published in 1863.

A memoir descriptive of the Geology of the Country between Folkestone and Rye, including the whole of Romney Marsh (Sheet 4), by F. Drew, was published in 1864. Various aspects of the geology were dealt with in the Geological Survey memoirs entitled The Geology of the London Basin, by W. Whitaker (1872); The Geology of the Weald, by W. Topley (1875); Pliocene Deposits of Britain, by C. Reid (1890);· Cretaceous Rocks of Britain, Volumes I to III, by A. J. Jukes-Browne (1900–4); The Water Supply of Kent, by W. Whitaker (1908); On the Mesozoic Rocks in some of the Coal Explorations in Kent, by G. W. Lamplugh and F. L. Kitchin (1911) and The Concealed Mesozoic Rocks in Kent, by G. W. Lamplugh, F. L. Kitchin and J. Pringle (1923). Catalogues of wells on these and neighbouring one-inch sheets were provided by parts vi and vii of Wartime Pamphlet No. 10, which were assembled by Messrs. S. Buchan, J. A. Robbie, S. C. A. Holmes, J. R. Earp, E. F. Bunt and L. S. O. Morris, and issued in 1940. A revised well-catalogue for the Canterbury (289) Sheet was compiled by Mrs. M. C. Davies and others, and was published in 1964.

In the primary six-inch survey a small tract was mapped by Dr. J. R. Earp in 1938; the remainder was surveyed between 1948 and 1956 by Messrs. J. G. O. Smart, G. Bisson and B. C. Worssam and the late F. H. Edmunds, under the supervision of the last-named as District Geologist.

This memoir has been written by Messrs. Smart, Bisson and Worssam, with palaeontological chapters by Dr. R. Casey and Mr. R. V. Melville. Fossils have also been named and palaeontological notes supplied by Dr. F. W. Anderson, Mr. M. A. Calver, Dr. W. G. Chaloner of University College, London, Dr. H. C. Ivimey-Cook, Mr. M. Mitchell, Dr. W. H. C. Ramsbottom, Mr. R. E. H. Reid of Queen's University, Belfast, the late L. F. Spath, Dr. A. J. Sutcliffe of the British Museum (Natural History), the late Sir Arthur Trueman, Mr. C. J. Wood and myself. Many of the fossils were collected by Mr. F. G. Dimes and Mrs. M. E. Evans. Petrographical contributions and descriptions have been provided by Dr. P. A. Sabine, Dr. R. Dearnley, Mr. R. W. Elliot and Miss H. A. H. Macdonald, X-ray work has been contributed by Mr. B. R. Young, chemical analyses have been made by Mr. G. A. Sergeant, spectrographic analyses by Mr. K. L. .H. Murray and particle-size analyses by Mr. J. Dangerfield. Mr. G. P. Jones has provided an account of the water supply of the area and electrical resistivity measurements in boreholes for water have been made by Messrs. D. A. Gray, T. K. Tate and K. H. Murray. Radiometric logs of boreholes in the district have been recorded by Messrs. D. Ostle and J. Taylor. The memoir has been edited by Mr. Bisson.

Our thanks are given to the National Coal Board and to British Railways for the ready co-operation and courtesy afforded to us by their staffs. We are also indebted to the Channel Tunnel Study Group for permission to publish details of their boreholes drilled in 1958, while the ages of certain shells and samples of wood are published with the consent of the Director of the National Physical Laboratory.

C. J. STUBBLEFIELD

Director

Geological Survey Office
Exhibition Road
South Kensington
London S.W.7.

21st March 1966.

R J Handiman

3/8?

CONTENTS

(References are listed at the end of the book)

PAGE

ILLUSTRATIONS

TEXT-FIGURES

viii

EXPLANATION OF PLATES

[1] Reference number of the photograph in the Geological Survey collection of geological
photographs.

LIST OF SIX-INCH MAPS

The following is a list of the National Grid six-inch geological maps included wholly or in part in the one-inch geological map sheets 289, 305 and 306, with the initials of the surveying officers and the dates of the survey for each six-inch map; the surveyors were G. Bisson, J. R. Earp, F. H. Edmunds, J. G. O. Smart and B. C. Worssam. Manuscript copies of these maps will be deposited for public reference in the Library of the Geological Survey.

CANTERBURY (289) ONE-INCH SHEET

TQ 94 SE	Great Chart	B.C.W.	1951, 1955
TQ 94 NE	Westwell	B.C.W.	1951
TQ 95 SE	Throwley	J.G.O.S., B.C.W. ...	... 1951–2
TQ 95 NE	Ospringe	J.G.O.S., B.C.W. ...	1948, 1951–2
TQ 96 SE	Teynham and Stone	B.C.W., J.G.O.S. ...	1948, 1951
TR 04 SW	Ashford	B.C.W., J.G.O.S. ...	1951, 1954–5
TR 04 SE	Brook and Brabourne ...	J.G.O.S., G.B.	1954–5
TR 04 NW	Eastwell	B.C.W., J.G.O.S. ...	1951, 1954–5
TR 04 NE	Wye	J.G.O.S., G.B.	... 1953–4
TR 05 SW	Molash	J.G.O.S., B.C.W.	1951–2, 1954–5
TR 05 SE	Chilham	J.G.O.S., F.H.E.	... 1952–4
TR 05 NW	Selling	J.G.O.S.	... 1951–2
TR 05 NE	Dunkirk	J.R.E., J.G.O.S., F.H.E.	1938, 1951–3
TR 06 SW	Faversham	J.G.O.S.	... 1951
TR 06 SE	Hernhill	J.R.E., J.G.O.S. ...	1938, 1951
TR 14 SW	Stowting	G.B., J.G.O.S.	... 1954–5
TR 14 SE	Elham	G.B.	1951, 1954–5
TR 14 NW	Waltham	J.G.O.S., G.B.	... 1953–4
TR 14 NE	Stelling	J.G.O.S., G.B.	... 1953–4
TR 15 SW	Petham and Lower Hardres...	F.H.E., J.G.O.S. ...	... 1953–4
TR 15 SE	Bridge and Bishopsbourne ...	F.H.E., J.G.O.S., G.B. ...	1951, 1953
TR 15 NW	Harbledown	J.G.O.S., F.H.E. ...	... 1951–4
TR 15 NE	Canterbury	J.G.O.S., F.H.E. ...	... 1951–4
TR 16 SW	St. Cosmus in the Blean ...	J.G.O.S.	1951, 1953
TR 16 SE	Sturry and Westbere ...	J.G.O.S.	... 1952–3
TR 24 SW	Acrise and Swingfield ...	G.B.	1951, 1954
TR 24 SE	Hougham Without	G.B.	1951, 1956
TR 24 NW	Barham and Denton ...	G.B.	1951, 1953–4
TR 24 NE	Sibertswold	G.B.	1951, 1953
TR 25 SW	Adisham	F.H.E., G.B. ...	1951, 1953
TR 25 SE	Nonington	F.H.E., G.B. ...	1951, 1953
TR 25 NW	Littlebourne and Wingham ...	J.G.O.S., F.H.E. ...	... 1951–3
TR 25 NE	Staple	J.G.O.S., F.H.E. ...	1951, 1953
TR 26 SW	Wickhambreux	J.G.O.S.	... 1952–3

FOLKESTONE (305) AND DOVER (306) ONE-INCH SHEETS

TQ 92 SE	Walland Marsh	J.G.O.S.	... 1956–7
TQ 92 NE	Snargate	J.G.O.S.	... 1956–7
TQ 93 SE	Warehorne	J.G.O.S.	... 1955–7
TQ 93 NE	Shadoxhurst	J.G.O.S., B.C.W. ...	... 1955–6
TQ 94 SE	Great Chart	B.C.W.	1951, 1955
TR 02 SW	Lydd	J.G.O.S.	... 1956–7

TR 02 SE	New Romney	J.G.O.S.	... 1956–7
TR 02 NW	Ivychurch	J.G.O.S.	... 1956–7
TR 02 NE TR 12 NW	}St. Mary in the Marsh	J.G.O.S., B.C.W. ...	... 1955–6
TR 03 SW	Ruckinge	J.G.O.S.	... 1955–6
TR 03 SE	Newchurch	J.G.O.S., B.C.W. ...	... 1955–6
TR 03 NW	Kingsnorth	J.G.O.S., B.C.W. ...	... 1954–6
TR 03 NE	Aldington	J.G.O.S., B.C.W. ...	... 1954–5
TR 04 SW	Ashford	B.C.W., J.G.O.S. ...	1951, 1954–5
TR 04 SE	Brook and Brabourne	J.G.O.S., G.B.	... 1954–5
TR 13 SW and part of TR 13 SE	}Burmarsh	B.C.W.	... 1955
TR 13 NW	Stanford	B.C.W., J.G.O.S., G.B. ...	... 1954–5
TR 13 NE and part of TR 13 SE	}Newington	B.C.W., J.G.O.S., G.B. ...	1951, 1955–6
TR 14 SW	Stowting	G.B., J.G.O.S.	... 1954–5
TR 14 SE	Elham	G.B.	1951, 1954–5
TR 23 NW	Folkestone	G.B., J.G.O.S.	1951, 1955–6
TR 23 NE	Capel-le-Ferne ...	G.B.	1951, 1956
TR 24 SW	Acrise and Swingfield	G.B.	1951, 1954
TR 24 SE	Hougham Without ...	G.B.	1951 1956
TR 34 SW and part of TR 33 NW	}Dover	G.B.	... 1956

Chapter I

INTRODUCTION

LOCATION AND PHYSICAL FEATURES

THE ONE INCH to one mile geological sheets Canterbury (289), Folkestone (305), and Dover A (306) depict a tract of East Kent at the eastern ends of the Weald and of the North Downs. The district extends from west of Ashford northwards to the outskirts of Faversham, north-eastwards to beyond Canterbury and south-eastwards to the Strait of Dover. In the south it includes the whole of Romney Marsh and parts of the adjoining marshland to the west and south. The Dungeness promontory lies three miles beyond the southern boundary.

Situated within the district are the City of Canterbury, the ancient Cinque Ports of Hythe and New Romney, the small but rapidly expanding industrial and market town of Ashford, and the seaside communities of Folkestone, Sandgate and Dymchurch. Elsewhere the district is essentially rural in character and is largely devoted to farming and forestry. Coal is mined in the east and other natural resources are worked locally for the production of sand and gravel, roadstone, bricks and lime.

The principal physical features of the district are illustrated in Plate I. The marshland is a virtually flat expanse, which has an attraction peculiar to itself ; much of its surface lies below high-tide level, and sea-walls hold back the sea ; the lowest area is in The Dowels and lies at about 6 ft above O.D. The marshland is bordered on the north by an escarpment composed of Wealden rocks in the west and Lower Greensand strata in the east, the latter mostly concealed by landslips. The Wealden country is dissected close to the scarp but farther north it is peneplanal. The Lower Greensand topography, on the other hand, is diverse, reflecting the varied nature of its component sediments ; its escarpment is weak, except where it overlooks the marsh, due to the comparative thinness of the resistant Hythe Beds.

North of the Lower Greensand outcrop a broad belt of Gault clay lies at the foot of the steep indented escarpment which bounds the North Downs on their south side. The Chalk hills cross the district from west-north-west to east-south-east ; they rise to 624 ft at Frittenfield and to 613 ft near both Hastingleigh and Paddlesworth, and excellent views of the country to the south can be obtained from the top of the escarpment. A drift-clad plateau slopes gently north-north-eastwards from the North Downs escarpment. The Chalk ridge has been breached north-east of Ashford by the Great Stour River ; elsewhere the Chalk plateau is dissected by numerous valleys, most of which are dry, and several have been truncated by the retreat of the Chalk escarpment. The Chalk country is scenically unspoiled almost throughout.

West of Canterbury a wooded upland known as the Blean is terminated on the west by an escarpment east of Boughton Street ; another scarp bounds the Eocene formations near Wingham.

The coast is low-lying adjacent to the marshland, where fine sandy beaches have encouraged the spread of building for holiday purposes. Elsewhere the beaches consist mostly of shingle. Landslips have affected almost the whole length of the Lower Greensand cliff, as well as the Chalk and Gault north-east of Folkestone, where The Warren is an area of wild beauty. East of The Warren the North Downs end in fine towering chalk cliffs.

DRAINAGE SYSTEMS

The Great Stour River carries the major part of the drainage of the district eastwards and the river enters the North Sea at Pegwell Bay on the East Kent coast. It rises west of the district near Lenham, and flows south of Ashford, where it is joined by two major tributaries. One, herein referred to as the South Stour, rises upon the Weald Clay north of Orlestone; the other, the East Stour, rises as a scarp spring near Postling and collects tributaries draining the Sellindge area, before traversing the Weald Clay towards Ashford. Below the confluence with these tributaries the Great Stour flows northwards into a trumpet-shaped chalk valley. Dry tributary valleys from west and north combine shortly before joining the system at Chilham, where the main stream turns north-eastwards towards Canterbury. It is joined by the Petham valley at Shalmsford Street; this valley has its upper reaches north-east of Hastingleigh and its course is northwards as far as Swarling Farm, where it turns abruptly west-north-west. On the north-west of this reach of the Great Stour tributary valleys draining Eocene and Chalk country are dry except for that joining south of Harbledown. The Stour eventually crosses the northern margin of the district near Fordwich.

In the north-western part of the area, west of Selling, a dendritic system of valleys, now mainly dry, passes northwards and finally reaches the north Kent coast in the Swale estuary. A small stream on the London Clay outcrop east of Dunkirk ultimately joins the Great Stour near Sarre to the north of this district. Another stream, here referred to as the Lampen Stream, rises near Hoath Farm, east of Canterbury; it collects the drainage of part of the Eocene outcrop before joining the Great Stour north of this district. The continuation of its valley through Lower Hardres, on the Chalk, is dry.

The nailbourne at Elham and Bekesbourne is also known as the Little Stour River. Its valley system occupies part of the Chalk outcrop; the main valley trends north-north-east from Etchinghill to Barham, whence it courses to the north-west as far as Bridge; there it turns sharply to the north-east, crosses Eocene strata and finally joins the Great Stour outside this district.

Valleys in the Chalk around Adisham join that of a stream (Wingham River) at Wingham, a mile south-east of its confluence with the Little Stour. A dendritic valley system originating about Womenswold and Shepherdswell passes north-eastwards to join the lower reaches of the Great Stour beyond this district.

Major valleys in the Chalk east of Hawkinge and Swingfield reach the sea at Dover. Small streams near Folkestone and Sandgate drain steeply southwards into the sea; others west of Sandgate pass into the complex artificial drainage system of the marshland. The Royal Military Canal was originally constructed for defence purposes, but now the drainage of the Dowels region is pumped into it.

Several of the small streams draining the Eocene beds are lost in swallow holes where the Chalk is approached or reached. The swallow holes occur mainly on the south side of the Blean, east of Lower Ensden (Prestwich 1854b). Others, on the west side of the Blean at Oversland and Southstreet, and on the borders of the Selling Tertiary outlier, were mentioned by Whitaker (1908, pp. 49–50).

Nailbournes are a feature of the Chalk drainage. They flow intermittently and at irregular time intervals in the Kennaways (Ospringe), Petham, Elham and Alkham valleys (Snell 1938). Whitaker (1908, p. 54) rejected a theory that such streams owed their origin to the periodical emptying of huge underground reservoirs by siphoning, and maintained that they were due to the gradual rise of an underground water-plane, after a wet season, to the level of the ground in certain low-lying parts, when the underground water emerged as surface water. Reynolds (1948, p. 78), however, contended that " the saturation of the chalk underlying the valley line does not take place in accordance with the rise of a uniformly sloping water table, but locally in those places where springs appear. The way in which the [Alkham] nailbourne flows only from a few well-defined places, and the large quantity of water sometimes discharged from them, suggest the existence in or near those places of large fissures close to the surface." Thus it seems that when the level of water in a fissure in the Chalk rises to ground surface a spring breaks out and flows until the supply of water dwindles and the level drops.

The district contains many interesting geomorphological features which have been fully discussed by Wooldridge. The characteristic tableland of the Chalk declines gently north-eastwards in an apparently unbroken manner, except where interrupted by dry valleys. However Wooldridge recognized that this surface was composed of two elements: the surface (or bench) on which Pliocene rocks were deposited and a steeper intersecting surface, cut by the Eocene sea (p. 175). He found (Wooldridge 1927, pp. 53–5) that the base of the Pliocene deposits was inclined not only to the north-east but also towards the main valleys, and he concluded that the Pliocene beds rested on a pre-existing peneplaned surface with minor undulations. He stated (1927, p. 77) that the Diestian (Pliocene) sea simply perfected the planation over limited tracts by bringing marine abrasion to bear on a surface already practically reduced to base-level and he considered that the main consequent streams followed virtually identical courses before and after the Diestian transgression.

The concept of a ' 200-ft platform ', put forward by Wooldridge (*in* Dewey and others 1925, pp. 275–8) in an account of the denudation history of the district, seems to represent an over-simplification of events in the early stages of that history. The phrase expresses the fact that summits to the south of the Chalk escarpment, around Ashford, are mostly around 200 ft above O.D., but its implication that this wide region has been ' bevelled ', as Kerney, Brown and Chandler (1964, p. 138) put it, has by no means been proved. The most that can be said is that at an early stage in the Pleistocene a sea level somewhat below 200 ft O.D. allowed the widespread formation of river gravels graded to that level. These gravels include those mapped in the present area and in the Maidstone district (Worssam 1963) as the 4th Terrace. The streams in 4th Terrace times evidently flowed much on the lines of the present rivers, through a landscape of low relief. Some summits on Wooldridge's ' 200-ft platform ', notably along the Hythe Beds escarpment, are and probably always have been interfluves, that have been progressively lowered in step with the down-cutting of valleys. The generally low altitude of the Hythe Beds around Ashford as compared with that farther west, near Maidstone, is more likely to be related to the absence of erosion-resistant chert in the beds than to any process of planation. The area shown by Kerney and others (1964, p. 137, fig. 1b) as a " 200–300 ft. planation surface " includes surfaces of various origins, one of which is an apron-like slope between 250 and 350 ft above O.D. fronting the Chalk escarpment in Eastwell Park and covered with Head deposits of a Clay-with-flints lithology (p. 208).

East of the Great Stour the line of intersection of the Pliocene and Eocene benches follows a line of the Tertiary escarpment during the period of forma-tion of the 4th River Terrace. The Chalk now forms a bold escarpment along the flank of the Little Stour River from Barham to Bridge (Wooldridge 1927, p. 80) and thence along much of the line to Shalmsford Street (Whitaker 1872, p. 360), and it is likely that in 4th Terrace times the Little Stour flowed south of this escarpment to Shalmsford Street. The retreat of the early Eocene escarpment was probably retarded long enough for the valley system to be established at its foot, by a very slight increase in the chalk dip which coincides with this line.

An extensive development of four river terraces occurs along the present drainage system. The majority of the terraces are occupied by gravels but in some of the Chalk valleys, notably the deeper and wider ones, the valley sides show discontinuous ' benches ', which for the most part are free from Drift deposits. Two such benches have been noted lying west of Throwley Valley farm, some 30 to 40 ft above the valley floor, at about 230 ft O.D. Others, in the valley west of Whatsole Street at about 430 ft O.D., around Lyminge at about 400 ft O.D., and on both sides of the valley between Barham and Kingston at about 200 ft O.D., are again 30 to 40 ft above the main valley floors, while the long spur north of Elham is at about 300 ft O.D., 35 ft above the valley floor. On the spur south-west of Bridge there is a high bench at approximately 200 ft O.D., and a lower bench is degraded and declines gently towards the valley. On the spur east of Wickham Bushes a bench is present some 75 ft above the valley floor. In the Alkham valley benches 40 to 50 ft above the present valley bottom range from 230 to 320 ft above O.D.

These benches may indicate an erosional pause, possibly coinciding in time with the deposition of the 3rd Terrace river gravels elsewhere. They should not be confused with the ' terrace-like ledges ' or ' lynchets ' mentioned by Whitaker (1872, p. 365) ; these have resulted from the accumulation against hedges or similar obstacles, of soil and other debris loosened by ploughing and washed downhill by rain (Scrope 1866 ; Wood 1961).

Some instances of river-capture are described in Chapter X. Other examples of capture have resulted from the cutting back of valleys on the Chalk escarpment by spring action and by ' run off ' from the Chalk. Several head-streams of the consequent Chalk drainage systems have been deflected southwards.

On Stowting Common (125432)[1] a small valley at the head of the broad embayment in the scarp now turns sharply eastwards at the elbow of capture with one of the head-streams of the valley (now dry) lying east of Stelling Minnis. The valley drifts of this system 'hang' in the wind gap. A more striking example is the double capture of a head-stream (also dry) of the Little Stour ¼ mile N. of The Pent, Postling (141398), and ½ mile E.N.E. of Postling itself (153393). Both valleys in the scarp display marked elbows of capture. The second was described by Fagg (1954), who also considered that the drainage of the trumpet-shaped entrance to the Little Stour gap, south of Etchinghill, had been reversed by the stream draining southwards to Hythe, but there is little evidence in the present topography to support this. A vestigial wind gap has been noted at Ashenfield Farm near Hassell Street (096471), where a tributary of the Crundale valley system has captured a head-stream of the Petham valley system. Another wind gap on the spur south-south-east of Petham (131505) reveals that a stronger tributary of the Petham system has captured a weaker. The capture of the stream at Harbledown (129580) by a tributary of the Great Stour has been noted by Coleman (1954).

The broad meander in the valley floor near Coombegrove Farm (Big Coombe Farm) (084466) appears to have been controlled by the resistant Melbourn Rock which crops out on the valley floor near Pett Street. The sweep of the river in the valley, north-east of the farm, has steepened and eroded the eastern side of the valley so that the tributary valley lying west of Hassell Street has almost been captured. The Hassell Street tributary valley, over part of its course, forms practically a ledge approximately 130 ft up the Crundale valley side. From the Hassell Street valley floor only a very slight rise of chalk separates its Dry Valley and Nailbourne Deposits from the steep slope of the Crundale valley.   J.G.O.S., G.B., B.C.W.

The Brockhill Stream, west of Hythe, has probably diverted to the coast a former tributary of the East Stour, on the course of which the 4th Terrace gravels at Pedlinge (p. 266) were laid down.

The V-shaped valley of a head-stream of the River Beult now reaches the southern corner of the wide alluvial tract, belonging to the Stour basin, ½ mile west of Court Lodge (993397). The head-stream, probably assisted by artificial diversion, receives drainage water from some ditches in this alluvium, and has captured a stream which rises north of Chilmington Green (983407) and formerly must have flowed into the northern corner of the alluvial area. With the accession of this latter, all the springs from the south side of the Great Chart outlier of Hythe Beds now drain into the Beult. If this capture were allowed to proceed to completion, by a reversal of drainage along the alluvial tract as far as Kingsnorth, about 1¾ miles to the east, the East Stour would be diverted into the Medway system. The Great Stour would then follow, turning south instead of north at Ashford, and its present valley through the Chalk escarpment would become a wind-gap.

    B.C.W.

---

[1] In this memoir the positions of localities are indicated by their National Grid reference numbers; those greater than (970216) lie within the 100-kilometre square TQ (or 51), while those less than (342406) lie within the square TR (or 61).

GEOLOGICAL SEQUENCE

The oldest rocks present at the surface in this district are of Tunbridge Wells Sand age. Details of the underlying formations are necessarily imperfectly known, but it is clear that if the cover of Cretaceous and later rocks could be stripped off a very interesting geological picture would be revealed.

Palaeozoic rocks, including Silurian, Devonian and Carboniferous formations, have been shown to constitute the basement upon which the younger sediments rest. The earliest Mesozoic strata are possibly of Triassic age, and these are succeeded by a diverse sequence of clays, sands and limestones which represent the Jurassic System. Upon the latter rest the deltaic sands and clays which make up the older rocks of the Wealden Series, namely the Ashdown Beds, the Wadhurst Clay and the Tunbridge Wells Sand.

The Tunbridge Wells Sand is followed by the clays with subordinate sands, limestones and clay ironstones which form the Weald Clay, and above them by the marine clays, sands and limestones of the Lower Greensand and Gault, which complete the Lower Cretaceous sequence. The outcrop of the overlying Chalk occupies the greater part of the area of Sheet 289, but gives way in the north to the succeeding Eocene clays, sands and pebble beds. The whole sequence of solid rocks is overlain by varied and widespread drift deposits. The formations represented on the maps and sections are given below.

SUPERFICIAL DEPOSITS (DRIFT)

RECENT AND PLEISTOCENE

Blown Sand	River Gravels
Marine Beach Deposits	Head
Storm Gravel Beach Deposits	Coombe Deposits
Marine Alluvium, clay	Head Brickearth
Marine Alluvium, sand	Head Gravel
Alluvium	Clay-with-flints
Dry Valley and Nailbourne Deposits	Sand in Clay-with-flints
Peat	

SOLID FORMATIONS

			Estimated thickness Ft
EOCENE	London Clay		250
	Oldhaven Beds		10 to 30
	Woolwich Beds		15 to 50
	Thanet Beds		95 to 120
CRETACEOUS	Upper Chalk		283 to ?400
	Middle Chalk		200 to 240
	Lower Chalk		180 to 262
	Gault		116 to ?191
	Lower Greensand	Folkestone Beds	up to 140
		Sandgate Beds	up to 120
		Hythe Beds	up to 60
		Atherfield Clay	up to 60
	Wealden	Weald Clay	up to 450+
		Tunbridge Wells Sand	up to 150

The following formations have been proved in boreholes or in mine workings:

CRETACEOUS ...	Wealden	Hastings Beds (including Tunbridge Wells Sand)	up to ?500
JURASSIC		Purbeck Beds	up to 68
		Portland Beds	up to 31
		Kimmeridge Clay	up to 262
		Corallian Beds	up to 342
		Oxford Clay and Kellaways Beds	up to ?198
		Cornbrash	up to 21
		Forest Marble	up to 18
		Great Oolite Limestone ...	up to 134
		Fuller's Earth	up to 23
		Inferior Oolite	up to ?54
		Upper Lias	up to 55
		Middle Lias	up to 45
		Lower Lias	up to 80
? TRIASSIC		... ...	up to 81
CARBONIFEROUS		Coal Measures	up to 2578
		Carboniferous Limestone ...	up to 451
DEVONIAN AND OLD RED SANDSTONE			up to 455
SILURIAN			up to ?182

G.B., J.G.O.S

BRIEF HISTORY OF GEOLOGICAL RESEARCH

The early geological writings on the geology of the Weald were outlined by Topley (1875, pp. 8–29), who was himself a major contributor. The interpretation of the sub-surface geology by Godwin-Austen (1856) led ultimately to the discovery of the Kent Coalfield, and the strata found in the boreholes and shafts made to explore and develop the coalfield were the subjects of papers by several authors; in particular the formations other than the Coal Measures were described in Geological Survey memoirs by Lamplugh and Kitchin (1911), and by these authors and Pringle (1923). The Coal Measures were described and correlated by Dines (1933, 1945), Stubblefield (1933) and Crookall (1933), and by Stubblefield and Trueman (1946).

The Wealden and Lower Greensand rocks at outcrop were described briefly by Fitton (1836) and in detail by Topley (1875). The Lower Greensand at Folkestone was described bed by bed by Price (1874b) and by Topley (1875). More recently Casey (1936–65) has thrown light on the palaeontology and conditions of deposition of the Lower Greensand; unlike most workers on the coastal sections he has carried his researches inland.

The type section of the Gault occurs at Folkestone and has been much featured in the literature, notably by De Rance (1868), Price 1874a), Jukes-Browne (1900) and latterly by Spath (1923 to 1943) and Casey (1950).

The Chalk away from the coast was largely neglected until the recent survey. It received only scant attention from Jukes-Browne (1903, 1904), who described the coast section fully, following on the works of Phillips (1818) and Price (1877), among others. Part of the zonal work by Rowe (1900) was carried out on the cliffs between Folkestone and Dover.

B

Prestwich's works (1850, 1852, 1854) formed the basis of English Eocene stratigraphy, and contained many references to the present district. Later work on the Eocene strata was done by Whitaker (1872). In the fields of geomorphology and structure the papers of Wooldridge are pre-eminent. Much of his information relating to the present area is recorded in Wooldridge and Linton (1955).   J.G.O.S.

Chapter II

STRUCTURE

PALAEOZOIC ROCKS

THE PALAEOZOIC formations of this district are wholly concealed beneath younger strata. Steeply-dipping Silurian shales were found in the Chilham Borehole (088545) (p. 16), and shales and mudstones probably of Silurian age, with dips between 20° and 30°, were present in the Brabourne Borehole (077423) (pp. 16–7). Old Red Sandstone rocks, including mudstones and sandstones with average dip of about 25°, were proved in the Harmansole Borehole (142529) (p. 17), beneath Carboniferous Limestone with a dip of 10° ; the junction was considered to be faulted (Lamplugh, Kitchin and Pringle 1923, p. 120).

The Coal Measures in Kent are folded into a trough or basin, the long axis of which trends about north-west–south-east, and the western part of the basin encroaches upon the present district. In this area the Coal Measures have everywhere been found to rest directly on an irregular surface of Carboniferous Limestone, Millstone Grit being absent, but no angular unconformity has been noticed. The western boundary of the Coal Measures basin lies east of Canterbury, between the Harmansole and Bishopsbourne (or Bourne) (192530) boreholes, and west of the Elham (180439) and Folkestone (239368) boreholes but east of the Brabourne Borehole. Arber (1914b, p. 693) and Dines (1933, p. 22) suggested that the western margin of the coalfield might be marked by a line of faulting ; Dines considered that the rise of the Carboniferous Limestone surface from about 3000 ft below O.D. at Bishopsbourne to about 1050 ft below O.D. at Harmansole was more sudden than in any other part of the basin. The Folkestone Borehole proved Coal Measures of the *Anthraconauta tenuis* Zone (pp. 29–30) between 1487 ft and 3400½ ft below surface, suggesting that this borehole was situated within a basin separate from the main one, and Dines (1933, p. 19) supposed that a ridge must be present between Dover and Folkestone, " in alignment with the anticlinal axis that separates the Carboniferous basins of the Pas de Calais and the Bas Boulonnais ".

Faults in the Coal Measures are nearly all normal, but some are reversed. The major faults strike north-west, roughly parallel to the axis of the main syncline.

DETAILS

In Chislet Colliery the Main West Fault was encountered 320 yd W.S.W. of the shafts, where it had a throw of 170 ft to the south-west ; it extends southeastwards on to the Canterbury (289) Sheet. The Stodmarsh Fault, which is believed to have a roughly parallel strike, passes between the Chislet Colliery underground boreholes No. 35 (218597) and No. 37 (216595) ; it is estimated that at this place the throw of the fault is of the order of 250 to 300 ft to the south-west (p. 28). The Coal Measures between the faults are gently folded and are affected by minor faults, most of which strike about east–west.

Dines (1933, p. 33) recorded that in Snowdown Colliery the Beresford Seam was affected by two parallel faults of north-west strike, which passed respectively 120 yd south-west and 150 yd north of the shafts ; their throws were 60 ft to the north-east and 50 ft to the south-west respectively. At the level of the Millyard Seam a fault of similar direction, with a downthrow of 65 ft to the north-east, was found between the shafts. This fault, with others of parallel strike, forms a zone of faulting which traverses the workings of the colliery. West of the shafts, faults develop with west-south-west strike and with throws of up to 70 ft to the north-north-west. The measures are also affected by numerous minor faults of diverse strike and by folds which modify the form of the major syncline.

MESOZOIC ROCKS

Mesozoic rocks rest uncomformably on the planed surface of the Palaeozoic formations. This surface may possibly be at 1706 ft below O.D. in the Brabourne Borehole (p. 30), and it rises towards the north-east, to 924 ft below O.D. at Littlebourne (196576) and perhaps 979 ft below O.D. at Stodmarsh (211600). It undulates slightly: two shallow troughs, with axes trending south-west, underlie Barham and Lower Standen respectively, and are separated by a gentle ridge.

The Lias is absent beneath the north-east corner of the district, where it is overlapped by rocks of the Great Oolite Series. Succeeding Jurassic formations generally dip to the south-west and have been bevelled to form the undulating plane of unconformity on which the Cretaceous strata rest ; this surface inclines to the north-east, so that each Jurassic subdivision extends a shorter distance to the north-east than its predecessor and is overstepped by Wealden and Lower Greensand strata (Lamplugh, Kitchin and Pringle 1923, figs. 2–4, pp. 5, 8, 9). Thus Wealden strata rest conformably on Purbeck Beds in the Ashford area, but overstep on to Portland Beds at Ottinge (169422), Kimmeridge Clay in the Elham–Folkestone–Dover area, and on to Corallian at Harmansole and Snowdown Colliery Shafts (247512). At Adisham Borehole (226543) Wealden rocks rest on Oxford Clay ; at Stodmarsh, if present, they rest on Great Oolite beds, while at Walmestone (261594) they probably directly overlie Coal Measures. In addition to the two great unconformities above and below the Jurassic sediments, several minor unconformities and non-sequences are present in the Jurassic succession (Chap. III).

Little is known concerning faults in the Jurassic rocks. At only one place, in the workings of Tilmanstone Colliery (Dover Sheet 290), has it been demonstrated that a fault in the Coal Measures also affects Jurassic strata, but this may be a common phenomenon.

The regional dip of the Cretaceous strata exposed at the surface is 1° to 2° to the north-north-east or north-east, varied by local folds and by normal faults. The folds are commonly monoclinal, with axes trending north of west and with northern limbs steeper than southern, but gentle anticlines also occur, with axial trends south of west. The faults are mostly of small throw.

G.B.

DETAILS

A weak anticline, the axis of which trends west-north-west, gives rise to an inlier of Tunbridge Wells Sand north of High Hockley (974339) and to the tongues of these beds extending northwards around Ham Street. A weak monocline passing through Swanton Farm (030358) has a similar axial trend ; the dip-slope of the Large-'*Paludina*' limestone near Rowling Street (020368) is inclined at $1\frac{1}{2}°$ to the north-east and must approximate to the dip on the northern limb of the fold ; from the extensive outcrops of the '*Paludina*' limestone it is evident that the dip on the southern limb is even flatter. J.G.O.S.

In the Goldwell Quarry (973426) (Worssam 1963, p. 6, fig. 3) a fault trending north-north-west is exposed ; to the west of the fault the strata are horizontal, while east of it the dip decreases from 30° to the E.N.E. at the fault to 10° to the E.N.E. at 50 yd from it. The presence here of a syncline with a west-north-west axis is suggested by the outlier of Sandgate Beds lying in a hollow. Beds on the northern limb of the syncline are no longer exposed, but were evidently seen by Drew, who showed opposed dip arrows at this place on the Old Series Geological Map (Sheet 3). At the south end of the Great Chart outlier another small syncline, with axis trending roughly north-west, is probably responsible for the presence of the outcrop of Sandgate Beds. Old quarries here show slight buckling of the Hythe Beds. The north-west face of one quarry (983408) showed a gentle anticline, with axis trending north-westwards. The north-east face showed two faults with throws of 3 ft to the north-west. A fault trending north-north-east across the south-east end of the outlier brings Hythe Beds against Weald Clay at its midpoint, where the downthrow is probably about 50 ft to the west. A similar fault borders the small outlier to the east. B.C.W.

The Lower Greensand outlier $\frac{3}{4}$ mile south of Kingsnorth church is bounded at its eastern end by a fault, the maximum throw of which is estimated to be 30 ft. A second fault is believed to branch north-eastwards from the first. The Old Series Geological Map (Sheet 3) shows contrary dips of 2° E. and 4° W. in this outlier. J.G.O.S.

A small carinate anticline with north–south axis, and possibly of superficial origin, was exposed in 1951 in a disused brickpit at Beaver Green (002412). A similar fold, but with an east–west axis, was described by Cornes (*in* Dewey and others 1925, pp. 259, 287, pl. 25A) from a brickpit at South Ashford (probably at 009414).

A strike fault between The Warren and Bybrook has a northerly downthrow which probably reaches 50 ft near the railway, where it cuts out the Sandgate Beds. A parallel fault $\frac{1}{4}$ mile to the south is shown on the map to account for the reported presence of ragstone in graves in the southern part of the cemetery (012433). A local east–west anticline also possibly occurs here.

Near Westwell and Eastwell the regional dip is about 2°, in direction N. 20° E. Along the narrow part of its outcrop between Westwell and Kennington the Gault probably has a locally steepened dip. A borehole at Westwell Waterworks (997463) proved the base of the Gault at 40 ft depth. The dip of the lower beds of the formation, cropping out south of the borehole, is calculated to be about 3°, and that of the upper beds about 10°, assuming that the thickness of 190 ft of Gault proved at Charing Pumping Station, $3\frac{1}{2}$ miles to the north-west, is maintained as far as this locality. A gentle syncline whose axis trends about N. 30° E. probably occurs along the Great Stour valley near Kennington. Evidence for this is the east-south-easterly fall in level of the outcrop of the base of the Chalk west of Kennington Hall (between points 016462 and 025456) and of that of the Folkestone Beds near Bybrook, implying that the dip near those places is to the north-east, whereas on the opposite side of the valley the base of the

Gault around the outlier south of Blackwall Farm dips to the north-north-west. The Stour Gap appears, therefore, to be developed on a syncline, rather than on an anticline as Kerney, Brown and Chandler (1964, p. 138, following Wooldridge 1926), suggested.

A monoclinal fold, with its steeper limb to the north, runs from Hinxhill church through Mersham-le-Hatch and to the north of Smeeth church, where it becomes a fault which dies out south-west of Lilyhole. The fold is mainly responsible for the narrow local outcrop of the Sandgate Beds. Small outliers of these beds in Hatch Park cling to the steep dip-slope of the Hythe Beds, and at Goodcheap Farm the Gault rests upon a similar slope of Folkestone Beds. In the boreholes at Hinxhill Pumping Station (p. 98) the thickness of Gault indicates that the dip on the north limb there is about 5°.   J.G.O.S.

South of Sevington the sandy basal beds of the Hythe Beds are cut out on the north-east side of each of two outliers by a strike fault with downthrow to the south-west, and on the east side of the more westerly outlier by a fault branching to the south.   B.C.W.

No evidence has been found of the faulting postulated by Worrall (1954, pp. 195-6) in the valley between Mersham and Aldington where displacements of the outcrops of the Hythe Beds are considered to be superficial. Such movements may have affected much of the weak syncline which lies south of the escarpment of the Hythe Beds at Mersham and extends south-eastwards through the outliers of Atherfield Clay and Hythe Beds east of Backhouse Wood (076378) and south of Partridge Farm (088370). A weak synclinal warp about a northerly axis is exposed in the quarry (063366) in the Hythe Beds at Aldington Corner.
                                                                        J.G.O.S.

No evidence was found for the faults north of Brook Farm (117386) and west of Otterpool Manor (100365) mentioned by Worrall (1954, p. 195). Around Sellindge, Stanford and Sandling Park the regional dip is about 1½ to 2°, in direction N. 30° E. East of the longitude of Hythe this direction changes to N. 45–50° E. A dip fault east of Perry Wood (140375) has a downthrow to the west-north-west of not more than 50 ft. A similar fault north-west of Stone Farm has a north-westerly downthrow probably of about 40 ft.

Between Summerhouse Hill (167377) and Scene Farm (175357) a line of faulted blocks trends approximately along the strike on the north-east, downthrow, side of a principal fault. The northernmost block contains Summerhouse Hill. The main fault is on the south-west side of the hill and brings the topmost part of the Gault against strata high in the Folkestone Beds, so that its throw must be about 150 ft; a fault with a downthrow to the south-east along the col north-west of Summerhouse Hill is indicated by an outlier of Middle Chalk capping the hill at less than 500 ft above O.D., at least 50 ft below the level of the base of the Middle Chalk on Brockman's Bushes. Two other down-faulted blocks, west of Bargrove and between Bargrove and Scene Farm respectively, are indicated by outliers of Gault occurring in topographic hollows on the dip-slope of the Folkestone Beds. A normal junction of Gault on Folkestone Beds is displayed at the south end of each outlier. The shape of the southern outlier suggests the presence of two intersecting fault blocks, each 200 to 250 yd wide and 600 to 800 yd long, one aligned north–south, the other N. 35° W. to S. 35° E.

Topley (1875, p. 235) mentioned a local disturbance in the valley at Horn Street; no trace of this is visible at the present day.   B.C.W.

In Sandgate, below Shorncliffe Camp (192351), the elevation of the top of the Hythe Beds beneath landslip debris indicates an inland dip greater than the average 1½° N.E., assuming that the thickness of the Sandgate Beds is uniform. The 'faulting' on the foreshore mentioned by Blake (1893, p. 468) is now considered to be landslip.

Mapping of the Chalk subdivisions shows that between the Stour valley and the valley east of Stelling Minnis the northerly regional dip of the Chalk is not constant, but is composed of two gentle folds separated by broad areas where the Chalk is horizontal or locally dips to the south. The structure is therefore a subdued continuation of the monoclinal type of fold present in the Wealden and Lower Greensand. The initial appearances of the Upper Chalk on the valley sides west of Pett Street (077472) and near Ashenfield Farm (093476), near North Leigh (128470) and near Fryarne Park (158470), coincide with the steeper limb of one of these folds. This same fold causes the Melbourn Rock to crop out in the valley floor at Pett Street (085471). The flat-lying chalk is best displayed by the outcrop of the base of Upper Chalk in the Petham valley. The dip appears to increase again at about the line of Chartham Downs. A weak southerly dip has been noted in the chalk-pit (164478) 930 yd E. 19° N. of Highchimney Farm.

Superimposed upon the structure described above are weak synclines with low dips of up to 1½°. One, roughly coincident with the Great Stour and Crundale valleys, is perhaps a continuation of the syncline at Kennington (p. 11). The structure contains a weak anticlinal warp along the upland 1¼ miles east of Wye, which dies out northwards. Another syncline coincides approximately with the Petham valley and a third embraces the valleys through both Lynsore Court (164488) and Dane Farm (176479).

Minor faults were noted in the Chalk at many places, but they could not be traced away from exposures. Such faults occurred at the following localities: Warren Quarry (029486) 800 yd W. 40° N. of Boughton Aluph church; the southern tip of Park Wood (042523) 1400 yd E. 13° N. of Coppins Farm; old pit (083443) 950 yd N. 38° E. of Beddlestone Farm; in the bank of Stone Street (133400) 620 yd E. 6° N. of Horton Park (p. 127); pit (167469) 200 yd E. 10° N. of Wildage Farm. J.G.O.S.

Reynolds (1948, p. 108) recorded the presence of faults in the chalk in adits driven from the Lower Standen, No. 3, Drellingore and Ottinge wells of the Folkestone Waterworks; these faults usually ran 40° south of east, but sometimes nearly due south. He also remarked that in the Terlingham Tunnel and at Drellingore the strata dipped to the east rather than to the north-east. He illustrated (*in* discussion of Wood and Viner-Brady 1955, fig. 39, p. 446) the courses of several faults striking south-east by east in the vicinity of Lower Standen and Drellingore.

In a chalk-pit (225380) 250 yd E.N.E. of Sugarloaf Hill a fault with a down-throw of 15 ft to the north is displayed, and 150 yd north-north-west of the quarry, on the north-east side of the Canterbury Road, the Lower Chalk–Middle Chalk boundary is again thrown down to the north. A fault with 17-ft down-throw to the west is seen in a pit (237377) west of Dover Hill, and on the cliff path above the western end of the Abbot's Cliff Tunnel (269385) another appears to throw 12 ft to the north-north-east (p. 135). A fault with a down-throw of about 5 ft to the north-east is probably at least partly responsible for the emergence of the Lydden Spout spring (281387). Another fault, with down-throw of about 5 ft to the west, is crossed by the Lydden Spout cliff path (283388) at the horizon of beds 16 and 17 of the section on page 135. Seven faults of various strikes and inclinations, and with throws of between 1 and 6 ft, were noticed in Shakespeare Cliff. G.B.

TERTIARY ROCKS

Within the present district the Eocene strata are relatively undisturbed and largely conform to the regional dip of $1\frac{1}{2}°$ to the north-east. It appears that a weak structural dome occurs in the vicinity of Rhodecourt Farm (057562) and that another is centred on the Chartham Hatch ridge. A weak syncline, which plunges towards the north-west, passes through the outlier of London Clay in Trenleypark Wood, while a broader weak syncline appears to occupy the Stour valley around Canterbury.

At Wingham the Eocene beds comprise part of the north-dipping limb of the broad syncline which occupies the lower part of the Great Stour valley.

J.G.O.S.

SUPERFICIAL STRUCTURES

In a few localities the Hythe Beds appear to extend farther down slopes than they would be carried by their dip, or are draped over spurs as if cambered. These localities are: the escarpment near Mersham; at and eastward of Park Wood, Smeeth; possibly on the north side of the Aldington ridge; the east side of the valley $\frac{1}{4}$ mile east of Harringe Court; and above Hythe, on the outcrop which slopes towards the valley south of Saltwood Castle. In a section (163352) 350 yd N. 10° E. of Hythe church the beds on the east side of a vertical joint are displaced upwards relatively to those on the west, as in dip-and-fault structure; a valley-ward dip of 17° to the east was shown by a 2-ft ragstone bed by the roadside (164353) 300 yd N.E. of this locality.

B.C.W.

The Eocene sands exposed in the pit at Shelford (161600), 750 yd N.W. of the Sewage Works near Fordwich, are affected by well-developed trough-faulting, which extends to about 50 ft below the surface and is the deepest effect of ground-ice within this district. The movements post-date the ferruginous staining of the sands. In that part of the face away from the Great Stour valley two single faults form a simple V, and there is little or no disturbance of the bedding or bands of ferruginous staining in the faulted masses. Towards the river the troughs are usually made up of four faults and the associated disturbance of the beds increases in intensity. The maximum throw of any one fault is approximately 4 ft and the total drop in the centre of a trough is 6 to 10 ft; breccias of laminated clay in fine sand occur in the fault planes. The overlying incompetent London Clay is not faulted, but sags into the troughs; one such sag, now quarried away, extended through the London Clay and let down 3rd Terrace gravel along the line of the trough, for a distance of at least 30 yd. Less well-developed trough faults, but probably of similar origin, are noted on p. 226.

These trough faults are thought to have been formed when wedges of ground-ice melted, at some time after the deposition of the 3rd Terrace. The freezing process, which resulted in the formation of the ice wedges, affected the Eocene sands differently from the surface of the London Clay (p. 269), due to their greater permeability and water content. In the top of the sands large wedge-shaped masses of ice, probably containing much sand and clay, appear to have formed and forced the mass of the sands apart. When the ice in the wedges melted, the surrounding sands, possibly partially frozen themselves, collapsed into the vacant spaces. The increasing degree of disturbance towards the Stour valley was probably due to the decreasing thickness of overburden, which permitted repeated, perhaps annual, thawing, and correspondingly greater disruption. The absence of disturbance in the trough faults away from the valley suggests that there the ice wedges were most impure, perhaps consisting simply of frozen wedges of Oldhaven Beds and Woolwich Beds, which collapsed on thawing.   J.G.O.S.

Chapter III

CONCEALED FORMATIONS

INTRODUCTION

IN 1855 GODWIN-AUSTEN (1856) read his historic paper to the Geological Society of London, entitled ' On the possible Extension of the Coal-Measures beneath the South-Eastern Part of England ', in which he wrote (p. 73) " . . . it will not be deemed too much to say, that we have strong *à-priori* reasons for supposing that *the course of a band of coal-measures coincides with, and may some day be reached, along the line of the valley of the Thames,* whilst some of the deeper-seated coal, as well as certain overlying and limited basins, may occur along and beneath some of the longitudinal folds of the Wealden denudation." This contention was not confirmed until 1890, when the Brady Borehole (295393) reached Coal Measures ; this borehole was made beside the railway at the foot of Shakespeare Cliff, 2 miles south-west of Dover, after work on the Channel Tunnel was stopped.

In the years that followed, deep boreholes and shafts were made at twenty-two sites within the district described in this memoir (Plate II), as well as at others to the north and east, to determine the extent and value of the new coalfield and of the iron-ore deposits that were found in the Corallian strata. Descriptions and classifications of the beds encountered in the various operations, often very widely at variance one from another, have been published by several writers, and the geological formations now known to be present are listed on p. 7.

SILURIAN

Silurian rocks were found beneath Lower Lias in the Chilham Borehole. They were described by Lamplugh, Kitchin and Pringle (1923, pp. 123, 129–30) as smooth, fine-textured hard shale or slate of dark bluish tint banded with dark grey, and with blue-black layers full of graptolites. The latter were identified by the late Miss G. L. Elles as an assemblage typical of the *Monograptus crispus* Zone of the Upper Llandovery. The shales were stained red and purple to a depth of 15 to 20 ft, they split readily along the bedding planes and were without secondary cleavage. The dip averaged about 80°, so that although the shales were penetrated to a depth of 182 ft,[1] the actual thickness of beds traversed was only about 30 ft.

Rocks probably of Silurian age occurred in the Brabourne Borehole between 1921 and 2004 ft. They comprised " Fine-grained homogeneous silty shale or massive mudstone, stained dull red and purple, with greenish streaks, in the upper 20 ft., but becoming dark sooty grey or nearly black below ;

---

[1] Some doubt exists as to the depth to which the shales were proved, for Lamplugh, Kitchin and Pringle listed 182 ft of shales between 1105 ft and 1187 ft (sic), while Dawkins (1913, p. 362) reported 51½ ft, with their base at 1154 ft. Specimens in the Geological Survey collection were recorded as from the range 1105 to 1200 ft, so that the final depth of the hole may have been in excess of 1200 ft, and was possibly 1287 ft.

showing traces of wavy bedding-stripe in places at an angle of 20° to 30° to the horizontal axis of the cores, and with a slight cleavage at about 60° . . ." (Lamplugh and Kitchin 1911, p. 54). They were taken to be of Devonian or Old Red Sandstone age by Dawkins (1905, p. 30), while Lamplugh and Kitchin (1911, pp. 54–6) recognized fragments of ribbed shell (possibly a Rhynchonellid) in cores from between 1973 and 1978 ft, and formed the opinion that the rocks were of marine origin and probably of Devonian age or older. However, after Silurian strata were found in the Bobbing Borehole, on the Chatham (272) Sheet (Dines and others 1954, p. 11), Lamplugh and his co-authors (1923, p. 194) noticed their resemblance to the lowest rocks in the Brabourne Borehole, and concluded that it was likely that the latter were also of Silurian age. That the fossils are marine has been confirmed by Sir James Stubblefield, who has identified fragments probably of brachiopod and trilobite, and polyzoans in material from 1930 ft.

OLD RED SANDSTONE

Within the present district, rocks of Old Red Sandstone age are known only from the Harmansole Borehole, where they were found between 1275 and 1730 ft below ground surface, beneath Carboniferous Limestone (pp. 9, 18). The Old Red Sandstone rocks consist of grey, greenish grey, mottled green and red, and dark red micaceous sandstone, with thin courses of conglomerate, the pebbles therein being mostly angular or subangular, and composed of sandstone. Some beds exhibit ripple-marks and rain-pitting. The average dip is about 25°. A Rhizodont fish scale and 'Estheria' of "a species distinct from E. membranacea of the Middle Old Red Sandstone" were found (Lamplugh, Kitchin and Pringle 1923, pp. 114, 120–1). 'E.' (Euestheria) cf. membranacea (Pacht) has been recognized by Sir James Stubblefield in a core from 1524 ft.

These strata were described by Baker (1920, p. 788) as "Grey, green, red, blue and mottled shales with subordinate sandstones and conglomeratic rocks". No detailed log of the beds has been preserved, but the few specimens in the Geological Survey collection include silty mudstone and siltstone. Sabine (in Smart, Sabine and Bullerwell 1964, p. 17) noted the resemblance of a dark reddish brown slightly calcareous ferruginous siltstone from 1447 ft to Palaeozoic siltstones from the Geological Survey's Canvey Island Borehole, and from 1614 ft he described a dark grey greywacke, rich in dolomite and containing plant remains.

CARBONIFEROUS

CARBONIFEROUS LIMESTONE

The zonal classification of the Carboniferous Limestone is as follows, based on Vaughan (1905) and Reynolds (1921, 1926) :

	Dibunophyllum Zone	...	$\begin{cases} D_2 \\ D_1 \end{cases}$
Viséan	Seminula Zone	...	S_2
	Upper Caninia Zone	...	$\begin{cases} S_1 \\ C_2 \end{cases}$

Tournaisian $\begin{cases} \text{Lower } \textit{Caninia} \text{ Zone} \quad \dots \quad \dots \quad \dots \quad \dots \left\{ \begin{array}{l} C_1 \\ \gamma \end{array} \right. \\ \textit{Zaphrentis} \text{ Zone} \qquad \dots \quad \dots \quad \dots \quad \dots \text{ Z} \\ \textit{Cleistopora} \text{ Zone} \qquad \dots \quad \dots \quad \dots \quad \dots \text{ K} \end{cases}$

In the Harmansole Borehole Carboniferous Limestone was found between 1249 and 1275 ft below surface (Lamplugh, Kitchin and Pringle 1923, p. 114), consisting of 2 ft of " Hard dark grey bituminous limestone, largely made up of crinoidal fragments ", on 11 ft of " Hard dark shale with thin dark grey limestones, varying from one to two inches in thickness ", on 1 ft of " Smooth dark shale ", on 5 ft of " Dark grey limestone with some shaly bands ", on 7 ft of " Smooth dark shales with *Modiola lata* Portl. At base the lowest 4 feet of beds are smashed and slickensided ". The dip is 10°.   G.B.

Fossils from these beds were originally named by J. Pringle, and on lithological and palaeontological grounds the rocks were taken to represent the Lower Limestone Shale (= K Zone). The faunal list, after revision by Sir James Stubblefield and the writer is as follows: crinoid columnals; Polyzoan (*Rhabdomeson?*) ; ' *Camarotoechia* ' *mitcheldeanensis* Vaughan, *Chonetes sp.*, *Eumetria sp.*, *Dielasma?*, *Productus sp.*, reticulate Spiriferid, Orthotetid, smooth Spiriferid, *Spirifer* cf. *ventricosus* de Koninck, *Syringothyris* cf. *elongata* North ; and *Promytilus? latus* (Portlock).

The fauna listed by Pringle (*in* Lamplugh, Kitchin and Pringle 1923, pp. 119–20) is substantially the same as that given above and the arguments for the Tournaisian age of these beds still hold good. The Orthotetids named as cf. *Schellwienella crenistria* (Phillips) by Pringle are represented only by dorsal valves and fragments of ventral valves, and are here considered to be indeterminate. The specimens of *Syringothyris* are closest to *S. elongata* and this species ranges up from the base of the Z Zone into the D Zone. One of the specimens of *Spirifer* is here compared with *S. ventricosus,* a species that has been recorded by Reynolds and Vaughan (1911, p. 364) from horizon β (base of Z Zone) in the Burrington Combe section.

The important fossil *Eumetria sp.* was recognized by Sir James Stubblefield and this brachiopod has been recorded only from the K Zone in the South-West Province. *Promytilus? latus* was found only in the shales between 1269 and 1271 ft ; although it is common in the K_m Subzone (base of K Zone) of the Bristol area, it is considered to be a facies fossil, indicating ' *Modiola* ' phase deposits, and not an indicator of horizon.

Polyzoa bands occur at several levels in the Lower Limestone Shale (K) of the Bristol area, but at present they have been insufficiently studied to be of any stratigraphical significance.

On the evidence of the *Spirifer* this Carboniferous Limestone might be assigned to the Z Zone. The higher parts of the Z Zone are usually characterized by the abundance of Zaphrentids and on the negative evidence of the absence of Zaphrentids, the age of the fauna is possibly either high K or low Z. The presence of *Eumetria sp.,* however, strongly supports the correlation of these beds with the K Zone.   M.M.

The Carboniferous Limestone in the Trapham Borehole (234571) was the subject of a controversy. Burr (1913, p. 732) recorded that Carboniferous

Limestone was entered at 2775 ft below surface, and that the bottom of the borehole was at 3226 ft. In a diagrammatic section he showed sandstone at the base of the Coal Measures, which persisted almost to the bottom of the borehole, where it was underlain by dark limestone, but in the text (p. 733) he described the Lower Carboniferous strata as "grey and carboniferous limestones with calcite veins, occasionally with shaly partings, slightly arenaceous in places, black shales with pyrites and carbonaceous matter, and black compact limestones with *Syringothyris cuspidata* (det. Woods), and in the lower portion some sandstone, free from lime." Arber (1914b, p. 687) agreed that 450 ft of Carboniferous Limestone were penetrated, and later (1915) described the strata involved, including limestone and black and greenish shales to 3192 ft, resting on hard sandstone and conglomerate to 3210 ft, on impure limestone to 3223 ft. *Productus sp.*, in limestone at 2900 ft, and *Syringothyris cuspidata* (J. Sowerby), in limestone interbedded with shale at 3150 ft, were identified by the late Henry Woods, and were held to indicate that the D Zone and at least part of the S Zone were absent, the C Zone being present. However, Bolton (1915, p. 655 ; 1916, p. 313) argued that Coal Measures continued to 3204 ft 2 in, on the grounds that the borer's log recorded sandstone and sandy bind, with some fireclay and conglomerate, to that depth, underlain by limestone.   G.B.

Specimens from the Trapham Borehole presented to the Geological Survey by the Kent Education Committee have been re-examined. At 2900 ft, *Productus* (*Linoproductus*) *corrugatohemisphericus* Garwood—a species considered diagnostic of the S_2 Zone—occurs in pale grey oolitic limestone. It is probably the fossil from this depth identified as *Productus sp.* by Henry Woods. A specimen of a similar pale oolitic lithology from 3015 ft is also present but contains only indeterminate shell fragments. Between 3150 ft and 3163 ft 6 in., fine-grained dark grey crinoidal limestone with thin dark grey mudstone partings contains : '*Zaphrentis*' *konincki* forma *typica* Carruthers towards intermediate form of Hudson and Mitchell 1937, p. 9 ; ? polyzoa ; '*Camarotoechia*' *mitcheldeanensis*, *Chonetes* (*Megachonetes*) *sp.*, *Leptagonia analoga* (Phillips), smooth Spiriferids, *Spirifer* cf. *suavis* de Koninck and *Syringothyris elongata*. H. Woods identified *Syringothyris cuspidata* at 3150 ft but the form is here referred to the species *S. elongata*. Smooth Spiriferids [juvs.] are the only fossils seen in dark grey fine-grained oolitic limestone between 3182 ft and 3182 ft 6 in. The lower group of dark-coloured limestone is assigned to the upper part of the γC_1 Zone. Arber's opinion (above) was thus more reliable than that of Bolton.   M.M.

Pringle (1917, pp. 36–8) described the Carboniferous Limestone in the Elham Borehole as "dark grey bituminous limestones with smooth fine-grained beds, or 'china-stones'", between 2289 and 2346 ft (base not seen). Specimens in the Geological Survey collection include fossiliferous Carboniferous Limestone labelled 2281 ft 6 in, 2282 ft and 2283 ft, coinciding with a bed of "fine-grained dark grey bituminous limestone with corals" recorded in the borer's log between 2281 and 2283 ft ; the underlying beds comprised 8 ft of "hard dark grey micaceous shaly sandstone with thin bed of chert", of Coal Measures aspect, and Pringle evidently considered that the core from between 2281 and 2291 ft had been inverted.   G.B.

The fossils in the limestone from the Elham Borehole were originally named by Pringle, who (1917, p. 38) gave their age as S₂ or D₁, and later (1928, p. 80) as S₁ or S₂. They were re-examined by Sir James Stubblefield and the writer, and the faunal list is as follows: *Carcinophyllum?*, *Diphyphyllum?*, *Lithostrotion martini* Edwards and Haime, *L.* cf. *sociale* (Phillips), *Syringopora* cf. *geniculata* Phillips ; *Athyris* cf. *expansa* (Phillips), *Chonetes* (*Megachonetes*) *sp.* (*papilionaceus* Phillips group), *Composita* cf. *ficoidea* (Vaughan), *Dielasma sp.*, Orthotetid fragments including fragments with *Orthotetes fascifera* Tornquist kind of costae, *P.* (*L.*) *corrugatohemisphericus*, *P.* (*L.*) aff. *corrugatohemisphericus*, *P.* (*Productus*) cf. *garwoodi* Muir-Wood ; and gastropod fragments. This assemblage is considered typical of the *Seminula* (S₂) Zone.

The coral fauna includes numerous well-preserved specimens of the S₂ form of *Lithostrotion martini* (Whittard and Smith 1943, pl. 15, fig. 4). A single specimen of a larger diameter which is close to *L. sociale* is also present.

P. (*L.*) *corrugatohemisphericus* is at present known only from beds of S₂ age. The fauna of the S₂ Zone is usually limited in the number of species, but the individuals of one species commonly occur abundantly, forming conspicuous fossil bands. *Chonetes* (*Megachonetes*) *sp.* (*papilionaceus* group) and *Composita* cf. *ficoidea* occur in this manner here.   M.M.

In a manuscript record of the strata found in the Adisham Boring, Pringle described the Carboniferous Limestone as follows: " Dark grey bituminous limestone, passing below into a fine-grained blackish limestone. The upper surface of the limestone appears to have undergone considerable sub-aerial weathering before the deposition of the Coal Measures. The Coal Measures are found filling the eroded hollows." The limestone was entered at 3237 ft 9 in and penetrated for 27 ft. Pringle remarked (1928, p. 80) that the limestones were unfossiliferous ; he likened their lithological characteristics to those of the Carboniferous Limestone found in the Elham Boring, to which he gave an S₁–S₂ age, but which is now considered to belong to the S₂ Zone above.

Carboniferous Limestone was reached at 2096 ft below Ordnance Datum in Chislet Colliery No. 35 Underground Borehole, where a cream oolitic limestone, with a stylolite and traces of pyrite, passed downwards into slightly oolitic calcite mudstone. Fossils in the oolitic limestone were identified by Dr. W. H. C. Ramsbottom as cf. *Composita sp.* [juv.], *Productus* (*Linoproductus*) *sp.*, a finely-costate Productid fragment and fragments of smooth Spiriferid ; no algae were seen. Dr. Ramsbottom remarked that this fauna, taken together with the oolitic nature of the limestone, suggested an S age.

In the Stodmarsh Borehole, Burr (1913, p. 734) put the point of entry into Carboniferous Limestone at 2145 ft below surface ; he wrote : " . . . the upper four-feet consists of black calcareous shales. Down to 2,220 ft we have white and grey massive limestone passing into shales, argillaceous limestone and black limestone interlaminated with black impure shales with calcite veins and some black shales with very little lime. At 2,257-61 the shales are disturbed and full of thick calcite veins ; the hole ended at 2,263 in typical limestone." According to the borer's log Carboniferous Limestone was entered at 2141 ft 7 in ; " clunchy fireclay and bind " were recorded between 2247 ft 9 in and 2257 ft 2 in.   G.B.

A specimen of *P. (L.) corrugatohemisphericus* from 2172 ft 10 in. in the Stodmarsh Borehole was presented to the Geological Survey by the Sedgwick Museum in 1961, and is probably the fossil identified as *Productus corrugatohemisphericus* by Henry Woods, on which evidence he put the beds in the *Seminula* Zone or the lower part of the *Dibunophyllum* Zone (Bolton 1915, p. 647). Additional material from this borehole, presented by the Kent Education Committee, has been re-examined and foraminifera and *P. (L.) corrugatohemisphericus* have been identified from pale grey brecciated oolitic limestone between 2157 ft 11 in and 2225 ft. This Productid is now taken to indicate an S_2 age.   M.M.

The Walmestone Borehole was discontinued at 2285 ft 6 in below surface, having entered Carboniferous Limestone at 2276 ft 6 in. Burr (1913, p. 732) recorded that the formation is represented by " grey limestone, containing only indeterminable mollusca, with some black shales intercalated." Arber (1914b, p. 708) commented that the only fossil plant found in the Carboniferous Limestone Series in Kent had been a leaf of *Cordaites* in the Walmestone cores. Little is known of the Carboniferous Limestone in the Bishopsbourne Borehole, where the formation was recorded from 3169 ft 2 in to 3235 ft below surface. According to the borer's log it consists of grey limestone, with pyrites and spar in places, and with beds of grey sandstone 1 ft 5 in and 8 in thick at 3182 ft 11 in and 3189 ft respectively. However, in an unpublished report, Arber stated that the Carboniferous Limestone in this borehole consists of dark limestone, containing only a few poorly-preserved fossils, and with thin shales near the top amounting to less than 3 inches in all.

No core of Carboniferous Limestone was taken from the Littlebourne Borehole, but the driller recorded that the formation was entered at 2613 ft 6 in below ground level, and the hole was discontinued at 2618 ft. Mudflush samples from 2617 ft and 2618 ft were examined by Sir James Stubblefield, who reported that they contained fragments of limestone, some of which were oolitic ; some contained traces of pyrite, and some foraminifera (*Endothyra*) and other micro-organisms such as were to be found in the Carboniferous Limestone. No determinable macrofossil was seen.

The West Court Farm Borehole (246481) did not reach the base of the Coal Measures, but pebbles and angular fragments of limestone (which resembled Carboniferous Limestone) were present in sandy mudstone and siltstone at intervals between 3910 ft and 3969 ft 7 in, and between 4014 ft and the bottom of the borehole at 4019 ft 6 in below surface. Thin sections of specimens from 3912 ft, 3935 ft and 3952 ft 6 in are described below.   G.B.

A specimen from 3912 ft in the West Court Farm Borehole (E 30136)[1] is a dark grey impure sandy pyritic carbonaceous limestone. It is composed of pale grey [Carboniferous] limestone fragments up to 0·5 cm across, and ooliths commonly 0·3–0·5 mm across, in a groundmass of finer angular quartz grains cemented by fine-grained calcite and clay materials including mica. There are abundant streaks of carbonaceous material, and granules and aggregates of pyrite in the groundmass and in the limestone fragments.

A heavy dark grey carbonaceous ankerite-siltstone from 3935 ft (E 30137) is nearly opaque in thin section. It is composed of a very fine-grained aggregate

---

[1] Numbers in parentheses preceded by the letter ' E ' refer to specimens and thin sections in the Petrographical Collections of the Geological Survey and Museum. Numbers preceded by ' MR ' denote specimens in the Museum Reserve collection.

of ankerite (ω showing a range near 1·773), clay minerals including mica, quartz silt, calcite, and carbonaceous matter, traversed by narrow calcite veinlets. There is abundant irregularly disseminated granular pyrite. The specific gravity of the rock is 3·27. Chemical determination by Mr. G. A. Sergeant showed the presence of 24 per cent CO_2. A heavy grey pyritic sandy limestone from 3952 ft 6 in (E 30138) is composed of abundant quartz grains 0·2 mm or less across closely set together and poikilitically enclosed in calcite which forms coarse plates many about 0·6 mm across. There is some flamboyant sparry calcite, perhaps stylolitic, and a little more finely granular limestone containing ooliths.   P.A.S.

COAL MEASURES

The Coal Measures in Kent consist of grey mudstones (sometimes called ' binds '), siltstones and sandstones, and rocks of intermediate grain-size, with seatearths, commonly, but not invariably, surmounted by coal. In common with the measures in other British coalfields they are of paralic type, at least in the lower part, having been formed in an " external basin ", so near the sea and " so nearly at sea-level that they were subject from time to time to marine incursions " (Trueman 1946, p. lii). They are considered to consist of a series of cyclic units or cyclothems, resting one upon another : an ideal example of such a unit was described by Trueman (1946, p. lvi) as : marine band, overlain by non-marine shale or mudstone, by sandstone, by rootlet bed, and by coal ; or, in the words of Edwards (*in* Eden and others 1957, p. 25) : " . . . the upward succession coal, detrital rocks with ironstone in the lower beds and sandier towards the top, seatearth with ironstone in the lower part and leached at the top, coal." The coal seams are thought to have been formed by the accumulation of vegetation, grown more or less *in situ,* in forest swamps which often extended over large areas, since some individual seams can be traced for great distances. At the close of the period of rest during which coal was deposited, the land surface was raised and the off-shore area depressed, with consequent inundation by the sea and deposition of dark grey marine mudstone. As the accumulation of sediments continued the sea was excluded by some means, and brackish-water conditions prevailed, when mudstones were deposited containing non-marine fossils and becoming progressively lighter grey in colour, followed by silty mudstones and siltstones with plant remains, and by sandstones. Finally, as the supply of sediment slackened its grain-size progressively decreased to that of mudstone, and the basin was filled to form a swamp, in which a new coal seam was deposited. Details of this process were discussed by Robertson (1932, pp. 87–9).

In practice the cycles are rarely ideally developed and many variants are found : in the Kent Coalfield many seatearths occur without superincumbent coal, marine bands are few and coals are commonly overlain by non-marine mudstones or by mudstones with plants, while sandstone roofs to the seams are not uncommon ; alternatively the sandstone phase may be missing.

A ' washout ' is an area in which a seam is locally wanting, and its place taken usually by sandstone with a conglomeratic base ; generally elongated in plan, washouts are thought to be due to bottom-scour by streams during the transport and deposition of the sandstone phase of the cycle.

Areas of comparatively thick coal, of small width but great length, are termed 'swilleys'; they may have been caused by the activities of contemporaneous streams traversing the forest swamps; they are generally adjacent to areas of thin coal, where the bedding in the coal can be seen to abut in places against humps or rolls in the floor of the seam. In such areas differential compaction of the sediments may have given rise to small-scale fractures and contortions within the coal seam. In addition a seam may have been affected by tectonic faults and folds.

Coal seams are usually composed of a characteristic sequence of layers of coal of different types, either hard or soft, dull or bright, interbedded with partings or bands of dirt; the latter can sometimes be shown to thicken and pass laterally into shales and so form a separate cycle, when the original seam is said to split; even so, the seam profile may be sufficiently constant to be traced for some distance. Other characters which assist in the recognition of a seam include its chemical properties and its contained spores.

Animal fossils, both marine and non-marine, are almost confined to the mudstones and shales, and their associated ironstone beds and nodules, but occasionally they may be found in the coarser sediments. Plant remains may occur throughout the cycle, but they are best preserved in the more argillaceous sediments. Fossils in seatearths are almost exclusively plant roots, although exceptionally non-marine shells have been found.

The classification of the Coal Measures into Lower, Middle and Upper divisions has been defined by Stubblefield and Trotter (1957): the base of the Lower Coal Measures is taken at the base of the marine band characterized by the goniatite *Gastrioceras subcrenatum,* the base of the Middle Coal Measures is at the base of the marine band with *Anthracoceras vanderbeckei,* and the base of the Upper Coal Measures is at the top of the marine band with *Anthracoceras cambriense.* On the basis of their contained non-marine lamellibranchs the Coal Measures have been further classified into zones (Trueman and Weir 1946), which, in order of upward succession, are named as follows: *Anthraconaia lenisulcata; Carbonicola communis; Anthraconaia modiolaris; Anthracosia similis* and *Anthraconaia pulchra; Anthraconauta phillipsii; Anthraconauta tenuis;* and *Anthraconaia prolifera.*

Bolton (1915, pp. 650–2) divided the Coal Measures in Kent on lithological grounds into an " Upper Series of sandstones, binds, and coals ", and a " Lower Series of shales, binds, and coals ". The latter he had never known to exceed 700 ft in thickness. The former he likened to the " Pennant Grits ", a comparison previously made by Dawkins (1907, p. 457), although palaeontologically he (Bolton) considered it to be equivalent to the Farrington Group, which overlies the Pennant Series in the Somerset Coalfield; the Lower Series he correlated palaeontologically with the Vobster, Ashton and Bedminster groups of the Bristol and Radstock areas. Arber (1909, 1912, 1914) investigated the palaeobotany of the Coal Measures and concluded that the Transition Coal Measures (= Staffordian Stage) and the " Middle

C

Coal Measures " (= Yorkian Stage) were represented, and that the " Upper " and " Lower Coal Measures " (= Radstockian and Lanarkian stages) were absent. Kidston (1919) proved the presence in the coalfield of the Keele or Farrington Group, the lower part of the Radstockian Stage, and Crookall (1933, p. 65) put the thickness of this division at 550 ft ; to the Staffordian Stage and the Yorkian Stage he allotted 1600 ft and 650 ft respectively.

Dines (1933) used the modified terms Lower or Shale Division and Upper or Sandstone Division, about 700 ft and 2100 ft thick respectively[1]. He remarked that in the lower division coals were often impossible to identify in sections due to splitting and changes in thickness ; sandstones were rare and generally thin, except in the upper part of the division in the north-western part of the coalfield. He put the base of the upper division at the bottom of a sandstone about 100 ft in thickness, which was surmounted by a persistent seam known as the Millyard Seam at Snowdown Colliery and the 'F' Seam at Betteshanger Colliery ; this coal was succeeded by some 200 ft of binds, with subordinate sandstones, thin coals and fireclays. The measures above were predominantly of sandstone, with some binds ; coals and fireclays were fewer than in the lower division ; near the top of the Sandstone Division shales again developed and coals included the Beresford Seam of Tilmanstone and Snowdown collieries. In 1945 Dines recognized fourteen more or less persistent coal seams in the coalfield, which he numbered in downward succession, Nos. 1 to 6 being within the Sandstone Division, and Nos. 7 to 14 within the Shale Division. No. 1 was the Beresford Seam, No. 3 the Snowdown Hard Seam, No. 6 the Millyard Seam, No. 7 and No. 9 the 'H' and 'J' seams respectively of Bettes-hanger Colliery. Dines also identified the Chislet No. 5 Seam with Kent No. 9, but this correlation was questioned by Stubblefield and Trueman (1946, p. 271), who preferred to relate this seam to Kent No. 7.

Trueman (1933) recorded the occurrence in the Kent Coal Measures of beds belonging to the *A. phillipsii* and *A. tenuis* zones and to the '*A. pulchra*' Zone. The faunal succession in the coalfield was described by Stubblefield (1933) and by Stubblefield and Trueman (1946). In the last-named work the authors showed the presence of the *communis* Zone as well as those of *similis–pulchra* and of *phillipsii* and *tenuis ;* they referred to Bolton's record (1915, pp. 664, 672) of *Lingula* at 2959 ft 3 in below surface in the Ripple Borehole (Dover Sheet 290), and suggested that this might represent a marine horizon near the base of the *similis–pulchra* Zone or in the underlying *modiolaris* Zone. They gave details of the fauna of two marine horizons, termed the Tilmanstone Marine Beds, near the top of the Shale Division, and probably within the *similis–pulchra* Zone. They discredited Bolton's record (1915, pp. 662, 675) of *Productus longispinus* from 363 ft above the Lower Tilmanstone Marine Bed in the Barfreston Borehole (Sheet 290), and they noted (1946, p. 278) the occurrence of a specimen of *Trigonoglossa* in Snowdown Colliery No. 1 Underground Bore-hole (243510) 273 ft below the Millyard Seam and 46 ft below a marine horizon with *Productus* ('*Pustula*') *rimberti ;* they tentatively took these two marine horizons to represent the two Tilmanstone Marine Beds.

[1] The Coal Measures trough is now thought to be deepest around Waldershare and St. Margaret's Bay, on the Dover (290) Sheet, where it reaches to about 3800 ft below O.D. and the Coal Measures attain a maximum thickness of some 2900 ft.

The Coal Measures strata penetrated in shafts and deep boreholes within the present area are displayed diagrammatically in Plate II. The suggested correlation of the coal seams in general follows the correlation adopted by Dines (1933 ; 1945), but differs from it in some respects.

In the detailed account of the Coal Measures below the majority of the fossils have been named by Mr. M. A. Calver, who has also brought up to date earlier identifications by Sir James Stubblefield, Dr. W. H. C. Ramsbottom and the late Sir Arthur Trueman.

DETAILS

LOWER COAL MEASURES

The *Gastrioceras subcrenatum* Marine Band has not been recognized in East Kent. The earliest Coal Measures sediment was described as sandstone in the Bishopsbourne, Trapham (pp. 18–9) and Littlebourne boreholes ; at Elham, Pringle (1917, p. 37) recorded " Grey micaceous shaly sandstone with *Stigmaria* and an irregular lump of black chert (pebble?) ", 8 ft, overlain by grey and black shales with thin bands of sandstone and a bed of black fine-grained bituminous limestone. In other boreholes within the present district the Carboniferous Limestone was overlain by silty or sandy mudstone, usually containing pyrite. In Chislet No. 35 Underground Borehole 3 ft 10 in of light grey pyritous silty mudstone rested on Carboniferous Limestone ; irregular unworn fragments of cream-coloured limestone were present in the lowest 1 ft 5 in.

Anthraconaia aff. *bellula* (Bolton) was found in Chislet No. 35 Borehole between 398 ft 11 in and 398 ft, 3 ft 9 in above a thin coal and 32 ft above the Carboniferous Limestone surface ; Mr. Calver comments that these shells appear to belong to the group of *Anthraconaia* species which occurs near the Millstone Grit–Coal Measures boundary. The thin coal is about 26 ft below a split seam which may be the equivalent of the Kent No. 14 Seam as correlated elsewhere in Chislet Colliery. In the immediate roof of the latter seam foraminifera and fish debris have been found in black shaly mudstone (Stubblefield 1953, p. 42), which is succeeded by silty mudstone and highly micaceous sandstone containing a *C. communis* Zone fauna (Stubblefield and Trueman 1946, p. 269). As marine strata are virtually unknown in the *C. communis* Zone the black mudstone may belong to the *A. lenisulcata* Zone.

The *C. communis* Zone fauna of Chislet Colliery occurs between the Kent No. 14 and No. 13 seams ; these coals can be traced from Stodmarsh to Adisham, and can be assumed to be represented among the four coals between 2207 ft 5 in and 2153 ft in the Elham Borehole. The fauna was probably represented by the " *Carbonicola acuta* " recorded by Bolton (1915, pp. 653, 683) from 2071 ft depth in the Stodmarsh Borehole, 69 ft above the Carboniferous Limestone surface. At about this horizon in Chislet No. 35 Borehole *Curvirimula sp.* was found in carbonaceous silty mudstone between 369 ft 6 in and 350 ft. An 8-ft bed of hard brownish grey calcite mudstone containing pyrite and sphaerosiderite occurred with its base at 328 ft 3 in, beneath a split seam, and above this seam *Carbonita spp.* were collected from between 292 ft 8 in and 291 ft 6 in.

In Chislet No. 29 Borehole (Faversham Sheet 273) *Anthraconaia sp.* and *Naiadites?* occurred separately above coals 38 ft and 47 ft respectively above the *communis* Zone fauna, and were thought likely to represent either the *modiolaris* Zone or the *similis–pulchra* Zone (Stubblefield and Trueman 1946, p. 270). The latter alternative has now been eliminated.

The Kent No. 12 Seam is taken to be the first seam below the Ripple Marine Band. It has no economic value.

MIDDLE COAL MEASURES

Ripple Marine Band. The Ripple Marine Band was so named (Stubblefield 1953, p. 42) because of Bolton's original record of *Lingula* at this horizon (2959 ft 3 in below surface) in the Ripple Borehole (p. 24). It has been shown (Calver *in* Anderson 1956) to be equivalent to the Clay Cross Marine Band of Yorkshire.

Marine fossils were found in the Chislet No. 35 Borehole in 3 ft of siltstone and fine-grained sandstone between a 3-inch coal and the base of an overlying seatearth. The fauna comprised *Lingula sp., Productus (Levipustula) sp., Euphemites sp., Parallelodon* cf. *geinitzi* de Koninck, *Polidevcia sp.* and *Hollinella sp.* In the West Court Farm Borehole marine fossils occurred as follows: Between 3872 ft and 3868 ft 10 in: sponge spicules; crinoid columnal [pentagonal]; *Lingula mytilloides* J. Sowerby, *Productus (P.) carbonarius* de Koninck, *P. (Levipustula)* cf. *piscariae* Waterlot, *P. (Linoproductus) sp.; Euphemites?,* cf. *Leptoptygma sp.* [juv.]; *Aviculopecten gentilis?* (J. de C. Sowerby), *Parallelodon?, Solemya* cf. *primaeva* Phillips; *Metacoceras sp. (cornutum?* Girty); platformed conodont. Between 3865 ft and 3861 ft 7 in: *L. mytilloides; Polidevcia acuta* (J. de C. Sowerby). Between 3859 ft and 3844 ft 2 in: *L. mytilloides; Euphemites?; Dunbarella sp., Nuculopsis sp.* [juv.], *Parallelodon* cf. *reticulatus* (M'Coy), *Polidevcia acuta;* and *Ephippioceras?* [juv.].

The seams Kent No. 11 to Kent No. 7 are of economic importance; they are, however, grouped in a small thickness of measures, and they tend to split or coalesce, and to vary in thickness, as do the intervening sediments. In the absence of faunal control the detailed correlation of these seams is thus very uncertain.

Spirorbis sp., Naiadites sp. intermediate between *productus* and *quadratus,* and *'Estheria' sp.* were found at 3803 ft at West Court Farm. In Chislet No. 35 Borehole four thin coals occurred at intervals of 10 to 12 ft below the Kent No. 11 (Chislet No. 11) Seam, and in the measures between them were found *Naiadites* cf. *productus* (Brown), *N. quadratus?* (J. de C. Sowerby), and species intermediate between the two. The beds above the Kent No. 11 Seam yielded only *Naiadites sp.* and *'Estheria' sp.,* but elsewhere in Chislet Colliery these measures have been found to contain a fauna of Lower *similis–pulchra* Zone age, confirming the suggestion of Stubblefield and Trueman (1946, p. 272).

In the Walmestone Borehole, cf. *Anthraconaia rubida* (Davies and Trueman), *Anthracosia sp. (ovum/phrygiana),* *?A. caledonica* Trueman and Weir, and *Naiadites sp. nov.* cf. *productus* (Brown) [cf. Dix and Trueman 1931, fig. 6] were found between 1937 ft 6 in and 1932 ft; *Naiadites sp.* [juv.] occurred at 1917 ft, 7 ft. below the coal here correlated with the Kent No. 9 Seam. In the Snowdown Colliery No. 2 Underground Borehole (250502) the lowest beds penetrated were some 42 ft below the Kent No. 9 Seam, and yielded *Naiadites sp.* [juv.] and *Palaeolimnadiopsis* cf. *pruvosti* (Raymond).

The Kent No. 8 Seam is thought to be merged with the Kent No. 9 Seam throughout the present district. In the Chislet No. 35 Borehole *Naiadites?* and *'Estheria' sp.* occurred above the Kent No. 8 Seam, while in Snowdown No. 2 Borehole this seam was succeeded by 4 ft of mudstone with plants, overlain by silty mudstone with *Naiadites sp.* [juv.] and *N.* cf. *productus* [juv.].

Kent No. 7 Seam. Most of the coal mined at Chislet Colliery has come from the Kent No. 7 or Chislet No. 5 Seam. The workings in the West District of the pit extend on to the Canterbury (289) Sheet, and in this area the seam is normally between 4½ and 6½ ft in thickness[1]. The coal is generally bright and friable, but a band of hard dull grey coal up to 10 in thick is usually present about 2 ft above the floor. Two partings of friable carbonaceous shale or 'rashes' are commonly present above the dull coal; in the West District the thickness of the upper parting is variable and may exceed 3 ft; the lower parting is thin and less than 1 in thick in places.

In the Snowdown No. 2 Borehole four coals, two of which were split seams, occurred within 75 ft above the top of the Kent No. 7 Seam; in the beds above the lowest coal were found fragments of *Naiadites?*, and above the third coal occurred *Spirorbis sp. Naiadites* cf. *productus, N.* cf. *obliquus* Dix and Trueman, and '*Estheria*' (*Palaeolimnadiopsis?*) *sp.*

Snowdown Marine Band. The fourth seam above the Kent No. 7 Seam in the Snowdown No. 2 Borehole was split, its upper leaf having an abnormally high sulphur content (4·23 per cent). In its roof aggregates of rod-like pellets about 0·2 mm long, some showing a suggestion of a longitudinal furrow, occurred in silty mudstone, and were identified as cf. *Tomaculum.* They were at the same horizon as that at which *Trigonoglossa sp.* was found in the Snowdown No. 1 Borehole (p. 24), and similar bodies have been collected at the same level in boreholes in Chislet Colliery and Betteshanger Colliery. In boreholes elsewhere in the coalfield this band has yielded other fossils, including *Planolites ophthalmoides* Jessen, *Lingula mytilloides, Orbiculoidea* cf. *nitida* (Phillips), *Pernopecten carboniferus* (Hind), cf. *Geisina sp.* and *Hollinella* cf. *bassleri* (Knight). This marine horizon is succeeded some 40 or 50 ft above by a strongly developed marine horizon which is taken to be the Lower Tilmanstone Marine Band, the intervening strata including several beds of seatearth. In the Snowdown No. 1 Borehole *Anthraconaia sp. nov.* cf. *pruvosti* (Chernyshev) and *Naiadites* cf. *obliquus* occurred 26 ft above the *Trigonoglossa* level, while in the Snowdown No. 2 Borehole *Anthraconaia?* and *Naiadites sp.* were found at about the same horizon. Comparison of borehole sections suggests that the marine horizon with *Trigonoglossa* is distinct from the Tilmanstone Marine Beds, and it is proposed to name it the Snowdown Marine Band.

Lower Tilmanstone Marine Band. Bolton (1915, pp. 654–5) recorded the presence of marine fossils in a 5-ft bed of shale in the Walmestone Borehole. Specimens in the Geological Survey collection from depths between 1724 ft 6 in and 1728 ft 6 in contain *Lingula mytilloides* and *Orbiculoidea* cf. *nitida.*

Marine fossils in mudstone between 227 and 220 ft below the Kent No. 6 Seam in Snowdown No. 1 Borehole include : sponge spicules* ; *Campylites* [*Sphenothallus*] *sp.* ; *Crurithyris carbonaria* (Hind), *L. mytilloides**, *O.* cf. *nitida**, *Productus* (*Levipustula*) *rimberti* Waterlot*, *P.* (*Linoproductus*) *sp.**, *P.* (*P.*) *carbonarius** ; *Coleolus carbonarius flenuensis* Demanet, cf. *Platyconcha hindi* Longstaff ; *Aviculopecten delepinei* Demanet, *Edmondia sp.*, cf. *Nuculopsis gibbosa* (Fleming) [juv.], *Polidevcia acuta, P. attenuata* (Fleming), *Schizodus?* [juv.] ; coiled Nautiloid ; *Hindeodella sp.* ; 'fucoids'. Fossils marked with an asterisk were also found in the Snowdown No. 2 Borehole between 268 ft 11 in and 259 ft 11 in below the Kent No. 6 Seam, in addition to *Planolites ophthalmoides* ; *Campylites* [*Sphenothallus*] *stubblefieldi* (Schmidt and Teichmuller) ; *Palaeoneilo?, Schizodus antiquus* Hind and '*Cypridina phillipsi*' Corsin.

In the West Court Farm Borehole, between 3374 ft 7 in and 3369 ft 8 in, sponge spicules, *Campylites sp.*, crinoid columnal [circular], *L. mytilloides, O.* cf. *nitida, Productus* (*Levipustula*) *sp.*, turreted gastropods (*Donaldina?*), ?*N.*

---

[1] Details of coal seams have been taken from the reports of officers of the Coal Survey, National Coal Board.

gibbosa, fish remains and pyrite-filled burrows occurred; between 3368 ft 3 in and 3366 ft 9 in *C. stubblefieldi,* crinoid columnals, *Chonetes (Neochonetes) granulifer?* Owen, *Chonetes (Chonetinella?) sp., Lingula sp.* [juv.], *O.* cf. *nitida, Phricodothyris sp., Productus (Linoproductus) sp.* and a Rhynchonellid fragment were present; and between 3361 ft 9 in and 3361 ft 1 in, *Lingula sp.,* Nuculid?, *Pernopecten sp.* and ' fucoids ' were found.

The Upper Tilmanstone Marine Band has not been detected within the present district.

Faunas found in the measures above the Lower Tilmanstone Marine Band in boreholes at Betteshanger Colliery and St. Margaret's Bay tend to confirm the suggestion of Stubblefield and Trueman (1946, p. 277) that the Tilmanstone Marine Beds may be equivalent to the Mansfield Marine Band of Yorkshire. There is also some palaeontological support for the idea of these authors that parts of the *similis–pulchra* Zone and of the *phillipsii* Zone may be missing in Kent.

Stubblefield and Trotter (1957, p. 3) put the junction between the Middle and Upper Coal Measures in Kent at the base of the Sandstone Division, although this horizon is not everywhere readily determinable due to the local appearance of sandstones in the upper part of the Shale Division.

Upper Coal Measures

Kent No. 6 Seam. This seam can be recognized from Stodmarsh to West Court Farm. It is probably also the 4-ft coal proved at 2225 ft 6 in in the Brady Borehole. It is worked at Snowdown Colliery, where it was named the Millyard Seam because it was encountered at about 1000 yd below surface, with its base at 3011 ft in No. 2 Shaft and at 2940 ft 6 in in No. 3 Shaft. Within the colliery workings its thickness varies between 3 and 5 ft; it consists predominantly of bright coal, but a band of dull coal 7 to 11 in. in thickness is present 10 to 16 in above the base of the seam. Two dirt partings are commonly present, the lower near the middle of the seam and the upper generally about 7 in from the top; above the upper dirt band the coal is usually of inferior quality. The lower dirt band thickens to the west and south-west in the colliery workings and it is likely that a split develops at this level, for in the Bishopsbourne and West Court Farm boreholes two seams are present at about this horizon, 15 and 19 ft apart respectively.

In the mudstone above the Kent No. 6 Seam in Snowdown Colliery *Anthraconaia pruvosti* (Chernyshev), *Anthraconauta phillipsii* (Williamson), *A.* aff. *phillipsii,* ' *Estheria* ' *sp.* and *Euestheria simoni* (Pruvost) have been recorded. *Anthraconauta?* [juv.] and ' *Estheria* ' *sp.* were found above the upper leaf of this seam in the West Court Farm Borehole ; 40 ft higher, *A. phillipsii* and *A.* aff. *tenuis* (Davies and Trueman) occurred, followed 20 ft above by *Spirorbis sp., A. phillipsii, E. simoni, Carbonita humilis* (Jones and Kirkby) and *C. salteriana* (Jones). In the corresponding strata in the Tilmanstone Colliery No. 3 Shaft *E. simoni* and *Anomalonema reumauxi* (Pruvost) were associated with *Anthraconauta* cf. *phillipsii, A.* aff. *tenuis, Anthraconaia* cf. *pringlei* (Dix and Trueman) and *A. pruvosti* (Stubblefield 1933). Mr. Calver comments that the assemblage above the Kent No. 6 Seam is likely to represent the *tenuis* Zone. The occurrence of *Anthraconaia* cf. *pringlei, Anthraconauta phillipsii, A.* cf. *phillipsii, A. tenuis?, Euestheria simoni* and fish remains in the Chislet No. 37 Borehole between 71 ft and 105 ft 4 in leads to the correlation of the 2 ft 2-in seam from which this borehole started, with the Kent No. 6 Seam, and to the conclusion that the Stodmarsh Fault has a downthrow of 250 to 300 ft to the south-west in this vicinity.

Bolton (1915, p. 654) recorded " indeterminable species of *Anthracomya* " from a level between the Kent No. 6 and No. 5 seams in the Walmestone Borehole ; from a similar horizon in the Stodmarsh Borehole he recorded " *Anthracomya minima* ", and a specimen from 1278 ft was subsequently renamed *A. phillipsii* by Sir Arthur Trueman.

The Kent No. 5 Seam can be correlated with reasonable confidence in the north of the area, but in the Bishopsbourne and Adisham boreholes a second seam some 30 to 60 ft above becomes of increasing importance (1 ft 11 in at 2070 ft 4 in at Bishopsbourne and 1 ft 7 in at 2171 ft 11 in at Adisham). This seam appears to persist as the 1 ft 9-in split seam at 2813 ft 9 in in Snowdown No. 2 Shaft and the 1 ft 6-in coal at 3006 ft at West Court Farm. The No. 5 Seam, on the other hand, seems to be reduced to a 1-ft split seam at 2851 ft in the shaft and to a 1 ft 2-in coal at 3050 ft 11 in at West Court Farm Borehole.

The Kent No. 4 Seam persists from Walmestone and Littlebourne boreholes to West Court Farm Borehole, but it does not exceed 2 ft in thickness in this part of the coalfield.

Kent No. 3 Seam. This seam can be traced from Adisham (2 ft 1 in split at 1544 ft 5 in) to Bishopsbourne (where two seams are close together at about this horizon), and to Snowdown Colliery ; in the No. 2 Shaft the seam was 4 ft 2 in thick, including at 1-ft parting 2 ft above the base, which was at 2241 ft 6 in ; in the No. 3 Shaft section the parting was described as " hard stone bind ", 8 in thick. The 1 ft 1-in seam at 1893 ft 11 in in the Ropersole Borehole (228487), which Dines took to be the Kent No. 3 Seam, would appear to be more reasonably correlated with one of the thin seams between Kent No. 1 and No. 2 seams.

The Kent No. 2 Seam was probably only represented in this area by the 1-ft coal at 1964 ft 9 in in the Snowdown No. 2 Shaft and by a thin split seam at 2165 ft at West Court Farm.

Kent No. 1 Seam. The Kent No. 1 or Beresford Seam was formerly worked at Snowdown Colliery ; its base was at 1500 ft 6 in below surface in the No. 2 Shaft, and its thickness was 7 ft 10 in, including 2 ft of fireclay 1 ft 9 in below the top, and 1 ft 2 in of ' dirt ' 8 in above the base. In the No. 3 Shaft the seam thickness was 4 ft 6 in, the partings being described as ' bind ', the upper 3 in and the lower 9 in in thickness. Dines (1933, p. 40) recorded that the average seam thickness in the workings was between 3 and 4 ft, consisting of 1 ft 6 in or so of good coal at the top, underlain by a fairly persistent parting of variable thickness ; the lower part of the seam was locally clean, but one or more partings were present at irregular intervals.

The 4 ft 4-in seam at 1510 ft 10 in in the Fredville Borehole (245512) is taken to be the Kent No. 1 Seam, and may be equivalent to a split seam at 1634 ft 6 in in the West Court Farm Borehole ; the 4 ft 2-in seam at 1519 ft in the latter borehole may equate with one of the thin coals above the Kent No. 1 Seam at Snowdown and Fredville.

In cores from the Fredville Borehole Bolton (1915, p. 660) named '*Leaia tricarinata* " at 1544 ft below surface, some 33 ft below the Kent No. 1 Seam ; and about 55 ft above this seam, at 1451 ft below surface, he recorded " *Anthracomya Phillipsii, A. pulchra,* and *Naiadites elongata*(?) ".

It has not been possible to relate with certainty the Coal Measures penetrated in the Folkestone Borehole to those found elsewhere in the coalfield. The 2-ft seam at 1676 ft 6 in below surface may be the Kent No. 1 Seam, although in this case the Lower Tilmanstone Marine Band should have been reached in the borehole. At 2473 ft *Spirorbis sp., A. tenuis?,* and *Carbonita sp.* were present ; at 2242 ft *Anthraconauta sp.* ; at 2019 ft *A. phillipsii, A. tenuis* and *Carbonita sp.* ; at 1987 ft *A. tenuis?* ; at 1886 ft *A. phillipsii* ; at 1868 ft *A.* cf. *phillipsii* [juv.] and *A.* cf. *tenuis* and at 1723 to 1725 ft *Spirorbis sp., A.* cf. *phillipsii* and *A. tenuis?.* Mr. Calver reports that it is likely that the whole 750 ft of measures in which these shells occur should be referred to the *A. tenuis* Zone. The plants from the borehole were named by Kidston (1919, pp. 47–9), who referred the strata between 1490 and 3298 ft to the Staffordian Series, and provisionally to the Etruria Marl Group.

?TRIASSIC

In the Brabourne Borehole the strata between 1840 and 1921 ft below surface were described by Lamplugh and Kitchin (1911, pp. 51–4): no core was recovered from the first 5 ft, but they considered that the colour of the washings left " little doubt that marls of Triassic character set in suddenly beneath the ferruginous Liassic rock-band and that transitional strata of the Rhaetic type were absent." Fragmentary cores of " hard lumpy yellowish-mottled marl " were recovered from 6 or 8 ft below the base of the Lias, and between 1853 and 1889 ft " Marls of variable colour, greenish- and yellow-mottled, greenish, and deep red " were found. Some bedding planes were marked by sandy and gritty streaks which were roughly horizontal and did not diverge noticeably from the bedding planes of the Jurassic strata. Lamplugh and Kitchin remarked that by " an increase in the thickness and coarseness of the gritty layers the marls passed gradually downward into the pebbly conglomerate forming the base of the Mesozoic sequence " ; they further described this conglomerate as: " Red and grey conglomerate of well-rounded and partly-rounded pebbles of limestones of several kinds, red and grey quartzites, limy sandstone, and red and black chert, in a matrix, sometimes scanty, sometimes abundant, of red and grey gritty marl, and occasional irregular intercalations of similar marl: yielding good cores: the pebbles mostly small, but ranging up to over two inches in diameter ", 32 ft thick, with its base at 1921 ft.

Etheridge (1900, p. 733) recorded 48 ft 4 in of " Triassic Conglomerates " in this borehole, while Dawkins (1905, p. 30) identified the beds between 1873 ft 5 in and 1921 ft 5 in as Dolomitic Conglomerate ; this correlation was accepted by Lamplugh and Kitchin. Pringle and Stubblefield (in an appendix to Lees and Cox 1937, p. 190) noted the resemblance of the Brabourne conglomerate to one found at the bottom of the Henfield Bore-hole, which they considered to be younger than Middle Devonian, and possibly of Lower Carboniferous age ; there is, however, no direct evidence of the age of the Brabourne conglomerate or of the overlying marls. Samples from 1891 ft and 1898 ft were examined for spores by Dr. W. G. Chaloner, but none was found.

JURASSIC

In the following account the greater part of the information is taken from Lamplugh and Kitchin (1911) and Lamplugh, Kitchin and Pringle (1923). Figures for the thicknesses of the Jurassic formations that occur in the present area are given in Table 1. The figures for Snowdown Colliery are from Ritchie (1920, pp. 162, 178), for Lamplugh, Kitchin and Pringle did not publish a separate account for Snowdown, but relied on the Fredville

TABLE 1

Jurassic formations proved in boreholes and shafts: Thicknesses in feet

Name of borehole or shaft	Height of ground surface above O.D. in feet	Thickness of post-Jurassic rocks	Purbeck Beds	Portland Beds	Kimmeridge Clay	Upper Corallian	Corallian Limestone	Lower Corallian	Oxford Clay	Kellaways Beds	Cornbrash	Forest Marble	Great Oolite Limestone and Fuller's Earth	Inferior Oolite	Upper Lias	Middle Lias	Lower Lias	Thickness of pre-Jurassic rocks penetrated
Abbot's Cliff	87	537	—	—	153	87+												
Adisham[5]	133	995	—	—	—	—	—	—	81	34⅓	19¾	8½	61½	—	4	5	3½	2052
Bishopsbourne	c. 185	1030	—	—	—	—	—	—	135		c. 54		72	—	12	8	?2	1922
Brabourne[2]	215	612	68	31	262	162	134	46	173	18		13	137	44	15	45	80	164
Chilham	c. 75	666	—	—	—	—	c. 76		c. 165	c. 16	c. 16	c. 10	c. 134	—	—	c. 22		182
Elham	275	719	—	—	199	78	129	45	130		68		123	8	41	20	38	748
Ellinge[1]	466	859½	67½	—	189¼		153		198½			100		54		54		129⅓
Farthingloe	c. 230	709	—	—	32	60+												
Folkestone	113	553	—	—	210	123	123	c. 81	c. 172		c. 38		c. 75		55	17	40	1913
Fredville	259	1035	—	—	—	—	—	54	128	43	21	14	c. 67	—	c. 9		c. 4½	459½
Harmansole	c. 200	814	—	—	—	—	87	43	164	14	15	8	86	—	—	14½	3½	481
Hothfield	200	724	68	17+														
Littlebourne[6]	135	1021½	—	—	—	—	—	—	—	—	—	—	38	—	—	—	—	1558½
Lower Standen	273	723	—	—	210	112	31+											
Ottinge	300	710¾	—	17¼	108½+													
Ropersole[3]	c. 400	1090	—	—	—		158½		189½			115				21½		554½
Shakespeare (or Dover) Colliery[2] ...	50	516	—	—	44	96	126	c. 73	c. 102	36	13	18	69	27	8¾	21¼	8	1150
Snowdown Colliery[4]	242	1010	—	—	—	—	—	43⅓	127		39		91½	—		16½		1698
Stodmarsh	87	1017½	—	—	—	—	—	—	—	—	—	—	?43	—	—	—	—	?1203
Trapham	59	c. 998	—	—	—	—	—	—	—		c. 40		c. 86	—	—	—	—	2102

Thicknesses of Mesozoic formations are taken from Lamplugh, Kitchin and Pringle 1923, with the following exceptions: [1] is from Dawkins 1905, p. 30; [2] are from Lamplugh and Kitchin 1911, the Shakespeare Colliery record being modified to accord with Lamplugh, Kitchin and Pringle 1923, pp. 21–2, 208, 211; [3] is from Dawkins 1913, pp. 372–3; [4] is from Ritchie 1920, pp. 162, 178; [5] is from Pringle 1928; [6] is from Geological Survey records. (See also pp. 30, 32.) c. signifies 'about'.

Borehole record, which they stated differed little from the shaft section; however, the Fredville figures do not accord with the thicknesses in the shaft as given by Ritchie, which are again different from those used by Brown (1923, fig. 7, pl. 1). The Ellinge Borehole (239428) figures are taken from Dawkins (1905, p. 30), whose account has been preferred to one by Ritchie (1920, p. 83), from which it differs, although both versions were based upon the log stated to have been kept by M. Ludovic Breton.

The records adopted for the Ellinge and Ropersole boreholes are couched in terms of Kimmeridgian, Corallian, Oxfordian, Callovian, Bathonian and Bajocian. These terms are evidently not European Stage-names strictly used, including as they do the term ' Corallian ', and it is here assumed that they are equivalent to the British formational names Kimmeridge Clay, Corallian, Oxford Clay, Kellaways Beds, Great Oolite Series and Inferior Oolite.

The fossil names have been revised by Dr. R. Casey and Dr. H. C. Ivimey-Cook.

LOWER JURASSIC OR LIAS

The thickest known development of the Lias in this area is at Brabourne, where Lamplugh and Kitchin (1911, pp. 49–51), with some uncertainty, put the thickness at 140 ft, against the 172 ft 9 in recorded by Etheridge (1900, p. 733). The formation thins to the north-east and is absent below the north-east corner of the Canterbury (289) Sheet area, where it is overstepped by younger Jurassic strata. At Stodmarsh " Tough dark blue clay " recorded between 1060 and 1066 ft was thought by Lamplugh, Kitchin and Pringle (1923, p. 175) possibly to belong to the Coal Measures, as Lias was unlikely to occur in the vicinity.

Lamplugh, Kitchin and Pringle found that elsewhere in the district the three major divisions of the Lias, Lower, Middle and Upper, were represented, except at Chilham and Harmansole, where the Upper Lias was thought to be absent, but parts of each division were shown palaeontologically to be missing in different places. The available information is necessarily imperfect, however, being based on meagre borehole cores. At Ellinge, Ropersole and Snowdown the Lias is not divided in the records available. It may be noted that Lamplugh and Kitchin (1911, p. 142) correlated the uppermost Lias bed at Shakespeare Colliery[1] (295393) with a bed at 1545 ft in the Ropersole Borehole, 8 ft above the top of the Lias as described by Dawkins.

Lower Lias

The Lower Lias consists mainly of grey to black smooth clay or shale, or silty shale, commonly calcareous and shelly, with thin beds of earthy or shelly limestone and occasional brown nodules. These clays belong to the *Prodactylioceras davoei* [" Capricornus "] Zone, including Brabourne, where the clay is 60 ft thick, the " Striatus " Zone of Buckman (1910, p. xvi). The zones of *Tragophylloceras ibex* (which includes Buckman's *valdani* Zone) and of *Uptonia jamesoni* may be represented by ferruginous sandy shale and muddy limestone at Brabourne, by iron-shot marl at Elham and by limestones at Folkestone and Shakespeare Colliery, but in the north-east these

---

[1] Shakespeare Colliery is also known as Dover Colliery.

zones seem to be missing. At the base of the formation a bed of pebbles, including many of Coal Measures sandstone, was found at Folkestone, Elham and Adisham, but the basal bed at Brabourne consists of brown iron-shot muddy sandstone, and at Shakespeare Colliery No. 3 Shaft it is a 6-inch bed of hard, rather sandy limestone. The Coal Measures surface is fresh and unstained.

The presence of Lower Lias at Bishopsbourne was not proved palaeontologically, but was suggested by the lithology of the 2 ft of shale immediately above the Coal Measures.

Middle Lias

In its thickest development, at Brabourne, the Middle Lias succession is briefly as follows (Lamplugh and Kitchin 1911, p. 50[1]): Dark green pyritous sandy rock with bitumen-stained joints (like Marlstone), 14 ft ; on greenish grey limestone, 17 ft ; on dark greenish blue micaceous shale, 9 ft ; on iron-shot ferruginous limestone, 5 ft. Lamplugh, Kitchin and Pringle (1923, p. 197) commented on the close lithological resemblance of the shales to some of the greenish micaceous shales of the *Amaltheus margaritatus* Zone in the Midland counties. They noted a broad division into a shaly facies below and a more pronounced limestone facies above, and suggested that, despite a lack of ammonites, it might be " permissible to allocate the shales below to the Margaritatus Zone, the limestone-series above to the Spinatus Zone, the boundary between them being as yet uncertain ". They thought that both zones are probably present at Brabourne, Folkestone and Shakespeare Colliery ; at Bishopsbourne, where the Middle Lias consists almost wholly of dark hard shelly limestones and calcareous shales, they considered (1923, p. 151) that the formation is probably incomplete, due either to non-deposition or to contemporaneous erosion ; at Harmansole (op. cit., p. 117) the limestones rest directly on Lower Lias clays of the *P. davoei* [" Capricornus "] Zone, suggesting to them that the *A. margaritatus* Zone is absent. However, these opinions were challenged by Ager (1954), who identified *Gibbirhynchia muirwoodae* Ager and *G.* cf. *amalthei* (Quenstedt) from borings and shafts in the Kent Coalfield. He concluded that only the *A. margaritatus* Zone is present and that a widespread break occurs at the top of the Middle Lias, due presumably to post-Domerian erosion.

Upper Lias

The Upper Lias is thickest at Folkestone. The uppermost 30 ft were not cored, but were described as dark blue tough clay with bands of hard rock ; below this, 6 inches of grey shelly limestone are underlain by 10½ ft of brownish shaly clay with silty seams, on 6 ft of greenish brown clay, on 8 ft of dark grey sandy limestone. Lamplugh, Kitchin and Pringle (1923, p. 33) assumed that the *Hildoceras bifrons* [" Communis "] Zone might

---

[1] Lamplugh, Kitchin and Pringle (1923, p. 51) considered that, due to lack of palaeontological evidence, part of the Upper Lias at Brabourne might have been included with the Middle Lias; there was no sharp lithological break at the Middle Lias–Upper Lias junction at Folkestone, Elham or Shakespeare Colliery.

be represented in the uppermost 30 ft; the first cores below were in the
Harpoceras falcifer [" Serpentinus "] Zone, and the impure limestone at the
base was referred with some doubt to the *Dactylioceras tenuicostatum*
[" Annulatus "] Zone.

At Elham 8½ ft of dark bluish grey clay and pale brownish clay, with
phosphatic pellets, (*H. bifrons* Zone), rest on 9½ ft of dark shaly clay with
thin silty streaks, on 8 ft of brownish green iron-shot shelly marl, on 7 ft of dark
shaly clay and greenish grey clay (*H. falcifer* Zone), underlain by 8 ft of dark
grey limestone with marly bands and bitumen-coated joints (*D. tenuicostatum*
Zone).

At Brabourne the *D. tenuicostatum* and *H. falcifer* zones were not detected,
but the *H. bifrons* Zone is represented, and the *Grammoceras striatulum* Subzone of the *Grammoceras thouarsense* [*Lytoceras jurense (pars)*] Zone is
also probably present. Only the *D. tenuicostatum* Zone and the *Harpoceras
exaratum* Subzone of the *H. falcifer* Zone were recorded from Shakespeare
Colliery. At Snowdown Colliery and Fredville, the *D. tenuicostatum* Zone alone
was found. The Bishopsbourne Borehole revealed the *H. falcifer* Subzone in 4 ft
of brown speckled marlstone at the top, with the *H. exaratum* Subzone below,
and possibly a thin development of the *D. tenuicostatum* Zone at the base. The
Chilham Borehole may have been devoid of Upper Lias, although this is
uncertain owing to the poor core recovery.

MIDDLE JURASSIC

INFERIOR OOLITE

The strongest faunal evidence of the occurrence of Inferior Oolite strata
in the present district was found at Shakespeare Colliery, where 27 ft of grey
shelly grit, with indurated calcareous bands, and with small phosphatic
nodules, pebbles and subangular fragments of vein-quartz near the base,
yielded no ammonites but a prolific lamellibranch fauna; this suggested
that the deposit was of Upper Inferior Oolite age, about the *Garantiana
garantiana* Zone or somewhat later (Lamplugh, Kitchin and Pringle 1923,
p. 200). Many of the shells had suffered abrasion, indicating that they
had been subjected to the action of strong, sand-laden currents.

Arkell (1933, p. 247) remarked that " The unconformable relations of
these sandy beds to the Lias, without the intervention of any Lower or
Middle Inferior Oolite strata, is in harmony with the palaeontological
evidence in pointing to their having been laid down during or after the
Vesulian Transgression."

At Brabourne, 44 ft of apparently unfossiliferous greyish blue muddy oolitic
limestone were taken to be possibly of Inferior Oolite age by Lamplugh and
Kitchin (1911, p. 48). This rock had a pebbly or brecciated layer at the base.
Again, at Elham, 7 ft of dark grey muddy limestone with seams of clay on 1 ft
of sandy marl and indurated calcareous sandrock with small quartz pebbles,
were allotted to the Inferior Oolite, without palaeontological support, while at
Folkestone no cores were taken from this part of the succession.

GREAT OOLITE SERIES

Fuller's Earth. In the Brabourne Borehole the Inferior Oolite is overlain by 23 ft of " Dark grey or bluish calcareous shales, with pyrites and

some ill-preserved shells, *Astarte, Pteria,* &c.; towards the base interbedded with thin brecciated bands of limestone . . ." These strata were uncertainly referred by Lamplugh and Kitchin (1911, p. 48) to the Fuller's Earth. At Shakespeare Colliery the basal beds of the Great Oolite Series consist of alternations of consolidated grey sand and sandy limestone, but in the absence of diagnostic fossils Lamplugh and Kitchin did not attempt to correlate them. Later, these authors and Pringle (1923, p. 47) found 23 ft 6 in of muddy and sandy limestone and marly clays at Elham, which were of Fuller's Earth aspect and yielded *Myophorella* [*Trigonia*] *signata* (Agassiz). They (1923, pp. 200–1) decided, however, to group all the rocks between the Inferior Oolite and the base of the Forest Marble under the term 'Great Oolite'. Arkell (1933, p. 322), on the other hand, accepted the resemblance of these shales at Brabourne to the Fuller's Earth facies of Dorset and Somerset, and correlated the sands and sandy limestones seen below the Great Oolite Limestone in sections north and east of Brabourne with the sandy facies of the Chipping Norton Limestone.

In the Fredville Borehole some 17 ft of sand and sandstone below the Great Oolite Limestone (Lamplugh, Kitchin and Pringle 1923, pp. 92, 103) were assumed to be contemporaneous with similar sediments in the Snowdown Colliery shafts which yielded a fauna probably of Lower Fuller's Earth age. At Bishopsbourne no sandy basement-beds are developed, although the lowest limestones are rather muddy, rubbly and streaked with clay. At Harmansole sandy limestone passes down into sandy marl, with a clayey and pebbly base, and rests on Middle Lias. The 'Great Oolite' section at Chilham shows an apparent lack of sandy basement beds, possibly due to imperfect core recovery.

Great Oolite Limestone. The Great Oolite Limestone consists of massively bedded cream or grey oolitic limestone, generally fine in grain, with occasional beds of clay or marl.

At Brabourne the succession is as follows: 49 ft of pale grey fine-grained oolitic limestone, sparingly mixed with grit and clay in places, on 7 ft of cream sandy limestone, on 25 ft of cream oolitic limestone, on 33 ft of greyish fine-grained rather sandy oolitic limestone, with small shells abundant in places.

At Chilham the limestones are about 134 ft thick, assuming that 10 ft of core missing in the lower part are not sandy basement beds. A 3-ft bed, consisting of dark greenish grey clay above, on a layer of dark grey shell marl, with blue-black clay with silty laminae and white shells below, has its base 41 ft below the top and a 1-ft bed of dark bluish grey loamy clay occurs 20 ft lower down. At Harmansole evenly-bedded creamy-white and bluish grey finely oolitic limestones, shelly in places, 79 ft thick, underlain by 7 ft of hard sandy, sparingly oolitic, grey limestone and sandy marl, make up the 'Great Oolite', and at Bishopsbourne the limestones were recorded as 72 ft thick, being dingy, shelly and streaked and mixed with clay and dark rubbly oolite in the basal 7 ft.

The Adisham Borehole showed $60\frac{1}{2}$ ft of limestone with thin marly bands, resting on 1 ft of black shelly clay and dark sandy clay with limestone pebbles. At Littlebourne, chip samples of grey clay with ooliths and of bluish grey oolitic limestones were taken to represent the Great Oolite Series, 38 ft thick, as against an estimated 86 ft of 'Great Oolite' at Trapham and 43 ft at Stodmarsh (neglecting the basal 6 ft of dark blue clay of uncertain age). The 'Great Oolite' is absent at Walmestone, having been removed by pre-Cretaceous planation.

At Fredville 50 ft of limestones were found, with a 6-inch bed of marlstone about the middle, while at Ropersole Dawkins (1913, p. 373) allotted 115 ft of shelly and oolitic limestones to the Bathonian and Bajocian, although the Inferior Oolite probably does not extend so far. The Elham section showed 99 ft 6 in of limestones above the supposed Fuller's Earth.

Forest Marble. " A somewhat abrupt change of physical conditions towards the close of Great Oolite time gave rise in this district, as in Oxfordshire and in Bas Boulonnais, to the accumulation of clayey, muddy and impure calcareous deposits of variable character constituting the Forest Marble " (Lamplugh, Kitchin and Pringle 1923, p. 203). These deposits consist predominantly of clay and marl, usually with a green tinge, with irregular bands of limestone, calcareous sandstone and oolitic rubble. Much of the limestone is peculiarly compact. Traces of lignite are commonly present.

In the area here considered this formation ranges in thickness from 8 to 18 ft, and Arkell (1933, p. 321) suggested that this uniformity indicated " that the north-easterly attenuation of the Great Oolite Series as a whole is not due to overstep by the Cornbrash. Rather is it a piecemeal overlap and thinning against a shore-line." Arkell further remarked that no trace of Bradfordian fossils was found in Kent, and that it was therefore likely that this Forest Marble was older than Bradford Clay, that it was of Kemble Beds age, and that in some of the borings part of the Forest Marble might belong to the *fimbriata–waltoni* Beds: there was thus a non-sequence below the Cornbrash in Kent.

The greatest recorded thickness of Forest Marble is at Shakespeare Colliery, where 4 ft of pale yellowish oolitic limestone rest on 3 ft of dark greenish grey clay marl, with shell-crumbs and patches of oolite-grains, on 3 ft of compact pale greenish calcareous claystone, interpenetrated by green clay in ramifying finger-like processes, on 8 ft of compact bluish grey calcareous claystone, with darker somewhat oolitic bands, containing fragments of lignite and small pale brown phosphatic nodules. At Brabourne, 7 ft of greyish white calcareous sandstone with bitumen-stained joints, and sandy limestone, rest on 6 ft of greenish blue sandy shale with thin intercalations of muddy oolitic limestone at the base.

At Adisham, the formation is represented by marly clay, greenish black and grey in colour, with one 6-inch bed of greenish clay and another of creamy oolitic limestone, the latter mixed with greenish clay. At Fredville, however, 6 ft of limestone are developed at the top of the 14-ft section, and at Chilham the formation consists predominantly of pale patchy limestone with dark blue nodular lumps, 9 ft, which rests on smooth grey marlstone, 1 ft. At Harmansole, the record showed: dirty creamy-white calcareous clay with greenish patches, $3\frac{1}{2}$ ft; on dark shelly clays, becoming black in places, with lignite, $1\frac{1}{2}$ ft; on soft oolitic marly limestone with bluish grey patches and bits of lignite, 3 ft.

CORNBRASH

The Cornbrash in East Kent was stated by Lamplugh, Kitchin and Pringle (1923) to consist generally of a variable impure limestone below, which yielded a rich assemblage of lamellibranchs, brachiopods and echinoids, and a thin argillaceous bed above, which contained lamellibranchs, but few brachiopods and no echinoids. Arkell (1933, p. 339) preferred to group this sandy clay with the Kellaways Beds, but for simplicity the records are here quoted as given by Lamplugh, Kitchin and Pringle. In the absence of fossils these authors found difficulty in placing the Forest Marble–Cornbrash contact.

At Brabourne the cores were incomplete at the Cornbrash horizon, but Lamplugh and Kitchin (1911, pp. 47, 200) considered that the formation is probably thin. At Shakespeare Colliery they doubtfully allocated to the Cornbrash 13 ft of bluish grey streaky sandy clay, of marly structure, with masses of indurated calcareous rock, small soft brown nodules and markings like borings.

At Fredville Borehole, Lamplugh, Kitchin and Pringle (1923, p. 91) recorded that the Cornbrash comprised 17 ft of fine-grained pale oolitic limestone, rather muddy in the lower part, resting on 4 ft of strata which were represented only by washings consisting mainly of medium-grained sand. This section is remarkable for the absence of the impure clay commonly present at the top of the Cornbrash, and for the comparative purity of the limestone, which resembled that of the Great Oolite Limestone, and failed to yield fossils characteristic of the Cornbrash. A similar limestone was found at Adisham Borehole by Pringle (1928, p. 79), who referred the following strata to the Cornbrash: grey clay, 8 in; on greyish white finely oolitic limestone, becoming marly below, 9 ft; on dull grey oolitic shelly limestone; 'casts' of shells filled with coarsely crystalline calcite; limestone mixed with dark marl at base, 4 ft; on hard cream-coloured limestone sprinkled with small brown grains of limonite, 5 ft; on dark grey muddy nodular limestone with dull brown specks of limonite, 1 ft.

Cornbrash is probably absent at Stodmarsh and Littlebourne, but it may be present at Bishopsbourne and at Trapham (Lamplugh, Kitchin and Pringle 1923, pp. 148, 174). At Harmansole the section is typical (Lamplugh, Kitchin and Pringle 1923, p. 113): thin grey sandy clay with *Meleagrinella* [*Pseudomonotis*], rests on hard greyish white calcareous sandrock, traversed by irregular wisps of dark marly clay, 5 ft; on brownish sandy marl with crushed *Pholadomya*, 4 ft; on greyish white nodular limestone with courses of pisolitic marl, *Obovothyris* [*Ornithella*] *obovata* (J. Sowerby) abundant, 6 ft.

The section at Chilham was marred by disturbance of the cores, but after some re-arrangement Lamplugh, Kitchin and Pringle (1923, p. 122) presented the following account: olive grey calcareous loam, about 3 ft; on pale greyish close-textured limestone, about 3 ft; on darkish grey oolite rubble, mixed and dappled with clay, with a firmer limestone band towards the top, 10 ft.

UPPER JURASSIC

KELLAWAYS BEDS

In the south and south-east of East Kent, Lamplugh, Kitchin and Pringle (1923) found the Kellaways Beds to consist mainly of impure marly sandstone with some ferruginous beds; farther north the formation was composed mainly of ferruginous marlstone, in places highly glauconitic, and in this area beds of ore-quality might occur locally. The iron was mainly in the form of polished coffee-coloured limonite ooliths termed 'millet-seed grains' or 'iron-shot'; in this respect the Kellaways Beds resembled the Upper Corallian iron ore, but their content of iron was lower and that of deleterious matter was higher. Lamplugh and his co-authors considered the Kellaways Beds readily identifiable, both lithologically and palaeontologically, but found the upper and lower limits sometimes difficult to define, which might account for some of the anomalous variations in thickness recorded for the formation.

Palaeontologically the Kellaways Beds in this district were found to be characterized by an abundance of *Gryphaea* of several species, whereas ammonites were rare. The greater part of the formation probably belongs to the *Sigaloceras calloviense* Zone, and some evidence was obtained that the *Proplanulites koenigi* Subzone is present at least locally at the base.

In the Brabourne Borehole, Lamplugh and Kitchin (1911, p. 46) allotted 18 ft of dark grey calcareous sandstone to the Kellaways Rock, with the reservation that the Cornbrash might also be represented. In the lower part the sandstone is interbedded with impure limestone, ferruginous and iron-shot layers, bands of fragmental-looking green loam, and thin shaly partings. The Kellaways Rock section at Shakespeare Colliery is as follows:

	Ft
Clayey rock with nodules, quartz grains and brown ferruginous oolitic grains; with 'casts' of shells so abundant as almost to form a lumachelle	1
Hard yellowish brown calcareous claystone with dark brown ferruginous or phosphatic grains, passing down into similar rock intermingled with blue-hearted concretionary clayey limestone; containing *Meleagrinella* and *Gryphaea,* but few other fossils ... ...5 ft to	6
Streaky dull grey and greenish loamy sand, dappled with dark clay, and thin tubular borings (?); with incipient concretionary structure, and irregular nodular hard bands in the lower portion. '*Belemnites*' and *Gryphaea* the most abundant fossils	24
Band of hard concretionary limestone in a matrix like the above. *Oxytoma, Myophorella,* etc.	2
Dull grey and greenish loamy sand, partly indurated, as above. No fossils seen	3

The uppermost bed described was assigned by Lamplugh and Kitchin (1911, pp. 27–8) to the Oxford Clay, but later, with Pringle (1923, p. 208), they re-classified the bed as Kellaways Rock, having concluded that the upward limit of the Kellaways Beds coincided with that of the conspicuous *Gryphaea* fauna. Specimens of *Myophorella* from near the base of the formation resembled those found at the Oxney Borehole (Dover Sheet 290) in association with an ammonite fragment which probably belonged to the *P. koenigi* Subzone; the presence of this zone at Dover is thus possible.

Dawkins (1913, p. 373) described the 'Oxfordian' and 'Callovian' succession in the Ropersole Borehole as grey clays and calcareous marls, with nodules containing fossils, 121 ft 6 in; on grey calcareous marlstone, with fossils, 8 ft; on grey sandy clay, 8 ft; on grey marly limestone, 2 ft; on sandy marl, 1 ft; on hard shelly ferruginous sandstone, 3 ft; on blue ferruginous oolite, 5 ft; on black sand, 6 ft 6 in; on calcareous sandy oolite, 3 ft 2 in; on sandy and ferruginous oolite, 31 ft 4 in. It is not clear what thickness should be allotted to the Kellaways Beds, but if sandiness be taken as a criterion they may account for 60 ft.

At Fredville, Lamplugh, Kitchin and Pringle (1923, p. 91) found the Kellaways Beds to consist of 2 ft of fine-textured muddy sandstone and sandy loam-rock, crowded with fossils, *Gryphaea bilobata* (J. de C. Sowerby) being particularly abundant; on 1 ft of hard shelly calcareous grit, sprinkled with yellow and brown iron-shot grains; on 13½ ft of hard olive-green glauconitic rock, with polished grit-grains and iron-shot; in places somewhat clayey and like green marlstone; on 1½ ft pale grey marlstone; on 2 ft glauconitic rock as above, but brighter green; on 8 ft rather coarse brownish sand; on 12 ft glauconitic rock as before, but mixed in places with dabs of black clay, and in places coarsely oolitic, with iron-shot and much polished grit ; ' *Pteria* ', ' *Rhynchonella* ', etc. ; on 2 ft hard green clay or marlstone crowded with coffee-coloured iron-shot ; dark green loamy ferruginous band at base, full of fossils ; on 1 ft of dull brown soft muddy sandstone. A selected specimen contained 34·59 per cent of iron ; the rock was described by Hallimond (*in* Lamplugh, Kitchin and Pringle 1923, p. 99) as a sandy oolitic iron ore, consisting of ooliths of ferric oxide, subangular sand-grains and small polished fragments of ' lydite ', with a few brown opaque particles and brown-stained fragments of crystalline siderite, all coated with chamosite and cemented with crystalline siderite.

In the Adisham Borehole the Kellaways Beds consist of brownish sandy shelly marl crowded with crushed shells of *Meleagrinella sp. nov.*, 1 ft ; on brownish grey calcareous sandrock with many iron-shot, 2 ft 3 in ; on soft brown sandrock with many iron-shot, 6 ft 8 in ; on greenish brown somewhat soft iron-shot sandrock, 12 ft 8 in ; no core, 6 ft 7 in ; on greenish brown sandy marl with plentifully scattered iron-shot, *Gryphaea*, etc. at base, 5 ft 2 in. The uppermost bed was included by Pringle (1928) in the Oxford Clay, but it has been re-classified on faunal grounds.

At Harmansole the Kellaways Rock is composed of 2 ft of brownish shelly iron-shot marl with *Gryphaea* on 12 ft of brownish and greenish brown sandy marl with iron-shot, passing down into soft brownish sandy marl ; while at Chilham, despite poor core recovery and some derangement of the cores, Lamplugh, Kitchin and Pringle (1923, p. 122) made out the following succession : olive-grey marlstone, streaked and dappled with rusty iron-shot grains, ?9 ft ; on dingy grey sandy limestone with ' pin-hole ' cavities and ' casts ' of shells, about 2 ft ; on brown iron-shot marlstone, greyish sparingly iron-shot calcareous loam and impure shaly limestone, about 5 ft.

OXFORD CLAY

The Oxford Clay is, according to Lamplugh, Kitchin and Pringle, the least variable of all the Jurassic formations in Kent ; it consists of greyish blue, fine-textured marly clay, without conspicuous bedding, but with some subordinate brown bands.

Lamplugh, Kitchin and Pringle recorded that the lowest division of the Oxford Clay, their " Ornatus Beds ", consists of brownish or greenish brown marly clay, characterized by *Kosmoceras* and a rich distinctive fauna of lamellibranchs, commonly preserved as brown-coated ' casts '. The fossils indicated that the *Kosmoceras jason* Zone is present at the base, succeeded by the *Erymnoceras reginaldi* horizon, with the *Kosmoceras duncani* horizon above.

The beds overlying those of the *K. duncani* horizon are less marly and of finer texture ; the basal part has a brownish tinge, the remainder being predominantly bluish grey. Here ammonites are preserved as ' casts ', commonly pyritized, indicating the zones of *Quenstedtoceras lamberti* and *Q. mariae* (*Creniceras renggeri* horizon) ; lamellibranchs are more scattered than in the preceding division.

The *Quenstedtoceras mariae* Zone, in the restricted sense used by Lamplugh, Kitchin and Pringle, is the uppermost Oxford Clay division ; it consists of smooth pale bluish grey clay, with ammonites preserved as pyritized ' casts '. Lamplugh, Kitchin and Pringle (1923, pp. 212–3) found

D

that at those localities in Kent where the nature of the beds below and above the Oxford Clay—Lower Corallian junction was ascertained, the pale bluish grey clay with fine brown filaments of the lower series was contrasted with the brown limestone and overlying ferruginous marlstone of the lowest Corallian beds. Below the junction *Quenstedtoceras* was found but not *Cardioceras* and *vice versa*. The strata of the *Cardioceras praecordatum* Subzone thus formed part of the Lower Corallian. Arkell (1933, p. 373) instead assigned the beds of this subzone to the Oxford Clay, although he remarked that they consisted largely of marls and marlstones full of iron-shot ooliths, very different from the typical clays in other parts of England. In order to avoid confusion the original classification is retained here.

At Brabourne 173 ft of pale bluish grey smooth marly clay, with semi-indurated claystone bands, represent the Oxford Clay. The thickness of the formation was estimated to be 135 ft at Bishopsbourne, where it was probably incomplete, being overlain by Cretaceous strata. At Elham, however, the full thickness is only 130 ft, according to the borer's record ; at Fredville, Lamplugh, Kitchin and Pringle put the thickness at 128 ft, contrasting with Ritchie's figure (1920, p. 178) of 127 ft for the combined Oxford Clay and Kellaways Beds at Snowdown Colliery. At Shakespeare Colliery the Oxford Clay was first allotted 88 ft of strata by Lamplugh and Kitchin, but later 1 ft was transferred to the Kellaway Beds (p. 38) and " some 15 ft. or 20 ft." were added at the top (Lamplugh, Kitchin and Pringle 1923, p. 211). The formation there consists of marly clay and marlstone, with two thin beds of glauconitic sandy calcareous rock.

CORALLIAN BEDS

Lamplugh, Kitchin and Pringle adopted a three-fold division of the Corallian in East Kent, namely Lower Corallian, Corallian Limestone and Upper Corallian. The Lower and Upper Corallian were composed mainly of clays and marls, marlstones and rubbly oolites, with subordinate bands of impure limestone ; in these beds there was often an admixture of small coffee-coloured granules of iron oxide, which in places were so abundant as to constitute beds of iron ore. The middle division consisted of pale coralline limestones.

Lower Corallian. The lowest 25 to 35 ft of strata consist largely of ferruginous olive-green and brown marlstones with iron-shot grains ; close to the base occurs a thin band of hard brown limestone with unevenly disseminated yellow ferruginous grains, while towards the top the presence of *Millericrinus* fragments suggested a correlation with some part of the ' Marnes à *Millericrinus* ' exposed near Boulogne (Lamplugh, Kitchin and Pringle 1923, p. 214).

Above the ferruginous beds the strata are mainly of pale grey calcareous marl and marlstone, with bands of limestone. A few feet above their base there is a well-marked thin bed of marly clay, slightly pisolitic and of a distinctive crumbling character. A few feet higher still, some 30 ft or more above the base of the formation, there occurs a thin bed of hard grey marlstone, with dark-coated ' casts ' of ammonites and lamellibranchs ; this bed, with its varied, abundant fauna, probably indicated a period of slow deposition.

The Lower Corallian yielded evidence of the *Cardioceras praecordatum* Subzone and of the *Perisphinctes plicatilis* Zone, but no trace of the *Cardioceras cordatum* Zone, indicating a possible non-sequence. Arkell (1933, pp. 438-9) considered that the beds above the *C. praecordatum* Subzone are equivalent to the Lower Calcareous Grit and part of the Berkshire Oolite Series.

At Brabourne the Lower Corallian comprises 11 ft of smooth pale slaty-grey or brownish calcareous clay and claystone, slightly sandy in places, with cuboidal marly fracture, resting on about 35 ft of thin bands of clay full of ferruginous oolitic grains, alternating with hard marl with similar grains, the clay increasing into preponderance downward. At Elham the thickness was put at 45 ft, not all represented by core, consisting of hard dark grey sandy limestone with dark grey marl below.

Corallian Limestone. This division is of uniform general character and thickness. Lamplugh, Kitchin and Pringle (1923, p. 217) stated that the upper part of the limestones might or might not comprise the whole of " Salfeld's Zone of *Perisphinctes wartae* " (see below), and might be surmised to reach to about the top of the Sandsfoot Clay of Dorset. Arkell (1933, p. 438), however, suggested that the limestones as a whole are the equivalent of part of the Berkshire Oolite Series, the Osmington Oolite Series and possibly part of the *Trigonia clavellata* Beds.

At Shakespeare Colliery the uppermost 20 ft are composed of " Creamy or greyish soft sandy limestone, with occasional layers of flaggy calcareous sandstone and of incoherent sandy shale ; and with rubbly bands mainly composed of rolled bits of shell, oolite grains, &c., containing many gasteropods, and *Pecten, Lima, Ostrea, &c.*" This rock passes down into " Coral Rag : irregular tabular masses of hard pale crystalline limestone, with hollows lined or filled with calcite crystals, usually in the interior of masses of coral or large gasteropods : around the hard masses, a softer rather sandy oolitic limestone full of fossils, and often with a strong bituminous odour ", about 33 ft ; on " Creamy-grey soft calcareous stone of sandy texture, containing few corals, but many shells and *Cidaris* spines ; with thin seams of very fossiliferous black shaly clay at the top and at the bottom ", about 12 ft ; on " Dark bluish muddy coral-limestone in hard tabular masses set in a softer calcareous matrix, with bluish-grey partings of calcareous silt : the coral masses frequently perforated by tubular borings and often converted in the interior into crystalline calcite : in limestone, the other fossils, few and poorly preserved, are mainly *Cidaris*-spines and the stem-ossicles of a large crinoid ; but in the clayey bands, *Terebratulae*, &c., are abundant ", about 61 ft (Lamplugh and Kitchin 1911, pp. 23-4).

Upper Corallian. The beds of this division are predominantly argillaceous, but at some levels the " clayey basis was mixed in varying and sometimes preponderant degrees with ferruginous, glauconitic, sandy, and calcareous ingredients " (Lamplugh and Kitchin 1911, p. 81). In certain beds the ferruginous content is so high as to constitute an iron ore.

Lamplugh, Kitchin and Pringle (1923, p. 191) found some suggestion that the lowest part of the Upper Corallian lies within " Salfeld's Zone of *Perisphinctes wartae* and *Cardioceras alternans* ' ". The iron ore they equated with the Westbury iron ore in Wiltshire. Arkell's opinion (1933, p. 437) was that the Upper Corallian is equivalent to the Glos Oolite Series and the Upper Calcareous Grit.

At Shakespeare Colliery, where the iron ore was first found in the Brady Borehole, a shaft provided the following Upper Corallian section (Lamplugh and
Kitchin 1911, p. 22) : oolitic limestone, 8 ft ; on alternations of calcareous claystone, clayey oolitic limestone, muddy grit, bands of marly clay and layers of
pisolitic rubble, 18 ft ; on grey clay with a few shining iron-shot grains, alternating
with brown gritty clay full of iron-shot, 7 ft ; on the ' millet-seed iron-ore ' :
small shining brown globules of iron carbonate crowded in a slightly clayey or
loamy matrix ; with some calcareous claystone concretions, 16 ft (base at 609 ft
below surface) ; on a hard concretionary rock-band with grains and rolled fragments of iron-ore, quartz-grit, lydites, etc., 3 in ; on sandy clay, glauconitic in
places, with streaks and bands of worn shell-fragments and pisolitic rubble,
7 ft 9 in ; on laminated clayey sandrock, showing current-bedding in places : much
perforated by boring organisms, f ft ; on calcareous claystone bands with thick
partings of grey and blue sandy clay, 12 ft ; on dark blue and grey clay with silty
layers and shelly layers, the latter particularly towards the base, 26 ft. The raw
iron ore contains on average 10·16 per cent water, 32·94 per cent metallic iron
and 14·96 per cent silica, while the dried ore contains 8·91 per cent lime
(Lamplugh, Wedd and Pringle 1920, p. 224).

At Brabourne, where the Upper Corallian is thickest, the succession is as
follows (Lamplugh and Kitchin 1911, p. 42): greenish grey glauconitic sandy
mudstone, with black specks, about 6 ft ; on blue-grey marly clay, about 14 ft ;
on ' millet-seed iron-ore ' : brown globules of iron carbonate in a marly matrix,
with ' casts ' of shells, and a few small lydites, 3 ft (base at 996 ft below surface) ;
on blue-grey marly clay, passing down by alternations into hard smooth pale
grey marlstone or argillaceous limestone, slightly pyritous in places ; fossils in a
few layers, but elsewhere scanty, 40 ft ; on layers of coarsely pisolitic pebble-like
rubble, alternating with bands of grey marlstone and clay partings, 4 ft ; on
banded pale grey marlstone or smooth argillaceous limestone, with few fossils,
38 ft ; on dark brownish grey fossiliferous marl, 31 ft ; passing down into greenish impure sandy oolitic limestone, becoming coarsely oolitic in a 3-ft band
toward base, 10 ft ; on dark brownish grey fossiliferous marl, as before, 16 ft.
The thickness of the iron ore has thus decreased to 3 ft.

Several boreholes were made to test the extent of the iron ore deposit about
Dover. At Farthingloe (287401) the Upper Corallian strata consist of pisolitic
greyish marl, blotchy whitish grey marl, dark iron-shot clay and pale smooth
claystone, together 34 ft thick, on 15 ft of ' millet-seed iron-ore ' (base at 790 ft),
on 11 ft of dark clay with thin sandy streaks. At Abbot's Cliff (266385) two
beds of iron ore were recorded, 3 ft 7 in thick at 715 ft 7 in below surface,
and 7 ft 10 in thick at 749 ft 6 in, but at Lower Standen (238399) only layers of
iron ore 2 to 3 in thick occur in 13 ft of brownish marly clay. In the coal
borehole at Folkestone the iron ore appeared to be represented by ferruginous
grains and lumps in 5 ft of hard brownish grey limestone, while at Elham iron
ore was absent from the Upper Corallian strata, which consist largely of clay
and marly clay, with 11 ft of pisolitic marl at the top. Assuming an average
thickness of 9 ft over an area of 6 square miles, reserves of the iron ore were
put at approximately 100 million tons (Lamplugh, Wedd and Pringle 1920,
p. 224).

KIMMERIDGE CLAY

The Kimmeridge Clay was divided by Lamplugh and others (1911, 1923)
into Lower Kimmeridge Clay, characterized by the presence of *Exogyra
virgula* (Defrance), and Upper Kimmeridge Clay, which was without *E.
virgula* but yielded *Modiolus autissiodorensis* (Cotteau) ; the Lower Clays

were more arenaceous and calcareous and of a more shallow-water facies than the Upper Clays.

A full thickness of the Kimmeridge Clay was proved only in the Brabourne Borehole within the present district. The Upper Clays are 64 ft thick, represented predominantly by smooth pale bluish grey or brownish calcareous mudstone, with darker clay and associated beds of argillaceous limestone both at top and bottom. The Lower Clays are 198 ft thick, consisting of dark grey to bluish grey clay, often calcareous or sandy, interbedded with sandy limestone and sandy marl. At Ottinge, as at Brabourne, Portland Beds rest non-sequentially on the Kimmeridge Clay. Elsewhere, the Kimmeridge Clay, where present, is overlain by Cretaceous strata, the stratigraphical horizon of the sub-Cretaceous unconformity becoming progressively lower from west to east ; it lies within the Upper Clays at Ottinge, but at Lower Standen it was thought to be near the top of the Lower Clays which are there 210 ft thick, and at Abbot's Cliff the thickness of the Kimmeridge Clay has decreased to 153 ft.

The Kimmeridge Clay section at Shakespeare Colliery is as follows (Lamplugh and Kitchin 1911, p. 22): blue clay with 'casts' of fossils and a 6-inch nodular band of limestone, 4 ft ; on green, greenish blue, and dingy brown glauconitic sandy loam and clay, with numerous concretionary hard bands of limestone and claystone ; the whole much dappled and streaked with coarse grit, lydites and glauconite, which often fill tubular borings, 10 ft ; on clayey greensand and hard calcareous greensand rock, 6 ft ; on dark and pale blue and brownish clay with hard calcareous rock-bands ; dappled and streaked with glauconite and polished grains of grit and lydite, 15 ft ; on firm smooth pale brown marly clay with thin layers of pisolitic rubble, shining 'millet-seed' ferruginous grains and hard concretionary rock, 9 ft. Lamplugh and Kitchin (1911, p. 153) concluded that these 44 ft of strata are stratigraphically equivalent to 138 ft of beds at Brabourne, demonstrating that the eastward attenuation of the formation is due to reduced sedimentation, as well as to erosion.

Portland Beds

Lithological and palaeontological evidence caused Lamplugh, Kitchin and Pringle (1923, pp. 228–9) to conclude that the Portland Beds in East Kent are equivalent to the Portland Stone of the Dorset succession, the Portland Sand being absent, and that the eastward thinning from Penshurst, in West Kent, where the formation is 131 ft thick, is due partly to less copious deposition and partly to the absence of some of the lower beds at the more easterly localities. Arkell (1933, p. 517) suggested that the transgressive stratum might be at the horizon of the Upper Lydite Bed and Glauconitic Beds of Wiltshire, which are equivalent to the upper part of the Portland Sand, thus indicating a reduced non-sequence in Kent. On the other hand Taitt and Kent (1958, p. 19) considered that the Portland Stone is absent in the Wealden and adjoining areas, and that this non-sequence is due to cessation of deposition rather than to subsequent erosion.

In the Hothfield Borehole (980460) 17 ft of Portland Beds were penetrated before the boring was stopped ; these consist of 7 ft of sandy calcareous grey and green limestone with particles of clay and green grains, on 10 ft of greenish grey calcareous sand with lignite.

At Brabourne the full thickness of the formation was proved to be 31 ft, made up of 11 ft of greyish yellow sandy limestone, hard crystalline nodular

ferruginous limestone and pale yellow oolitic limestone, with calcareous sandstone at the base, resting on 16 ft of greenish grey sandy mudstone and semi-
indurated dirty bituminous calcareous sandstone and 4 ft of calcareous bituminous
conglomeratic-looking hard rock, resting with a sharp junction on the Kimmeridge
Clay.

At Ottinge Borehole, where Purbeck strata were not found, the Portland
Beds are represented by hard grey calcareous sandstone with green grains,
1 ft 3 in, on grey sand with green grains, 10 ft, on hard grey shelly calcareous
sandstone, 6 ft. In the Ropersole Borehole, Dawkins (1905, p. 29) originally
classified the strata between 1081 and 1090 ft below surface as "Portland", but
later (1913, p. 372) included these beds in the Wealden (pp. 61, 65).

PURBECK BEDS

Purbeck strata are probably confined to the south-west corner of the
present district, and these beds were recognized by Lamplugh, Kitchin and
Pringle only at Brabourne and Hothfield; it is possible, however, that the
Purbeck was also reached in a borehole at a brewery in Ashford (p. 59).
The records of Purbeck strata at Ellinge, by Dawkins (1905, p. 30) and by
Ritchie (1920, p. 83), were probably erroneous, while at Ropersole, Dawkins
(1905, p. 29) referred first to Purbeck–Wealden beds between 1025 and 1081
ft, but later (1913, p. 372) to Wealden only, between 1025 and 1090 ft.

The Purbeck strata were of fresh-water origin, except about the middle
of the series, where marine beds were thought to be "comparable with the
strata of similar nature in the Middle Purbeck series of Dorset" (Lamplugh,
Kitchin and Pringle 1923, p. 230; Casey 1955, p. 220). The marine strata
referred to include equivalents of the 'Cinder Bed', which is the horizon
taken by Casey (1963) to mark the Jurassic–Cretaceous boundary in southern
England.

At Brabourne (Lamplugh and Kitchin 1911, pp. 38–40), 12 ft of mottled
green and yellow marly clay below the Hastings Beds were doubtfully referred
to the Purbeck because of their predominantly green tint; they rest on an
earthy breccia of claystone fragments in a matrix of green sandy loam, 3 ft,
which overlies hard green marly shale and compact bluish grey claystone flags,
interbedded with bands of similar breccia, 13 ft; below this, mottled green and
yellow marly clay, 10 ft, overlie hard smooth flaggy claystones with conchoidal
fracture, and subordinate green marl and black shale, together 30 ft thick. The
Purbeck–Portland junction was not clearly seen. The probable thickness of
Purbeck strata at Brabourne is thus 68 ft, just equal to that at Hothfield. At
the latter locality the formation comprises green and grey clay and marlstone,
with beds of white breccia towards the top and a 1-ft bed of compact white
limestone 9½ ft below the top, 52 ft 9 in, resting on blue sandy shale interbedded
with grey laminated earthy limestone, 15 ft 3 in. The mention in the record
of beds of breccia towards the top of the succession suggested to Lamplugh,
Kitchin and Pringle (1923, p. 58) that these might be equivalent to those
found at Brabourne.   G.B.

Chapter IV

LOWER CRETACEOUS

GENERAL ACCOUNT

WEALDEN

THE LOWER PART of the Wealden formation, the Hastings Beds, consists mainly of fresh-water sediments and is made up in ascending order of Ashdown Beds (including Fairlight Clays at the base to the south-west of the present district), Wadhurst Clay and Tunbridge Wells Sand. The upper part, the Weald Clay, is of fresh to brackish-water origin. Though virtually the whole of the Wealden is present in the district covered by this memoir, only the Weald Clay and part of the Tunbridge Wells Sand are exposed at the surface. Wealden strata have been shown by boreholes to occur extensively beneath younger rocks in the district, but they may be absent in the Chilham, Littlebourne and Stodmarsh boreholes, and to the north-west of them.

In boreholes north of the outcrop the subdivisions of the Hastings Beds lose their identities; Lamplugh, Kitchin and Pringle (1923, p. 17) remarked that even where the Wealden beds are thinnest there is a change in character from sandy and silty beds below to clayey and shaly beds above, although they conceded that these thin subdivisions could hardly be regarded as strictly equivalent to the Hastings Beds and Weald Clay of the Weald. They also commented (1923, p. 231) that where the Hastings Beds lie conformably upon the Purbeck Beds, as at Brabourne and Penshurst, there is difficulty in fixing a dividing line between the two formations.

The formational thicknesses given below are taken from Lamplugh and Kitchin (1911) and Lamplugh, Kitchin and Pringle (1923) unless otherwise stated. The thickest recorded development of Wealden within this district is based upon a reinterpretation of the log of a borehole at Ashford (p. 62), where 713 ft are believed to be present. The formation is, however, probably somewhat thicker, perhaps over 1000 ft, at the outcrop at the western margin of the Folkestone (305) Sheet, if the rate of thickening of about 100 ft per mile between the Brabourne Borehole (309 ft) and Ashford is maintained, although it is only 544 ft thick at Hothfield.

In the Ottinge and Elham boreholes 195¾ ft and 153 ft of Wealden beds were proved respectively, so that the amount of expansion between the area of these boreholes and that at Brabourne is only about 20 ft per mile. Isopachytes on the Wealden trend south-eastwards from the Ottinge–Elham area and there is only a slight thickening between there and the Folkestone Borehole, where about 218 ft were recorded. To the north-east of Folkestone the Wealden becomes thinner, to 82 ft in the Farthingloe Borehole and ?85 ft in the Shakespeare Colliery Shafts, at the rate of about 35 ft per mile. North of Folkestone the formation thins to 183 ft at Lower Standen and perhaps 129 ft at Ellinge (Ritchie 1920, p. 83), including 67 ft of supposed Purbeck Beds (p. 44). In the Ropersole Borehole 65 ft of Wealden were encountered (Dawkins 1913, p. 372), in the Harmansole Borehole 46 ft, at Fredville 35 ft and at Adisham 30¼ ft. To the north of Adisham the borehole records are unreliable, but ?13 ft were logged in the Walmestone Borehole, although only 2 to 3 ft of Wealden core were seen. In the Chilham Borehole the absence of Wealden was inferred but not proved.

Associated with the thinning of the Wealden, differences in relative development of the upper clayey and lower sandy divisions occur, but these do not form a regular pattern. Some of the variation in the thickness of the Weald Clay is due to erosion which preceded the deposition of the Lower Greensand ; erosion at this level was noted in the Shakespeare Colliery Shafts by Lamplugh and Kitchin (1911, p. 17), who found no trace of the brackish-water fauna as present at Hythe. They considered that the Atherfield Clay did not exhibit a sufficiently inshore facies for the erosion to have been considerable. In the north-east of the Canterbury (289) Sheet the sandy division decreases to perhaps 1½ ft and 1 ft in thickness at Fredville and Adisham respectively, and beyond it is probably overlapped by the clay division.

Allen (1954 ; 1959) visualized today's Weald as a morass in mid-Hastings Beds times, bordered on the north and north-east by low-lying Jurassic strata, with an upland region of Palaeozoic rocks beyond. Two major rivers carried sediment into the morass, from the north and north-east respectively. He connected Ashdown Beds with Wadhurst Clay, and Tunbridge Wells Sand with Weald Clay, as facies of major depositional cycles, in which the sands were deposited as normal deltas during periods of increased transport by the rivers, followed by deposition of the clays during periods of reduced transport. Minor sand–clay cycles were superimposed upon the main cyclothems.

Hastings Beds: Tunbridge Wells Sand

The Tunbridge Wells Sand is composed of incoherent, fine, well-graded, fawn or brown sand, with seams of grey to brown or red clay occurring intermittently ; the latter are responsible for small springs breaking out at the surface. An estimated thickness of 80 ft crops out in this district, out of a total thickness of about 150 ft.

Weald Clay

The Weald Clay is composed of light brown to grey, often mottled, heavy clays and silty clays in the weathered 5 or 6 ft below surface ; at greater depth it becomes shaly and dark grey to brown in colour. On the basis of its contained ostracods Anderson (in Worssam 1963, pp. 16–9) proposed the following upward sequence of zones: *Cypridea dorsispinata* Zone, *C. tuberculata* Zone, *C. clavata* Zone and *C. valdensis* Zone.

Within the clays Topley (1875, p. 102) recognized a sequence of subordinate beds which can be traced over wide areas of the outcrop; starting with the oldest these beds are as follows: No. 1, Horsham Stone; No. 2, Sand and sandstone; No. 3, Limestone (Small-'*Paludina*'); No. 4, Limestone (Large-'*Paludina*'); No. 5, Sand and sandstone with calcareous grit; No. 6, Limestone "Sussex marble" (Large-'*Paludina*'); No. 7, Sand. Of these beds, Nos. 1, 2 and 4 do not occur in the present district.

Thin seams and lenses of grey siltstone and brown ironstone are common in the clay, and thin beds of bright red to crimson clay also occur, usually in association with seams of sand. Concentric nodular ironstones and boxstones may be of secondary origin where they are associated with sand beds, having been deposited by percolating ferruginous ground-water. The ironstones are locally known as ' crowstones '.

The presence of limestone " with Paludina " was first recorded in this district by Fitton (1836, p. 161), at Hurst, now Falconhurst, near Aldington Knoll. This is the Large-'*Paludina*' limestone, also known as Bethersden Marble, Petworth Marble, Sussex Marble or ' snailshell '. It is made almost entirely of shells of the fresh-water snail *Viviparus sussexiensis* (J. de C. Sowerby) in a grey to greenish grey calcareous matrix. In places the matrix is ferruginous and has a brownish colour, and there is meagre and at present inconclusive field evidence that the limestones pass laterally into ironstones. Small-'*Paludina*' limestone, composed of the smaller gastropod *Viviparus elongatus* (J. de C. Sowerby), has been found in this district as thin wafers only (p. 61).

Large-'*Paludina*' limestone (Topley's Bed 6) has been mapped in the present area at only one horizon, usually about 100 ft below the top of the Weald Clay. It has a maximum thickness of about 6 inches, but it is usually only 2 or 3 inches thick. The clays near the limestone are commonly green-coloured for a thickness of several feet. The Hothfield Borehole proved two beds of the limestone 20 ft apart. Locally at the outcrop two beds appear to be present, separated by about 10 ft of clay, but both are probably impersistent within the mapped outcrops.

Fitton (1836, p. 162) noted the presence of sand within the Weald Clay near Falconhurst, and correctly recorded other localities where the sand occurs inland, although he confused it with the Tunbridge Wells Sand at Warehorne. The sand is similar to the Tunbridge Wells Sand except that in two exposures it is rather coarser in grain. It occurs at two main horizons, of which the lower is the more important, as well as in small, rather indefinite and apparently lenticular beds. The sand at the upper main horizon is slightly above the Large-'*Paludina*' limestone and represents Topley's Bed 7; it is patchily distributed and near Bilsington Priory the bed is lenticular and there is interdigitation with the surrounding clays. The lower main horizon was thought by Topley to represent his Bed 2, but it is in fact rather higher in the succession, and Bed 2 is not

present as such in this district. In the eastern part of its outcrop the top
of the lower sand horizon lies an estimated 55 to 60 ft below the Large-
'*Paludina*' limestone, but near Ruckinge it displaces nearly the whole of
this thickness of clay, occurring locally in two beds, of which the upper
is the more important and variable and ranges up to nearly 50 ft in
estimated thickness ; the lower bed is only about 5 ft thick and may be
impersistent. These beds apparently reunite north-west of Noakes Farm
(017344) and from there can be traced to near Bromley Green, where the
sand passes into silty clay with lenses of fine sand in places. Its lower levels
pass into clay with seams of silt which crop out near Capel Farm (998356)
and form a rise in the ground at Orlestone.

At Shadoxhurst, sandy clay (p. 61), which there represents the lower
sand horizon, is about 300 ft above the Tunbridge Wells Sand ; north-east
of Ham Street the interval between the sand and the Tunbridge Wells
Sand is nearly 100 ft and at Ruckinge it is an estimated 50 ft. Farther
east exposures are unreliable in the landslipped ground bordering the
northern edge of Romney Marsh, but if the eastward attenuation continues
the Ruckinge–Bilsington sand beds and the Tunbridge Wells Sand must
coalesce in the vicinity of Hythe. Thus it seems that almost the whole
of the Weald Clay up to the level of the Large-'*Paludina*' limestone
passes laterally into sand. This hypothesis is supported by (*a*) the presence
of appreciable amounts of silt within the lowest 100 ft of Weald Clay near
Ham Street, (*b*) the thickness of 165 ft assigned to the Weald Clay in the
Folkestone Borehole, which is much the same as the thickness of Weald
Clay usually present above the Ruckinge–Bilsington sand beds at outcrop,
and (*c*) the Folkestone area is considered by Allen (1954, pp. 502–3) to lie
near one of his main river estuaries which passed sediments into the
Wealden marsh, and towards which a facies change to coarser sediments
could be expected to occur.

The Ruckinge–Bilsington sand beds occupy a horizon which received
coarse sediments locally ; the lateral passages into clay which can be
traced, particularly that about 200 yd south of Flint Farm (017352), suggest
that the sands occur along the course of a Wealden river. It was probably
not large enough to be the main feeder mentioned by Allen, although it
may have been a tributary or distributary of that river.

The uppermost Weald Clay, here as elsewhere, includes brackish-water
deposits. Drew (1864, p. 6), quoting in part from the notebook of R.
Etheridge, recorded that " bands of *Ostrea distorta* associated with *Cyrena* "
had been found " by Mr. Mackeson about 30 feet below the top of the
Weald Clay behind the School of Musketry at Hythe." Associated with
these were " other fossils characteristic of the Wealden age, viz., *Cypris
valdensis*, *C. tuberculata*, *C. spinigera*, *C. granulosa*, *Paludina fluviorum*,
and spines of fishes ". Dr. F. W. Anderson comments that this may well
have been the horizon recorded at Woodhatch in Surrey (Kitchin 1934,
p. 77), where *Nemocardium* (*Pratulum*) *ibbetsoni* (Forbes) was recorded
25 ft from the top of the Weald Clay and was associated with *Viviparus*
and '*Cyrena*'. Dr. Anderson adds that the ostracods listed by Drew are
almost certainly wrongly named : *Cypridea valdensis* (J. de C. Sowerby)

and *C. spinigera* (J. de C. Sowerby) are possible, but *C. tuberculata* (J. de C. Sowerby) is a species found much lower in the Weald Clay, and *C. granulosa* (J. de C. Sowerby) is not found above the Cinder Bed of the Middle Purbeck. The non-marine ostracods characteristic of the uppermost 30 ft of the Weald Clay, are *Cypridea spinigera*, together with three undescribed species of *Cypridea*, the brackish-water form *Theriosynoecum fittoni* (Mantell) and the marine species *Sternbergella cornigera* (Jones).

As Drew pointed out, the plane of separation between the Weald Clay and the Atherfield Clay is definite despite the development of the brackish-water beds in the upper part of the former. Black clays, which are locally and thinly developed at the top of the Weald Clay, were believed by Cornes (*in* Dewey and others 1925, p. 259) to be these brackish-water deposits, but he over-stressed their thickness and importance. Also in the top of the Weald Clay near Mersham and less frequently near the Lower Greensand outlier south of Kingsnorth church, chocolate-brown clay closely similar to that found within the Atherfield Clay, has been noted. This clay is thin and local and perhaps represents a weathered surface associated with the unconformity postulated near Kingsnorth (p. 62).   J.G.O.S., G.B.

LOWER GREENSAND

The Lower Greensand consists of clays, sands, sandstones and sandy limestones, of shallow-water marine origin, grouped in ascending order into Atherfield Clay, Hythe Beds, Sandgate Beds and Folkestone Beds. The last three of these names were applied by Drew (1864, p. 7) to divisions of the Lower Greensand already established by Fitton (1836), whose classification was based on observations along the whole of the outcrop in Kent and Surrey. Hythe, Sandgate and Folkestone are therefore not, strictly speaking, type localities. Within the district described here the Lower Greensand is thickest at Folkestone, where it attains about 250 ft, while at Brabourne it is 237 ft thick. It thins to about 160 ft at Elham, 130 ft at Shakespeare Colliery and at Farthingloe, 72 ft at Ropersole, 18 ft 9 in at Adisham, where only Sandgate Beds and Atherfield Clay are preserved (Pringle 1928), $12\frac{1}{2}$ ft at Littlebourne and 12 ft at Walmestone. At the last-mentioned locality only Folkestone Beds and Sandgate Beds were thought to be present (Lamplugh, Kitchin and Pringle 1923, pp. 181–2). The thickness of the Lower Greensand at Chilham was put at 42 ft, and at Stodmarsh ? Lower Greensand and Wealden were allotted 56 ft (op. cit., pp. 122–4, 175).

At outcrop the Atherfield Clay rests with a sharp junction on the Weald Clay, and south of Ashford there is possibly a slight unconformity at this horizon. The top of the Lower Greensand is less easy to define than is the base. Casey (1950) has emphasized that there is no palaeontological break between the Folkestone Beds and the Gault, despite the intervals of restricted deposition that are indicated by layers of phosphatic nodules at the junction of these two formations. Following Casey, the Lower Greensand in the present district is taken as corresponding to the Aptian and Lower Albian, for which stages he has proposed the following scheme of ammonite zones and subzones (Casey 1961, p. 497):

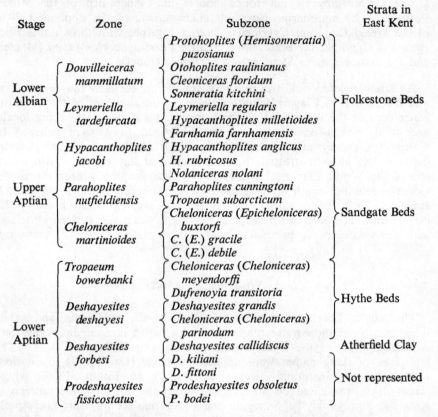

Stage	Zone	Subzone	Strata in East Kent
Lower Albian	*Douvilleiceras mammillatum*	*Protohoplites (Hemisonneratia) puzosianus* *Otohoplites raulinianus* *Cleoniceras floridum* *Sonneratia kitchini*	Folkestone Beds
	Leymeriella tardefurcata	*Leymeriella regularis* *Hypacanthoplites milletioides* *Farnhamia farnhamensis*	
Upper Aptian	*Hypacanthoplites jacobi*	*Hypacanthoplites anglicus* *H. rubricosus* *Nolaniceras nolani*	Sandgate Beds
	Parahoplites nutfieldiensis	*Parahoplites cunningtoni* *Tropaeum subarcticum* *Cheloniceras (Epicheloniceras) buxtorfi*	
	Cheloniceras martinioides	*C. (E.) gracile* *C. (E.) debile*	
Lower Aptian	*Tropaeum bowerbanki*	*Cheloniceras (Cheloniceras) meyendorffi* *Dufrenoyia transitoria*	Hythe Beds
	Deshayesites deshayesi	*Deshayesites grandis* *Cheloniceras (Cheloniceras) parinodum*	
	Deshayesites forbesi	*Deshayesites callidiscus* *D. kiliani* *D. fittoni*	Atherfield Clay
	Prodeshayesites fissicostatus	*Prodeshayesites obsoletus* *P. bodei*	Not represented

Worrall (1954) found differences in heavy mineral composition between the Hythe Beds and higher beds of the Lower Greensand. His subsequent suggestion (1957) that the calcium carbonate content of the Hythe Beds is of secondary origin and due to recent weathering processes has been adequately refuted by Casey (1961, p. 520).

Locally interrupted sedimentation in the Lower Greensand is explained by Casey (1961, p. 520) as resulting from earth movements. These were effective in the present district at the commencement of the Upper Aptian, and again at the change from Aptian to Albian time, which was marked by a break in sedimentation in the Folkestone Beds. The sandstone and sand beds of the *tardefurcata* and *mammillatum* zones are localized in their distribution, indicating a period of instability which closed with the development of the widespread nodule bed of the *puzosianus* Subzone.   B.C.W., G.B.

ATHERFIELD CLAY

The outcrop of the Atherfield Clay forms a narrow band, for the most part along the face of the escarpment formed by the overlying Hythe Beds. The outcrop on the escarpment facing south over Romney Marsh is obscured by landslipping; the Atherfield Clay here has been mapped on the basis of an assumed thickness of 40 ft. The formation was formerly seen at lowest tides on the foreshore opposite Shorncliffe Battery (Drew 1864, p. 7), where it may have been brought down by landslipping, but this most easterly outcrop is now buried beneath shingle.

The Atherfield Clay in this district consists mainly of bluish grey, in places brown-mottled, sandy clay and pale grey slightly glauconitic clay; its lower part includes reddish brown or chocolate-brown clay. In the region of Kingsnorth, Aldington and Sellindge a thin development of stiff chocolate-brown clay at the junction of the Atherfield Clay and Weald Clay has been included on the map with the latter formation, with which it appears lithologically more akin. Numerous springs issue along the junction with the Hythe Beds, and in consequence the ground on the Atherfield Clay outcrop is commonly wet or boggy.

Around the Great Chart outlier the thickness of the Atherfield Clay is estimated to be 10 to 15 ft, and near Ashford it is about 20 ft. In the East Stour valley the formation shows considerable and locally abrupt changes in thickness. Though generally about 40 ft thick near Aldington (075363), it is only 20 ft north of Clap Hill (060370). At the Collier's Hill outlier (038380) it attains 60 ft. While some of these variations may be depositional, others appear to be due to superficial movements affecting the Hythe Beds, which tend to be very rubbly near their junction with the clay and to be carried down over the clay below their true stratigraphical level. Thicknesses found in shafts and boreholes near the outcrop include $66\frac{1}{2}$ ft at Brabourne, $49\frac{1}{2}$ ft at Hythe and 28 or 38 ft (there are two versions of the record) at Folkestone. At Shakespeare Colliery the thickness totals 43 ft and the section recorded by Lamplugh and Kitchin (1911) provides the most complete record of the formation available from this district (p. 68).

Casey (1961, p. 518) remarked that the Atherfield Clay of this district has no 'Perna' Bed at the base and is the correlative of the Crackers and Upper Lobster Beds of the Isle of Wight, not of the Atherfield Clay *sensu stricto*.

B.C.W., J.G.O.S.

HYTHE BEDS

At and west of Ashford the Hythe Beds crop out on the south-facing hillside north of the Great Stour, and form outliers near Great Chart. Two important outliers occur respectively ¾ mile south of Kingsnorth church and at Collier's Hill. South-eastwards from Willesborough the main outcrop forms an escarpment which reaches its highest point, 350 ft above O.D., at Lympne. Eastward of Hythe the beds crop out in a cliff crowned by the Folkestone Beds. Their north-easterly dip brings them down to sea level at the point (Mill Point) (221351) 1¼ miles east of Sandgate Coastguard Station.

The Hythe Beds consist of alternate layers, generally 6 in to 2 ft thick, of ragstone and hassock, the former a hard, greyish blue, glauconitic sandy limestone and the latter a grey to brownish grey, glauconitic, argillaceous, calcareous sand or soft sandstone. Worrall (1954, p. 187; table I) estimated the average percentage of calcium carbonate in ragstone from localities between Maidstone and Sandgate as about 85 per cent. The rock is notably dense, with a specific gravity about 2·7. It is quarried for roadstone.

At outcrop the Hythe Beds become thinner from south to north and boreholes have proved that the formation wedges out altogether under cover of younger rocks over the whole of the present district. Thus on the Great Chart outlier the thickness of the formation is estimated as 55 to 60 ft ; on the main outcrop at Ashford it is about 30 ft ; while in the waterworks borehole at Westwell (997463), only $1\frac{1}{2}$ miles north-east of the nearest outcrop, it is 6 ft or less. Quarries at Aldington suggest a total thickness of about 40 ft, but in the Brabourne Borehole rag and hassock were absent. Old descriptions of the quarries at Hythe suggest that the thickness there may be as much as 60 ft. Boreholes at Folkestone (Whitaker 1908, pp. 136–8 ; Lamplugh, Kitchin and Pringle 1923, p. 29) give thicknesses between 23 and 34 ft, and in the shafts at Dover the Hythe Beds are absent. Lamplugh and Kitchin (1911, p. 12) believed that although the Dover area remained beneath the sea during the accumulation of the Hythe Beds, conditions there were such that permanent sedimentation was not possible.

At outcrop between Ashford and Sandgate the rag and hassock beds show a general uniformity of sequence which suggests they were laid down in a steadily subsiding basin. They are in continuation with the lower division of the Hythe Beds of the adjoining Maidstone district (Worssam 1963), and the *Exogyra* Bed which marks the top of that division is well exposed at the Goldwell Quarry (973426), $\frac{1}{2}$ mile north-west of Great Chart, although it is not traceable east of that locality. Casey (1961, p. 520) described the beds as belonging to the *deshayesi* and *bowerbanki* zones of the Lower Aptian. The upper division of the Hythe Beds of Maidstone, only 5 ft 8 in thick at the Goldwell Quarry, is represented in a phosphatic nodule bed of the *martinioides* Zone at the base of the Sandgate Beds at Sellindge.

In quarries at Chilmington Green (Plate IIIA), $\frac{1}{2}$ mile south of Great Chart, the ragstone beds display a fineness of grain and constancy in development which, together with the absence of shelly beds, suggests deposition in tranquil off-shore waters. Around Ashford are signs of a more shallow-water facies. Loosely consolidated sand, in which ragstone beds are few and widely spaced, is well developed in the lower part of the Hythe Beds there. Sand is also traceable at the base of the Hythe Beds eastward beyond Ashford, as far as the coast.

At the Otterpool Quarry (112366) near Sellindge, the topmost part of the Hythe Beds consists of 7 ft of highly glauconitic green hassock and greenish grey sandy limestone beds, belonging to the *meyendorffi* Subzone. Equivalent beds are more than 16 ft thick at Hythe. They thin markedly northwards from there to Saltwood, where they total only $3\frac{1}{2}$ ft in a quarry (164360) near the Castle, and eastwards towards Sandgate, but the exposures are not clear enough to show whether this thinning was primary, or due to erosion at the base of the Sandgate Beds. Sections on the foreshore near Mill Point show phosphatic and ferruginous boxstones near the top of the Hythe Beds, which indicate a period of non-deposition and erosion before deposition of the Sandgate Beds (pp. 78–9).   · B.C.W.

SANDGATE BEDS

The Sandgate Beds consist predominantly of glauconitic silty clays and silts. The outcrop extends as a ragged band from Ashford to Folkestone,

A 9220)

A.   HYTHE BEDS IN A QUARRY AT CHILMINGTON GREEN

B.   FOLKESTONE BEDS AND GAULT IN GRANARY COURT SAND-PIT

(*A* 9137)

ranging in width from 100 yd or less near Mersham, to nearly a mile at Pedlinge (140355), with small outliers scattered on the Hythe Beds dip-slope. In Sandgate and in the West Cliff at Folkestone the formation is extensively affected by landslips. The thickness of the Sandgate Beds at outcrop in the Ashford area, between Godinton and Willesborough, is about 50 or 60 ft. A total of 51 ft is indicated by a borehole at Westwell (997463) and of 55 ft by a trial borehole at Henwood (024430). More than 70 ft occur in a borehole (055431) at Hinxhill, 53 ft at Brabourne and 75 ft at Sandling Park (142367). The beds are possibly no more than 41 ft at Saltwood (Whitaker 1908, p. 188), but appear to be 50 to 70 ft thick at outcrop between Saltwood and Seabrook. A water borehole at the Metropole Hotel, Folkestone, is recorded (Whitaker 1908, p. 36) as proving 118 ft, but this seems excessive. The Shakespeare Colliery Shafts proved 75 to 82 ft.

The base of the Sandgate Beds is marked at some places by a layer of phosphatic nodules, which Casey (1961, p. 526) regarded as continuous, discounting records which indicate that the nodules are locally absent. The lowest 10 to 20 ft of the formation everywhere consist of dark green, highly glauconitic loam or clayey sand. The silty clays above this are dark grey or bluish grey at depth and weather to a pale or medium grey mottled with orange. Among them beds of fine-grained grey to yellow sand occur locally.

The boundary between the Sandgate Beds and Folkestone Beds along the inland outcrop is marked by a spring line where the clays give place to overlying sands. The junction thus traced agrees with Fitton's definition of the beds now called Sandgate Beds (1836, p. 115) as occupying a " flat and marshy tract ". In a section on the West Cliff between Sandgate and Folkestone, Fitton (1836, p. 123) classified as Sandgate Beds the strata up to the base of the lowest stone band of the Folkestone Beds, though in the Seabrook valley just west of Sandgate the stone beds are at least 40 ft above the spring line that marks the oncoming of sand above clay. In the present memoir the top of the Sandgate Beds in the West Cliff section is taken 1 ft 6 in below the position Fitton indicated, and at the East Cliff, Folkestone, the Sandgate Beds are taken to extend upwards to a nodule bed defined as the base of the Folkestone Beds by Price (1874b, p. 138).

Above the basal nodule bed of the *martinioides* Zone the bulk of the Sandgate Beds appears to belong to the *nutfieldiensis* Zone. *Parahoplites nutfieldiensis* (J. Sowerby) has been found by Dr. Casey at Sandgate, while clayey sands in temporary exposures at Folkestone, apparently from near the top of the formation, have yielded ammonites indicating the *nolani* Subzone of the *jacobi* Zone (Casey 1961, p. 524).   B.C.W.

FOLKESTONE BEDS

The Folkestone Beds, in Fitton's words, consist " principally of sand, white, yellowish, or ferruginous, with concretions of limestone and of chert, frequently in false stratification ". Westward of Stanford (130380) the Folkestone Beds form an irregular hilly tract. East of Stanford stone beds at the top of the formation give rise to an escarpment dissected by small valleys draining to the coast ; the dip-slope is particularly well marked south and east of Newington. At Sandgate and on the West Cliff, Folkestone, the escarpment borders the sea coast and its face is extensively landslipped.

The thickness of the formation appears to be between 100 and 140 ft throughout the tract from Ashford to Sandgate. A borehole at Westwell indicates 118 ft, one at Hinxhill 115 ft and the Brabourne Borehole 118 ft. At the Saltwood Tunnel (155369) the thickness is at least 100 ft (p. 90), while at Shorncliffe the beds are about 100 ft thick. In the East Cliff (237363) the total thickness, down to a nodule bed just below the base of the stone beds, is 65 ft (Casey 1950). In the Shakespeare Colliery Shafts the Folkestone Beds are condensed to $4\frac{1}{2}$ to $5\frac{1}{2}$ ft (p. 97).

Since the time of Fitton (1836) varied opinions have been expressed on the position of the junction between the Folkestone Beds and the Gault. Now, following Casey (1961), it is taken at the top of the ' Sulphur Band ' and at the equivalent horizon inland. The ' Sulphur Band ' is a bed of phosphatic nodules in the East Cliff section, at the top of the *mammillatum* Zone, and owes its name to the yellow alteration products of pyrite which encrust the nodules.

North-west of Ashford fine-grained sands form the lowest 20 ft or so of the Folkestone Beds, and above these are clean current-bedded sands in which masses of ferruginous sandstone or ' carstone ' are well developed locally. Near the top come the phosphatic nodules of the *mammillatum* Zone.

Much clay occurs in the middle and upper parts of the Folkestone Beds between Hinxhill and Brabourne Lees. The local succession here consists of fine-grained brown, grey and green sand forming the lower half of the formation, rather coarser sand overlying this, and coarse green to brown sand in the top 10 to 20 ft, capped by the *mammillatum* nodule bed. East of the wood north of Quarrington the middle division wedges out and the top division thickens. A similar succession, of fine-grained sands forming the lower and middle parts of the formation and coarse gritty sands its upper part, has been traced between Horton Priory (106393) and Hillhurst, south of Stanford. Sands with cherty layers and massive doggers of a hard, pale grey calcareous sandstone known as Folkestone stone come in at the top of the Folkestone Beds at the fault line $\frac{3}{4}$ mile east of Stanford.

Non-sequences occurred in the Folkestone Beds at different times in different parts of the district, and are a notable feature of the succession (Casey 1939 ; 1950 ; 1961). It is believed that most of these represent intervals of non-deposition and winnowing of unconsolidated sediments on a shallow sea floor, rather than of subaerial erosion. The direction of movement of successive centres of maximum deposition during Folkestone Beds times appears to have been eastward (Fig. 1). The *H. rubricosus* Subzone is possibly represented in the fine sands that, grading downwards into the Sandgate Beds clays, form the lower part of the Folkestone Beds inland from Folkestone. The *anglicus* Subzone forms the bulk of the Folkestone Beds west of Sandling Junction (148368), but thins to the east. A pause occurred at the commencement of the Albian : the *F. farnhamensis*

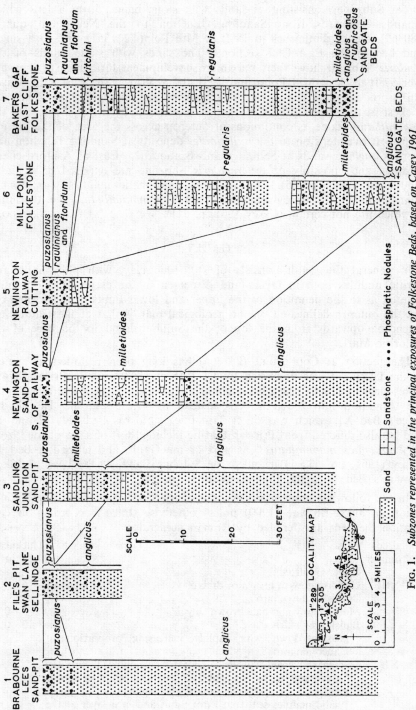

Fig. 1. Subzones represented in the principal exposures of Folkestone Beds, based on Casey 1961

E

Subzone has not been found in the present district. Beds of the *mille-tioides* Subzone, consisting of sands with stone bands, form a lens which expands eastwards from Sandling Junction to the Newington quarry (180369), thence diminishes to 18 ft at Mill Point (221351), and to a single sand bed in the East Cliff section. The sands with stone bands of the *regularis* Subzone have their maximum development farther east ; they are over 57 ft thick at Mill Point and total about 50 ft at the East Cliff (Casey 1961, pp. 531–2). The lower subzones of the *mammillatum* Zone are thickest at the East Cliff. Only the *puzosianus* Subzone, at the top of the *mammillatum* Zone, is continuous at outcrop across the area of the present maps. It oversteps on to the *milletioides* deposits at Sandling Junction and overlies *anglicus* sands at Sellindge and Brabourne. East of Ashford chert-bearing sands possibly of the *regularis* Subzone are exposed in a pit in Sandyhurst Lane (p. 85). Above them, phosphatic nodules with *Sonne-ratia kitchini* Spath are evidence of a trough of *mammillatum* age, the limits of which are not certain (Casey 1961, p. 541).   B.C.W., J.G.O.S.

GAULT

In general the Gault consists of stiff blue clay, with phosphatic and pyritic nodules both in layers and dispersed in the clay. Opinions have differed as to the definition of the upper and lower limits of the formation, but the outcrop delineated on the geological map is that of the Gault clay, extending upwards from the top of the Sulphur Band to the base of the Chloritic Marl.

The section at Copt Point (243365) has long been regarded as typical of the Gault ; it was described by De Rance (1868) and by Price (1874a). Jukes-Browne (1900) based his account mainly on that of Price, but divided the formation into fourteen beds instead of Price's eleven. He divided Price's Bed XI, which was 56 ft 3 in thick, into his Beds 11, 12 and 13, and he also differed from Price in that he included 6 ft of sand of the " zone of *Ammonites mammillatus*" as part of the Gault, and termed it Bed 1a. Casey (1961, pp. 530–1) included the Sulphur Band, at the base of Jukes-Browne's Bed 1, in the Folkestone Beds (p. 54).

In the following summarized account of the Copt Point section quotations are from Jukes-Browne (1900) unless otherwise stated ; according to convention the beds are denoted by Roman numerals :

Bed Number		Thickness Ft in	
	Upper Gault:		
XIII	Uniform greyish brown or drab clay ... ... seen for	6	0
	passing down into		
	Mottled grey and pale brown clay	3	0
	Bluish grey marly clay	13	0
	passing down with rapidly increasing proportion of glauconite into		
XII	' Greensand seam ' of Price (not to be confused with the ' greensand seam ' in Bed I): " dark green argillaceous greensand or marly clay full of glauconite grains "; phosphatic nodules scattered throughout and in a layer at the base	3	3

Bed Number		Thickness Ft in
XI	Pale grey marly clay with a seam of phosphatic nodules at the base	35 6
X	Rather hard homogeneous pale grey marly clay, with layers of phosphatic nodules at the top and 2 ft 5 in down; the clay between the nodule bands is mottled	5 1
IX	Pale grey marly clay with numerous " silvery impressions of *Inoceramus sulcatus* ". At the top a layer of phosphatic nodules with " vast quantities of crushed forms of *Inoceramus sulcatus* " (Price)	9 4½
	Lower Gault:	
VIII	' Junction Bed ' or ' nodule bed ': mottled grey clay between two layers of phosphatic nodules ... ... about	– 10
VII	Uniform dark grey, very fossiliferous clay	6 2
VI	' Mottled bed ': dark grey clay mottled with light grey. The upper limit is at the top of a 6-inch seam of " hard [?calcareous] ragstone "	1 0
V	' Coral bed ' of Price: dark grey clay, rather lighter than Bed VI and mottled with smaller light buff spots and markings. Occasional " Lumps of fossil resin " and " pieces of ironstone "	1 6
IV	Light grey clay, surmounted by a layer of phosphatic nodules and rolled ' casts ' of fossils	– 4
III	' Light bed ' or ' crab bed ': light buff or fawn clay, with a layer of phosphatic nodules and fossils at the top, and here and there small lenses and layers of clay ironstone	4 6
II	Dark grey clay, with a layer of crushed fossils and phosphatic nodules at the base and a layer of phosphatic nodules 15 inches above the base. Fossils characterized by their " deep rich colour " or nacreous lustre	4 2
I	Dark grey clay with a layer of phosphatic nodules near the base	7 4
	' Greensand seam ': dark green glauconitic clayey sand or sandy clay, with two layers of phosphatic nodules	1 6
	' Sulphur Band ' (p. 54)	1 3

Jukes-Browne found that the total thickness of Bed XIII exceeded 22 ft, as against the 17½ ft of Price. In Bed XII he noted that the phosphatic nodules were either black, water-worn and irregular, or light brown or buff coloured, and that phosphate of the latter type in places cemented together nodules of the first type and also enclosed glauconite grains, being evidently formed *in situ,* while the brown nodules seemed " to have been derived from some other bed ". The ' Junction Bed ' (Bed VIII) was taken by De Rance and Price to mark the passage from Lower to Upper Gault.

The borings put down to investigate the Folkestone Warren landslips (p. 296) were studied by Mr. A. M. Muir Wood and Dr. R. Casey, who concluded that the thicknesses of the several Gault subdivisions in the vicinity lie between the following limits: Bed I, 9½ ft to 17 ft (including the Sulphur Band which attains a thickness of up to 2½ ft); Bed II, 4 to 5 ft; Bed III, about 4 ft; Bed IV, 4 inches to over 4 ft; Bed V, about 1½ ft; Bed VI, 1 to 2½ ft; Bed VII, 6 to 7 ft; Bed VIII, 10 inches to 2½ ft; Bed IX, 9 to 11 ft; Bed X, 5 to 8 ft; Bed XI, 35 to 50 ft; Bed XII, about 3 ft; Bed XIII, 22 to at least 45 ft. The total thickness of the Gault is here about 140 ft.

In boreholes it is normally possible to recognize the base of the Gault with precision; on the other hand the Gault–Lower Chalk boundary is often difficult to establish, so that thicknesses recorded for the Gault may be misleading, particularly where cores were not taken. This uncertainty may account for the considerable local changes of thickness implied by the figures listed on p. 101.   G.B.

Zoning of the Gault. Nowhere is the Gault more readily accessible, its fossils more abundant or more perfectly preserved than in the cliffs and shore of Folkestone. The descriptions of the Gault cliff at Copt Point by De Rance (1868) and Price (1874–9) constituted one of the earliest uses of palaeontology for stratal subdivision on modern lines. Added to the more recent researches of L. F. Spath, these investigations have raised the Gault succession at Folkestone to the status of an international yard-stick for Middle and Upper Albian times.

The 'zones' of the Gault of Folkestone recognized by De Rance, like Price's 'beds', were not intended to have more than local application. One of the results of Spath's intensive study of the Ammonoidea of this formation was a refined zonal classification based exclusively on this Order of Cephalopoda. The table below is based mainly on the work of Spath (1923a; 1923b; 1942), but incorporates some modifications introduced by Breistroffer (1947) and Casey (1961). This scheme is applicable in detail to England and the Boulonnais and may be followed in outline across Europe to the fringe of Asia.

Stage	Zone	Subzone	Beds
Upper Albian (Upper Gault)	Stoliczkaia dispar	Stoliczkaia dispar & Mortoniceras perinflatum / Arrhaphoceras substuderi	XIII
	Mortoniceras inflatum	Mortoniceras altonense	XII
		Callihoplites auritus	XI
		Hysteroceras varicosum	X
		Hysteroceras orbignyi	IX
		Dipoloceras cristatum	VIII
Middle Albian (Lower Gault)	Euhoplites lautus	Anahoplites daviesi	VII
		Euhoplites lautus & E. nitidus	V–VI
		Mojsisovicsia subdelaruei	IV
	Hoplites dentatus	Dimorphoplites niobe	III
		Anahoplites intermedius	I (part), II
		Hoplites dentatus & H. spathi	I (part)
		Hoplites benettianus	
		Hoplites eodentatus	I (part)

R.C.

DETAILS

WEALDEN

HASTINGS BEDS: TUNBRIDGE WELLS SAND

The Tunbridge Wells Sand is exposed in the roadside (974322) 1000 yd N. 30° W. of Higham Farm, where 10 ft of hard brown and grey fine laminated sand with seams of dark brown ironstone up to about ⅛ inch thick pass up into 4 ft of greyish brown sandy loam at the surface. The road cutting (000335) 200 yd S. 20° W. of Ham Street Station shows 2 to 3 ft of laminated, gently current-bedded fine khaki sand with some thin ferruginous laminae and occasional clay seams ½ inch thick. There are poorly preserved impressions of wood on some bedding planes. The exposure is very close to the top of the formation. A well hereabouts, 150 ft deep, encountered "a great quantity of gas" at 65 ft and several seams of carbonaceous material; clay was proved between 110 and 122 ft (Whitaker 1908, p. 151).

HASTINGS BEDS (UNDIVIDED)

In boreholes in the marshland area a thick development of the Wadhurst Clay appears to be a feature of the Hastings Beds. A borehole at Poplar Hall (976257) is reported to have encountered 92 ft of drift (p. 253), resting on dense dark grey clay, 112 ft; on light grey sand, 7 ft; on hard dark blue clay, 45 ft. These strata probably represent the Wadhurst Clay, though the thickness is somewhat greater than that at outcrop on the marshland edge westwards. Strata in another borehole (032312), 566½ ft deep, at Langdon, 630 yd S. 9° E. of Oxpound, were classified by Whitaker (1908, pp. 235-7) as follows: "Turf and mould", ½ ft; Drift, 65½ ft; ?Weald Clay, 88 ft; ?Hastings Beds, 412½ ft. The position of the mapped base of the Weald Clay between Ham Street and Ruckinge suggests that the strata underlying the drift are in fact Hastings Beds. The lowest part of the borehole may have passed into rocks older than Hastings Beds, but no correlation can be attempted on the lithological descriptions available.   J.G.O.S.

In his account of the Hothfield Borehole section, based on the manuscript of Boyd Dawkins, Whitaker (1908, p. 225) did not separate Purbeck from Wealden. Lamplugh, Kitchin and Pringle (1923, pp. 56-8), however, allotted 144 ft of strata to the Hastings Beds, comprising 17 ft of white hard sand, on 62 ft of pale grey clay with pyrite, on 41 ft of alternations of sand, sandy clay and mottled loam with plants and brown and purple clay, on 24 ft of dark clay with pyrite and calcareous bands.   G.B.

Two boreholes, 722½ ft and 782 ft deep, were sunk in 1874 and 1901 respectively, at a brewery (013423) 200 yd N. of Ashford Station. Tentative classifications of the strata they encountered were made by Whitaker (1908, pp. 74-5). The 1874 borehole seems to be the more accurately recorded; in this the lowest 9½ ft penetrated, comprising green clay, green sandy clay and 3 ft of "Hard white stuff", are so essentially similar to the uppermost Purbeck Beds in the Hothfield Borehole (p. 44) that they are here tentatively re-classified as Purbeck Beds. On this basis the Hastings Beds are 259 ft thick.

In a borehole (021413) at a railway yard 880 yd W. 10° S. of Willesborough church dark sandstones between 371½ and 512 ft below surface were regarded by Whitaker (1908, p. 76) as ? Tunbridge Wells Sand. The sandstones contained a seam of light-coloured tough clay between 445 and 452 ft below surface. A further 21 ft of "Tough clay" below 512 ft, at the bottom of the borehole, were assigned by Whitaker to the Wadhurst Clay.   J.G.O.S.

In the Brabourne Borehole, Lamplugh and Kitchin (1911, p. 38) recorded 200 ft of Hastings Beds, with their base at 612 ft depth, as follows:

	Ft
Grey sandy silt alternating with tough mud and compact marly clay, in part dark and streaky	19
Mottled red and green hard clay or marl and pale green marl with semi-nodular structure, with two or three bands of calcareous ironstone up to 4 inches thick	26

Ft

Darker muddy clays with stone-bands, and beds of greenish marl:
fish-remains and cyprids abundant in a gritty phosphatic layer,
and fresh-water shells ('*Cyrena*', *Viviparus*) in other layers   18
Banded light and dark grey silts with a nodular band; and brown
silts with lignite and plant-remains, including *Onychiopsis
mantelli* towards the base   35
Pale rather coarse sand, with green grains and masses of lignite
near the top (no cores obtained from the bottom part)   ...   22
Muddy silts, brown, dirty grey, and paler below, with fragments
of plants   34
Band of hard calcareous grit with crystalline structure   1
Pale silts   23
Mottled red and yellow clay with marly structure ...   .   ⎫
Brownish grey muddy silts ...   .   ⎬ 22
                                                                        ⎭

The Ottinge Borehole record (Lamplugh, Kitchin and Pringle 1923, pp. 54–5)
was based on the manuscript of Boyd Dawkins. It showed 35¾ ft of Hastings
Beds, predominantly of grey clay, with hard calcareous beds or nodules, accompanied by sandy shale and sandstone. At Elham, Lamplugh, Kitchin and
Pringle (1923, p. 45) gave the formation 73 ft, based on the driller's log.
At Lower Standen Borehole, coring started in Hastings Beds, which had a
thickness in excess of 40 ft, consisting of silty and sandy clay interbedded
with quartz sand.

The thickness of Hastings Beds in the Folkestone Borehole was given as
53 ft by Lamplugh, Kitchin and Pringle (1923, p. 29); they did not divide
the 117 ft of Wealden at Abbot's Cliff Borehole (1923, p. 36), but commented
that, the absence of coarse sand apart, the lower Wealden beds were similar
to those of the Shakespeare Colliery shaft section; at the latter locality the
Hastings Beds were about 35 ft thick, with their base resting on an eroded
surface of Kimmeridge Clay at 516 ft; they were described by Lamplugh and
Kitchin (1911, p. 16) as follows:

Ft

Pale grey silt interbedded with thin layers of tough blue-black
pyritous mud or clay, containing ostracods and fish fragments ...   4
Pale greenish and yellowish grey silt and clay as above, with
seams of coarse sand passing into pebbly sand in No. 3 Pit   3 to 7
Mottled red and yellow ('catsbrain') gritty clay and silt ... about   3
Brown and grey loamy clay and silt, with plant remains and
streaks of black mud with ostracods as above ...   ... about   2
Pale greenish grey and white sandy clay and silt, with streaks
of lignite in places ...   .   6 to 8
Coarse white sand and silt, with pebbles of quartz and quartzite
up to 3 inches diameter in No. 2 Pit   1
Grey and brown clayey silt, with lignite, passing into ...   ...   1 to 2
Coarse white sand and silt with very variable pebbly bands or
lenticles, and lignite, saurian bones, etc.; thickness varying
from a few inches in No. 3 Pit to 9 ft in No. 1 Pit   ... about   8

In the Farthingloe section, 49 ft of Hastings Beds were thought to be present,
with 13 ft 6 in of conglomerate at the base, although cores were not obtained.
Ritchie (1920, p. 176) recorded 24 ft 6 in of Hastings Beds at Snowdown
Colliery, but Brown (1923, pl. 1, fig. 7) showed only about 3 ft of "Silty
clay and gravel". At Fredville, Lamplugh, Kitchin and Pringle (1923, pp. 90–3)
were non-committal, but presumed on lithological grounds that the lower
beds allotted to the Wealden (p. 46) were equivalent to some or all of the

Hastings Beds: the lowest 1½ ft were represented by sandy washings only. Similarly at Ropersole Borehole the basal 10 ft of Wealden strata, consisting of dark grey sandy clay, may represent the Hastings Beds.

At Adisham (Pringle 1928, p. 78) the basal foot of the Wealden consisted of hard pale grey fine-grained sandstone with fragments of lignite; and at Harmansole, 7 ft 6 in of dirty white clayey sand, with specks of lignite, were thought by Lamplugh, Kitchin and Pringle (1923, p. 112) to be Hastings Beds.

G.B.

WEALD CLAY

Warehorne—Orlestone—Shadoxhurst—Ashford—Willesborough.

At High Hockley (973339) a well and borehole passed through the lowest 133½ ft of Weald Clay (recorded as blue clay and marl) and entered the Tunbridge Wells Sand for 29 ft. These thicknesses are based on a reinterpretation of the records of the borehole (Whitaker 1908, p. 155), in which the junction between the Weald Clay and Tunbridge Wells Sand is not sharply defined.

Sheets of hard ironstone up to approximately 1 ft in thickness occur at or near the base of the Weald Clay between Hockley and Warehorne. The ironstone in the banks of a pond (986328) 175 yd S. 15° W. of the cross-roads at The Leacon is 2 to 3 inches thick and partly developed as nodules set in stiff brown and grey clay. It was composed of nodules up to 5 inches in diameter, some with a brown siltstone core, in a temporary exposure 150 yd W. 40° N. of Warehorne church.

North of Warehorne and west of Orlestone there is a long dip-slope with much greyish green clay at the surface. Wafers of grey or green Small-'Paludina' limestone ¼ to ½ inch thick and showing infillings of suncracks on their lower surfaces, were noted in debris from ditches at two localities on this slope.

The even dip-slope is broken at Orlestone by a slight rise formed by the occurrence of silt seams 1 to 3 ft thick within the clays, and at Shadoxhurst by a prominent double escarpment feature. These escarpments are due to thin nodular ironstone beds which occupy parts of their dip-slopes. Their scarp faces comprise brown weathered silty clay and silt, with two lenticular masses of sand on the more southerly feature ½ mile south-east of Shadoxhurst church. The scarp features form the north-westerly continuation of the beds of sand found in the Ruckinge and Bilsington districts.

Along the roadside at Hornash, north-east of Shadoxhurst church, temporary diggings showed 6 to 7 ft of grey and brown laminated clays with ¼-inch thick siltstone bands.

J.G.O.S.

Near Chilmington Green, a lens-shaped outcrop (975407) of fine-grained grey to yellow sand is an estimated 60 ft below the top of the Weald Clay. Probably at the same horizon, red clay is seen (989397) 400 yd west of Court Lodge and again in small temporary exposures around the cross-roads at Kingsnorth. In the bank of a stream (000397) 400 yd N. 10° W. of the cross-roads occurs whitish silty clay with a 2-ft bed of fine sand; a little farther north occur 4 ft of red clay.

In the Hothfield Borehole (Whitaker 1908, p. 225), the Weald Clay comprised 305 ft of stiff blue sandy clay with two 3-inch bands of 'Paludina' limestone 267 and 287 ft respectively from the top; overlying 28 ft of red and green mottled clay, sandy in the lowest 13 ft; overlying 67 ft of blue and grey clay. "Cypridea, Cyrena media, Melania and Paludina" and "hard calcareous nodules" occurred in the top 267 ft.

The record of a borehole at a brewery in Ashford (p. 59), starting just above the top of the Weald Clay, indicated that the base of the formation was 454 ft below surface. In the record of a second hole at the same site no definite junction between Weald Clay and Tunbridge Wells Sand can be recognized. About $\frac{7}{8}$ mile to the south-east a borehole for the railway works $\frac{1}{2}$ mile W. by S. of Willesborough church, starting some 20 ft below the Atherfield Clay, proved clays to $371\frac{1}{2}$ ft depth, on sandstone beds (p. 59), indicating a total thickness of less than 400 ft of Weald Clay.

Cornes (*in* Dewey and others 1925, pp. 259, 287) described an exposure at South Ashford as showing shaly clay interstratified with lenticular masses of silt, and containing fossiliferous bands of ironstone and pale grey argillaceous limestone, yielding *Unio*, '*Cyrena*', '*Paludina*', *Cypridea*, fragmentary reptilian and fish remains, and a single large specimen of "*Exogyra sinuata*". A brick-pit at Beaver Green (002412), mentioned by Cornes (op. cit., p. 260) as recapitulating the features of the South Ashford pit, exposed in 1951, 25 ft of bluish grey shaly clay with five bands of clay ironstone 2 to 4 inches thick, the lowest 8 ft up and the others at 3 to 4 ft intervals above it. Most of the ironstone bands were deeply weathered. The top one, 2 inches thick, was composed of coalesced thick-walled boxstones of 1 to 2 inches diameter, with hollow centres, some containing loose pellets of black clay. The band next below this had a $\frac{1}{2}$-inch layer at its base crowded with *Cypridea valdensis* and specifically indeterminate remains of *Viviparus*.

An old brickpit (026409) just under $\frac{1}{4}$ mile S. 35° W. of Willesborough church showed small exposures of bluish grey to dull brown mottled stiff clay, black finely silty clay and reddish brown clay. The horizon is at the top of the Weald Clay, and Atherfield Clay could formerly have occurred as an outlier here, which might account for the numerous specimens of "*Exogyra sinuata*" obtained at the pit by Cornes (op. cit., p. 260).   B.C.W.

Kingsnorth—Bromley Green—Ham Street—Bilsington—Brabourne Borehole.

Adjacent to the faulted outlier of Hythe Beds and Atherfield Clay $\frac{1}{4}$ mile south of Kingsnorth church there are only about 10 ft of Weald Clay between the base of the Atherfield Clay and the Large-'*Paludina*' limestone. This proximity of the Large-'*Paludina*' limestone to the Atherfield Clay at Kingsnorth was mentioned by Drew in his field notes, but was never published. It is apparently the result of an unconformity at the base of the Atherfield Clay. This unconformity seems to be extremely local and abrupt, for the Weald Clay ridge at Shipley Hatch, only about $\frac{1}{4}$ mile north-west of the outlier, rises 45 to 50 ft above the Large-'*Paludina*' limestone and is thought to be composed of clays overlying it. Furthermore, south-eastwards of the outlier the limestone returns to its normal horizon, about 100 ft below the Atherfield Clay, within a short distance.

The precise structure in the vicinity is in doubt, for it was impossible to trace the Large-'*Paludina*' limestone with any certainty for 1000 yd south-south-east of the boundary fault of the outlier, although several auger-holes between 8 and 15 ft deep were made for this purpose. Direct evidence for the presence of the limestone in the faulted segment to the north-east of the outlier was also lacking, but an indefinite seam of fine sand found hereabouts was regarded as the sand commonly found overlying the limestone. It was apparently faulted out at each end of its outcrop. A record of a 120-ft well (003385) at the road junction 700 yd N. 43° E. of Halfway House shows only clays and is no help in interpreting the structure.

South and south-west of the Lower Greensand outlier the Large-'*Paludina*' limestone forms a low escarpment. From the appearance of the ground it has been deduced that two bands of the limestone, less than 10 ft apart, are present along parts of the outcrop. Westwards the limestone bands apparently die out. A tiny outlier of the limestone caps a slight rise 650 yd north of Bromley Green.

The more northerly sand at Bromley Green can only be traced beside the roadway for about 900 yd east of the cross-roads. It seems to have no representative in the eastern part of the Weald Clay outcrop. This sand is very fine, fawn or grey in colour, and is associated with small quantities of red clay.

A temporary section (990370) 120 yd W. 35° N. of the cross-roads at Bromley Green showed 7 to 8 ft of grey and brown mottled laminated clays with a ¼-in siltstone band at the bottom of the section. The same thickness of clays of similar lithology was seen at a locality 500 yd N. 35° E. of these cross-roads.

In a dredging for a ditch (018372) 830 yd E. of Lone Barn Farm the Large-'Paludina' limestone is found cropping out on the banks. It ranges from 2½ to 6 inches in thickness and has a partial ferruginous staining. At Rowling Street it again forms a double escarpment and apparently two bands of the limestone are present here.

The limestone was proved by augering to be 5¼ ft below ground surface (020359) 720 yd N. 30° E. of Flint Farm and to be overlain by 2¾ ft of green clay with 3 ft of brown and grey clay above.

The sand beds are exposed for 4 ft at Hollybush (012350), where they comprise current-bedded medium-grained sand and friable sandstone with thin light grey clayey seams. Numerous carbonaceous fragments occur on the bedding planes. At the surface the friable sandstone is well bedded and contains some small boxstones. Some 3 ft of slightly clayey sands with many carbonaceous fragments and fossil wood are exposed in the railway cutting north-west of Hollybush. Much of the sand cropping out locally is finer in grain than that seen in these exposures.

The base of the Weald Clay is intermittently exposed in the roadside south of the railway bridge at Ham Street Station. To the east of Ham Street it follows the edge of the marshland and passes below the marsh deposits near Ruckinge church.

In the woodland north-east of Ham Street many silt seams are present in the 100 ft of strata between the lowest sand bed mapped and the base of the Weald Clay.

In the outlier of Large-'Paludina' limestone (018347) to the east of Gill Farm and at a Rectory (021342) 300 yd north of Noake's Farm the limestone rests upon or very close to the sand beds, which have replaced the 50 to 60 ft of clay that normally separate the sand and the limestone. There are, however, only an estimated 40 ft of sands and silts here at maximum, for the base of the sand appears to be developed at a slightly higher horizon than to the east. This is a normal trend for the Weald Clay at outcrop, for more and more clay occurs below the sand beds as one proceeds westwards.

At Noake's Farm, Herne House and Bilsington the sand crops out at two distinct levels. The lower appears to be impersistent and only a few feet in thickness and at an horizon only 50 ft above the base of the Weald Clay. The upper bed, with its base about 20 ft above the lower, forms a prominent bench feature, particularly at Bilsington village. At Herne House the upper sand bed lies an estimated 50 to 60 ft below the Large-'Paludina' limestone found in a small outlier, the intervening strata being clay. However, a larger outlier of limestone to the north (027350) has sand underlying it on its western side and clays on its eastern side, the facies change associated with the watercourse occurring below the outlier (see p. 48).

The Large-'*Paludina*' limestone caps the ridge running south towards Bilsington, where locally the double development seems to be present. Fine loamy sand, usually grey in colour, overlying the limestone, has a very patchy distribution and seems to be impersistent. . This sand occurs at localities: (028355) 500 yd W. 10° S. of Fagg's Farm, (031356) 150 yd W. of this farm, (030360) on the main outcrop of limestone 500 yd N. of Swanton Farm, and (036357) capping a ridge 400 yd E. of Fagg's Farm ; there are two patches near the Priory (042357), respectively 350 yd and 600 yd S. 30° E. of Home Farm, and three patches (050357) north-west of Bonnington.

East of Bilsington, the sand beds flank the marshland and pass into the landslips of Aldington Knoll. Drew (1864, p. 6) mentioned two sections at Falconhurst, showing 8 ft of " moderately coarse rock-sand " within these landslips. The Large-'*Paludina*' limestone can be traced by the occurrence of rubble and old diggings on the slopes of Aldington Knoll. It is exposed in the bed of a stream 620 yd W. 32° N. of the summit of the Knoll. From an ironstone near the horizon of the Large-'*Paludina*' limestone recorded by Fitton close to Aldington Knoll, Dr. F. W. Anderson has recorded a fauna typical of the *Cypridea valdensis* Zone, i.e. *C. valdensis* and *Theriosynoecum fittoni*.

An isolated mass of coarse brown ferruginous sand caps a mound (057351) 300 yd E. of Pinn Farm. About Collier's Hill the Weald Clay forms a plain broken only by a low escarpment of clay ironstone (crowstone) to the south of the hill.   J.G.O.S.

Pale bluish grey clays crop out on the low hills within the alluvial tract of the East Stour River north-east of Kingsnorth. In a ditch section (039394) just north of Swanton Court, finely silty bluish grey to brown clay was overlain by bluish black stiff clay, the latter bed, probably 3 to 5 ft thick, being overlain by Atherfield Clay.   B.C.W.

The Weald Clay thickness was only 109 ft in the Brabourne Borehole, made up of 12 ft of alternating tough bluish black clay and streaky grey clay and silt, with few fossils apart from fish scales, resting on about 5 ft of grey laminated shaly clay, with '*Cyrena*', *Unio* and ostracods, on 92 ft of alternating dark clay and paler silty clay, as before, with a few bands of grey, calcareous claystone from 2 to 4 in thick (Lamplugh and Kitchin 1911, p. 37).   G.B.

Lympne—Hythe—borehole and shaft records.

Disturbed beds in a landslip scar (100344) some 40 ft above the marsh level 700 yd S.S.W. of Port Lympne, comprised whitish silty clay or silt with blocks of yellow siltstone, overlain by 6 ft of red and white mottled slightly silty clay ; an adjacent section showed $3\frac{1}{2}$ ft of massive fine-grained sandrock. On the landslipped ground between here and Hythe are numerous small exposures of stiff dark grey to fawn clay.

In a bank on the west side of the Seabrook to Horn Street road a temporary section (186353) at the top of the Weald Clay showed about 10 ft of light grey clay with some thin seams of bluish black clay, and one or two lines of clay ironstone nodules, the latter measuring up to 1 ft across. Some larger nodules possessed an unweathered core of pale grey siderite mudstone and outer coats of brown limonitic stone ; others were composed throughout of a soft buff claystone crowded with ostracods. One nodule contained a $\frac{1}{2}$-inch seam full of *Viviparus sp.* Dr. Anderson reports that the fauna is dominated by *Theriosynoecum fittoni* and a new species of *Cypridea,* and that the assemblage is very characteristic of the highest part of the *Cypridea valdensis* Zone. He also remarks that the specimens used by Sowerby (*in* Fitton 1836) for his description of *C. valdensis* are said to have come from the Hastings Beds near Hythe, but were almost certainly from above Topley's Bed 7, at a somewhat lower horizon than that of the section described above.   B.C.W.

In the Ottinge Borehole 160 ft of Weald Clay were found, as follows: 24 ft of grey clay with 'Cyrena' and plants, on 1 ft of fine white shale, on 135 ft of dark grey clay, brown sandy clay and brown clay, with lignite and plants. At Elham the borer's log indicated only some 80 ft of Weald Clay.

In the Folkestone and Abbot's Cliff boreholes the Weald Clay was not cored, but at the former locality Lamplugh, Kitchin and Pringle (1923, p. 29) doubtfully allotted the formation a thickness of 165 ft. At Farthingloe, however, where cores were taken, the Weald Clay was found to be 33 ft thick, with its base at 660 ft below surface; it comprises 15 ft of grey, greenish and dark shaly clays with ostracods, on 8 ft of pale grey clay with white silty streaks, on 8 ft of bluish grey clay with ostracods, on 2 ft of greenish silty clay (Lamplugh, Kitchin and Pringle 1923, p. 39). Dr. Anderson has recognized *Cypridea valdensis* and *Theriosynoecum fittoni* in specimens from 630 to 637 ft.

Lamplugh and Kitchin (1911, pp. 15–6) described the Weald Clay seen in the Shakespeare Colliery Shafts as follows:

	Ft
Greyish blue clay with laminae of pale silt, and films of ostracods, etc. ... . Dark blue and black clay laminated with silt and with layers of fresh-water shells ('Cyrena', Viviparus and Unio), ostracods and fish-remains	5
Laminated dark blue, brown and greenish clay, dull black elastic mud, and pale silt, with two bands of clay ironstone 3–6 inches thick, above and below, associated with fossiliferous pyritous layers. Ostracod and fish bands numerous, with *Viviparus*, 'Cyrena', etc.	8
Dark blue, grey and black clay and mud, with pale silty laminae, and curiously mixed blue and grey bands mottled as if by the burrowings of small organisms and laced with tubular markings like *Sabella*-tubes, charged with comminuted shell-fragments. Shells rare, but many partings crowded with ostracods, bits of fish, etc., and some containing small rounded pellets of grey clay; at base, two clay ironstone bands, 7 inches and 3 inches thick, with a shaly parting full of fish-remains, etc.	21
Laminated blue, black, brown and grey clay or cyprid shale and silt, similar to the above, with 'Cyrena', Viviparus, etc., in some bands, and strongly marked layers of black and brown fragments of fish, with ostracods: towards the base one of these layers has a sprinkling of coarse grit grains (seen in No. 1 Pit only); at the bottom is a thin band of brown nodular claystone	11
Paler silty clay, intermingled blue and grey; tough greenish blue mud; and brownish loam with plant remains, lignite and pyritized wood; less shaly than the beds above, and forming a passage into the more sandy Hastings Beds below; no shells seen	5

Dawkins (1913, p. 372) recorded 65 ft of Wealden strata at Ropersole, of which 55 ft may have represented the Weald Clay, consisting of black shale, 8 ft, on dark grey marl with fine sand, 13 ft, on dark grey clay, 34 ft.

It is not clear whether clay or sand dominate the Wealden strata at Fredville Borehole and Snowdown Colliery, for Lamplugh, Kitchin and Pringle (1923, p. 90) did not divide the Series at the former locality, while in the colliery shafts

Ritchie (1920, pp. 162, 178) put the Weald Clay thickness at only 8 ft, but Brown (1923, pl. 1, fig. 7) gave it some 35 ft.

At Adisham Borehole, Pringle (1928, p. 78) recorded 29 ft of slaty grey clay and bluish black and grey silty clays with ostracods and fish remains, surmounted by 3 inches of bluish grey limestone, the latter showing worm-burrows infilled by the basal Atherfield Clay, the Wealden surface being 964 ft 9 in below ground level. *Cypridea valdensis* and *Theriosynoecum fittoni* are present in great abundance between 964 ft 9 in and 965 ft, according to Dr. Anderson, who also recognized a few examples of *T. fittoni* at 990 ft 6 in.

Positive evidence of the presence of Wealden strata was not found at Trapham or Stodmarsh boreholes, but at Walmestone the Weald Clay may have been as much as 13 ft thick. At Harmansole the sequence of the Wealden cores was incomplete, but the Weald Clay is probably 38 ft 6 in thick, with its base at 806 ft 6 in ; cores seen by Lamplugh, Kitchin and Pringle (1923, p. 112) consisted of slaty grey and white streaky clays. Dr. Anderson reports that *C. valdensis* and *T. fittoni* occur in large numbers in specimens from 770 ft 6 in.

G.B.

LOWER GREENSAND

ATHERFIELD CLAY

Great Chart—Kingsnorth—Mersham—Brabourne Borehole.

Topley (1875, p. 112) recorded that a road section (987416) just west of Singleton Manor, showed stiff blue probably Wealden clay, below brown clay overlain by blue sandy clay with *Exogyra* and '*Rhynchonella*'. The projecting tongue of the outcrop 1 mile west of Ashford is probably due to a local increase in thickness of the formation, coupled with a slight dip that nearly coincides with the slope of the ground surface. The bank of a railway siding (994434) ⅝ mile N. 30° E. of Bucksford showed 6 ft of pale bluish grey plastic clay with *Plicatula placunea* Lamarck and *Exogyra tuberculifera* Koch and Dunker[1].

The log of a trial borehole (997463) made in 1933 at the Waterworks 1450 yd S. 28° E. of Westwell church records " light grey sandy clay " between 209 ft and the base of the hole at 255 ft. The upper 6 ft of this possibly belongs to the Hythe Beds (p. 71), leaving 40 ft to be assigned to the Atherfield Clay.

Just north of Swanton Court a ditch section (039395) showed, on Weald Clay, reddish brown clay with estimated thickness of 5 ft, overlain by about 15 ft of greyish blue to brown finely glauconitic clay with beds of greenish fine sandy clay midway and near the top, overlain by Hythe Beds.
B.C.W.

At the outlier ¼ mile south of Kingsnorth church the Atherfield Clay consists of soft grey silty or sandy clay between 20 and 30 ft in thickness and rests apparently unconformably on the Weald Clay (p. 62). Ditches on the east side of Collier's Hill (038380) showed sections of about 1 ft depth in soft buttery grey and orange mottled silty clay, partially speckled by glauconite, and darker grey silty micaceous clays.

Near Mersham the sinuous form of the base of the Hythe Beds is not reflected in the base of the Atherfield Clay and is suggestive of superficial movement of the former beds over part of the outcrop of the latter. Springs are plentiful in the vicinity and much of the outcrop is marshy. Worrall (1954, p. 186) recorded a section showing a passage upwards from the Atherfield Clay to the Hythe Beds in the railway cutting (051393) 190 yd S. 40° W. of Mersham

---

[1] The Lower Greensand and Gault fossils have been named by Dr. R. Casey and Mr. R. V. Melville.

church: grey clay which crumbles into cubical lumps, 2 ft thick, was overlain in upward succession by very pale, almost white, silty clay, becoming heavier at depth, 10 ft; pale grey silt, mottled orange, 4 ft; a black and brown mottled horizon, 6 in; greenish hassock, 1 ft 6 in; reddish brown coarse sandy loam with blackish iron oxide at the base, 1 ft 2 in; soil, 10 in.

Topley (1875, p. 112) also stated "Atherfield Clay, somewhat sandy and of a bluish and greyish colour is seen below the Hythe Beds near Smeeth Station", but Cornes (*in* Dewey and others 1925, p. 260) wrote that the formation was absent and that "Hythe Sands" rested on Weald Clay in a slip scar at about the same locality. No section is now visible (1955): however, Casey (1961, p. 519) confirmed the presence of Atherfield Clay here, recognizing *Deshayesites forbesi* Casey in specimens collected in 1925.

In the Brabourne Borehole, Lamplugh and Kitchin (1911, pp. 36–7) described the Atherfield Clay as hard brown clay with marly structure, 18 ft, resting upon hard blue clay, 3½ ft, basing their description upon chisel samples from depths between 282 and 303½ ft. However a fragmentary ammonite recovered from between 240 and 250 ft, in greyish green loam, and recognized by Casey (1961, p. 519) as "*Deshayesites, apparently D. forbesi*", shows that the Atherfield Clay is at least 53½ to 63½ ft thick and possibly 66½ ft (p. 81). It appears therefore that the Atherfield Clay has expanded when compared with its thickness at the Mersham outcrop, where it is about 25 ft thick: a complementary behaviour to that of the Hythe Beds (p. 52).   J.G.O.S.

Hythe—Folkestone—borehole and shaft records.

The bed and banks of the Brockhill stream (149356), ⅞ mile W. 25° N. of Hythe church, expose pale grey to fawn slightly silty and micaceous clay in a brecciated state, consisting of fragments up to 6 inches across in a pasty clay matrix. The fragments have yielded *Panopea gurgitis* (Brongniart) *plicata* J. de C. Sowerby, *P. mandibula* J. Sowerby, *Resatrix* (*Vectorbis*) *vectensis* (Forbes), *Thracia rotundata* (J. de C. Sowerby) and *Deshayesites forbesi*.

Simms (1843, p. 207) described a trial shaft put down "from the bottom of the quarries at Hythe". It proved, beneath Hythe Beds, 49 ft 6 in of clay with marine fossils, separated from the uppermost beds of the Wealden by a layer of soft sand 1 inch thick. A collection of fossils acquired by the Geological Survey from the Geological Society in 1911 contains a number of Lower Greensand specimens from Hythe which were contributed by Simms. Dr. Casey reports that some of these, in an Atherfield Clay matrix, bear the label "Hythe, shaft" or figures agreeing with the depths in the shaft given by Simms. These specimens are determined as follows: *Pinna robinaldina* d'Orbigny, *Cuneolus lanceolatus* (J. de C. Sowerby), *Nemocardium* (*Pratulum*) *ibbetsoni* (Forbes), *Resatrix sp., Ensigervillia forbesiana* (d'Orbigny); *Deshayesites forbesi* and *D. sp. nov.* The *E. forbesiana* is the basis of the record of "*Perna mulleti*" from the base of the Lower Greensand in this shaft.

A temporary excavation in 1955 (159349) 300 yd W. of Hythe church showed, on a face 10 yd long, 6 ft of bluish grey to fawn very finely glauconitic stiff clay, with two or three slipped blocks of ragstone embedded in it.

A bank on the west side of the Seabrook to Horn Street road (186353), showed, above the Weald Clay, about 3 ft of reddish brown stiff clay, passing up to dark grey to greyish brown stiff clay, about 1½ ft, overlain in turn by 4 ft of pale grey finely glauconitic sticky clay with nodules of white race of about one inch diameter.   B.C.W.

Whitaker (1908, p. 136) described the strata in two boreholes at the Metropole Hotel (216354), 1100 yd S. 28° W. of the principal station in Folkestone. In this memoir the record of the later borehole is used: this penetrated the topmost 8 ft of the Atherfield Clay, recorded as "Blue marl". Whitaker (1908, pp. 137–8) also quoted two records of a borehole in Tontine Street (231360) in which the thicknesses assigned to the beds are 28 ft and 38 ft. J.G.O.S.

The Atherfield Clay section at Shakespeare Colliery was described by Lamplugh and Kitchin (1911, p. 13) as follows:

	Thickness Ft	Depth from Surface Ft
Pale tough greenish grey clay, probably decolorized, perforated by molluscan borings . . ., with casts of ammonites and other fossils, and small pale brown nodules	1	389
Banded marly clays, in alternating tints of pale and dark bluish grey, brownish grey, and brown, mostly smooth and unctuous-feeling, but with some darker layers mottled with silty and sandy streaks. Fossils abundant	23	
Firm brown banded clay or marl with small pale brown concretions and many fossils, particularly ammonites ('*Hoplites deshayesi*')	7	
Smooth blue-black unctuous clay, crowded with *Pinna robinaldina*; and with many small *Exogyra* at base: '*Pinna*-bed'	1 to 2	
Firm marly clay of rich chocolate colour, with pale brown nodules, fauna varied and abundant, included a band of small crushed echinoderms, but cephalopods rarer	9	
Unctuous dark blue clay (like that of the '*Pinna*-bed') with a thin irregular seam of sandy grit with lydites and brown phosphatic fragments like crumbs of bone, etc.	1	
Resting tightly on, and interlocked with, a corroded deeply pitted surface of Weald Clay at		431

The Dover No. 1 (Aycliff) Borehole (295396), drilled in 1958, entered the formation to a depth of 16 ft 9 in, in dark grey and brownish grey silty clay.

In the Ottinge Borehole Dawkins gave a thickness of 111 ft to the Atherfield Clay (Whitaker 1908, p. 224), but Lamplugh, Kitchin and Pringle (1923, pp. 54–5) suggested that only the lowest 59 ft of blue and dark brown clays with *Pinna* represented the Atherfield Clay (pp. 79–80). The thickness of the formation was stated to be 19 ft 8 in at Ellinge (Dawkins 1905, p. 30; Ritchie 1920, p. 83), but farther north, at Ropersole, Dawkins (1913, p. 372) recorded 21 ft of brown clay. The thickness was further reduced at Fredville, where Lamplugh, Kitchin and Pringle (1923, pp. 90–2) allotted 7 ft of chocolate-brown clay to the Atherfield Clay; Burr (1909, p. 390) also classified the overlying 10 ft of silty clay and silt as Atherfield Clay, but this was included with the Sandgate Beds by Lamplugh and his co-authors, on the grounds that the surface of the chocolate-brown clay was seen to be worn and pitted by molluscan borings when exposed in the nearby Snowdown shafts.

The Adisham Borehole showed 10 ft 4 in of "Firm marly clay of a rich chocolate brown colour", with numerous phosphatic grains and rounded fragments of bone in the lowest 3 ft 9 in (Pringle 1928, p. 78), while at Harmansole the Atherfield Clay was represented by only 2 ft of brown sandy clay, full of small smooth brown phosphatic fragments (Lamplugh, Kitchin and Pringle 1923, pp. 112–4). G.B.

HYTHE BEDS

Great Chart—Ashford—Westwell—Sevington.

The disused Goldwell Quarry (973426), ½ mile N.W. of Great Chart (Worssam 1963, p. 44), has an upper and a lower working separated by the plane of a reversed fault trending between south and south-south-east. The correlation of the beds in the two workings is not certain, but the total of Hythe Beds exposed appears to be about 40 ft. The lowest bed seen is a massive 2-ft fawn sandy limestone. Above this come some 19 ft of hard, fine-grained ragstones with interbedded hassock, then 8 ft 6 in of grey sandy ragstones and hassock, including the *Exogyra* Bed at the top. The latter is a pale grey, chalky, sandy limestone occurring in doggers up to 8 inches thick, with abundant small *Exogyra latissima* (J. Sowerby) and other fossils. The 8-inch hassock above it also contains *Exogyra,* and scattered small phosphatic nodules. Above this the topmost 5 ft 8 in of the Hythe Beds comprise two massive grey ragstone beds with an intervening hassock. The upper of these ragstone beds contains some chert and has a planed top surface.

At Chilmington Green the following section (981411) was exposed in 1955 in a quarry (Plate IIIA) west of the road that crosses the Great Chart outlier, ½ mile S. 5° E. of Great Chart church; the quarrymen's names for certain distinctive ragstone beds are given in brackets:

		Ft	in
	Soil	1	0
20.	Massive, light grey fine sandy limestone, broken by weathering (Large Rock)	2	0
19.	Hassock, even-grained, compact	1	0
18.	Hard, fine-grained light grey limestone (E 28163) 5 in to –		6
17.	Hassock	–	6
16.	Limestone (as bed 18, the two together named the Little Diamonds) ... ... 5 in to –		6
15.	Hassock	–	6
14.	Massive limestone. At the north end of the face is a parting 4 to 6 inches from the top, and the top part consists of oval doggers; towards the south end of the face the top splits off to form a separate thin bed, with 2 to 3 in of hassock beneath it. Base of main bed sharp but wavy (Leathercoat) (E 28162) ...	1	6
13.	Hassock, containing oval doggers of limestone, which in places are stuck on to bed below ... 6 in to –		9
12.	Massive limestone, top uneven (E 28161) ... 9 in to 1		0
11.	Hassock ... ... 4 in to –		6
10.	Limestone, hard, bluish grey, blue-hearted, fairly constant in thickness (Nos. 10 and 12 are the Lord and Lady)	–	9
9.	Hassock ... ... 4 in to –		6
8.	Limestone, hard, fine-grained, blue-hearted, speckled with glauconitic grains. A prominent bed with flat top, broken into massive blocks 4 ft square by joints trending roughly N.N.E. and E.S.E. (Square Rock) (E 28160)	1	4
7.	Hassock, hard ... ... 1 ft to 1		2
6.	Limestone, very massive, with in places a 3 to 4-in. doggery basal layer separated from the main bed by a hassock parting (Hook and Eye) 1 ft 6 in to 2		0

						Ft	in
5.	Hassock	... ...				4 in to –	6
4.	Massive limestone, with very irregular, knobbly top and bottom surfaces					10 in to 1	2
3.	Hassock with thin doggers of limestone, ... about					1	8
2.	Limestone, massive					1	2
1.	Hassock ... ... seen for					–	4

The largest of the Chilmington Green quarries (983408) east of the road, ⅛ mile E. of New Street Farm, disused in 1955, exposed beds from the hassock underlying the Square Rock upwards. The lower of the two beds named Lord and Lady in the quarry just described (bed 10) seems to be represented here only by a thin band of limestone on the west face and by occasional flattened doggers on the north face. The section is as follows, the beds being numbered in sequence with those of the previously-described quarry :

		Ft	in
Sandgate Beds: (p. 80) ... ... seen for		3	0
Hythe Beds:			
27.	Hassock	3	0
26.	Massive, tough, dark greenish grey glauconitic limestone, in places in large, flat-bottomed doggers	2	0
25.	Hassock	–	9
24.	Massive limestone (E 28168)	1 ft to 1	3
23.	Hassock	–	4
22.	Limestone (E 28167)	9 in to 1	0
21.	Hassock with scattered rounded lumps of limestone	1 ft to 1	6
20.	Massive, irregularly fracturing limestone, with a parting midway (Large Rock) (E 28166) ... 1 ft 8 in to 2		0
19.	Hassock, laminated	9 in to 1	0
7–18.	(Square Rock to Little Diamonds, beds 16 and 18 sectioned in E 28164 and E 28165) ... about	9	0

A third quarry (984410), now filled in, lies adjacent and to the north of that last described, and ⅛ mile S. 22° E. of Great Chart church. In this quarry beds from the Little Diamonds upwards were exposed. Bed 27 was missing, the nodule bed at the base of the Sandgate Beds being noted in 1951 to rest directly on a line of massive doggers of greenish grey glauconitic limestone, similar to those forming bed 26 of the section given above.

A disused small quarry (984406) at the extreme southern end of the Great Chart outlier showed five or six beds of pale grey, soft sandy limestone (E 28170) with thin intervening layers of hassock or soft flaky stone, in all about 6 to 10 ft, near the base of the Hythe Beds. ·Crenella bella (J. de C. Sowerby) and Cheloniceras sp. were obtained. The thickness worked, about 15 ft, was probably all below the 27 ft or so exposed in the larger quarries, which suggests a total of at least 40 to 50 ft for the Hythe Beds.

Dr. R. Dearnley reports as follows on ragstone beds from the quarries near Chilmington Green :

The rocks consist of a granular mosaic of calcite grains in which coarser and finer grained portions are irregularly distributed, apparently bearing no relation to the original bedding. The finer portions average about 0·01 mm in grain size (0·005–0·02 mm variation) and the coarser areas average about 0·20 mm (0·05–0·30 mm variation) (E 28162–8, E 28170). Distributed throughout the calcite mosaic are organic fragments, including foraminifera and shells, the latter usually replaced by coarsely crystallized calcite, especially in the larger fragments. Glauconite occurs as scattered grains (about 0·20 mm diameter) in each specimen but is more abundant in some specimens (E 28160–2) than in others (e.g. E 28163–4). It locally replaces organic debris and especially foraminifera fragments (E 28166–7).

Quartz grains, some rounded but more usually subangular to angular and commonly showing embayments, are present in all specimens, but in widely varying amounts. Irregular contacts between the quartz grains and the surrounding calcite grains may be indicative of secondary growth. The quartz usually shows undulose extinction. Grains of granulated or sutured quartz crystals are common and are taken to represent detrital (? meta-morphic) grains. Other grains, or irregular patches, consist of chalcedony, and some (E 28160–3) may be due to secondary silicification. Local silicification of shell debris occurs. Small fragments (0·05 mm), perhaps of chert, occur in microcrystalline aggregates.

On the main outcrop at Ashford the Hythe Beds are estimated to total about 30 ft. A former estimate of 15 ft (Worssam *in* discussion of Worrall 1954, p. 201) did not take account of the sands developed in the lower part of the formation. These sands were seen in a deep drainage trench (999435) 100 yd E. of Repton Manor, where rag and hassock beds above them included shelly seams. Excavations for a bridge to carry the Ashford by-pass over a minor road (003436) 300 yd E. of the railway cutting, were noted by Mr. Smart to show a small anticline with axis approximately east–west, in strata as follows, in ascending order : greenish grey glauconitic sand, 12 ft ; greyish brown ragstone with oysters, Trigoniids and Terebratulids, 1 ft 6 in ; grey and brown sand, 2 ft 6 in ; nodular and impersistent grey limestone, up to 8 in ; brown hassocky sand, 4 ft, capped by weathered and broken ragstone, about 2 ft.

At the Henwood Pumping Station (021429) ¼ mile W. 30° S. of Conningbrook (Whitaker 1908, p. 77) four shafts are connected by an adit 350 yd long, 30 to 40 ft below the ground surface. The shafts passed through the following strata : sandy clay (Sandgate Beds), 6 to 35 ft ; ragstone, 7½ to 15 ft ; sand, 15 to 17 ft ; ragstone, 2½ ft ; on blue clay (presumably Atherfield Clay), proved for 6½ ft. A trial borehole (024430), just over ½ mile W. 30° S. of Conningbrook, proved Hythe Beds, consisting of " rock and veins of loam " at 68 ft to 87 ft 6 in depth, resting on 8 ft 6 in of " clay ", presumably Atherfield Clay.

In a trial borehole at the Waterworks at Westwell (p. 66) the Hythe Beds are 6 ft or less in thickness. The log (Buchan and others 1940, p. 45) gives 1 ft 3 in of " Hard limestone " with base at 207 ft depth, overlying 2 ft of " Dark grey clay ", overlying in turn, to the bottom of the hole, 46 ft of " Light grey sandy clay ". Specimens kept by the Ashford Borough Engineer showed the " dark grey clay " to be a greenish black, non-calcareous glauconitic silty clay, of Sandgate Beds type ; while others from 210 ft to 212 ft 6 in were of grey calcareous silty sand, possibly a hassock or pulverized soft ragstone. Clays occurred at 215 ft and below. The nature of the " hard limestone " is thus in doubt, while the Hythe Beds would appear to have occurred between 209 and 215 ft.

Between Willesborough and Sevington the sands in the lower part of the Hythe Beds have a wide outcrop east-south-east of Boys Hall. The ground is poorly drained. Small springs issue at the junction with the overlying rag and hassock beds, the soil on which, by contrast, has good natural drainage. Debris from a ditch (040395) 250 yd N. by E. of Swanton Court indicated a limestone bed at the junction of the sands with the underlying Atherfield Clay.   B.C.W.

F

Kingsnorth—Mersham—Smeeth—Aldington.

The beds have been worked extensively in the outlier $\frac{1}{4}$ mile south of Kingsnorth church. There are no sections (1955), but Drew (in MS.) recorded a $6\frac{1}{2}$ ft section showing five ragstone beds 5 to 9 inches thick alternating with hassock in a pit probably on the north side of the outlier (006380). He also noted "The hassock is a drab-coloured clayey sand containing some green grains ; this must be very nearly the bottom of the rag."

A few feet of the grey and brown clayey sand at the base of the formation cap Collier's Hill and the outlier of Atherfield Clay $\frac{2}{3}$ mile west-south-west of Mersham church.

The formation is worked in quarries (046411) alongside the main road 1050 yd E. 20° N. of Sevington church. On the south side of the road the section is :

		Ft	in
17.	Soil	2	0
16.	Broken ragstone	1	8
15.	Hassock	–	8
14.	Broken weathered grey ragstone with slight nodular tendency	2	0
13.	Brown hassock with seam of ragstone of variable thickness in centre	1	4
12.	Coarse grey ragstone	–	10
11.	Hassock with dark bands and numerous small ragstone nodules	2	3
10.	Light grey massive ragstone with slight nodular tendency	2	4
9.	Greyish brown hassock	1	3
8.	Light grey medium-grained ragstone	–	8
7.	Greyish brown hassock with small nodules of ragstone	–	6
6.	Light grey medium-grained ragstone	–	8
5.	Hassock, bluish grey at base passing up into greyish brown colour at top	–	8
4.	Greyish blue coarse crystalline ragstone	–	5
3.	Dark grey hassock	1	8
2.	Greyish blue coarse crystalline ragstone	–	9
1.	Dark grey hassock ... ... seen for	2	0

Fossils include *Ceriopora ramulosa?* Michelin, *Cheloniceras sp.* and *Dufrenoyia truncata* Spath. North of the roadway the following section was noted :

		Ft	in
14.	Soil : sandy loam with ragstone pieces, clayey at base	2	0
13.	Broken grey ragstone, perhaps originally in two bands	4	0
12.	Hassock with small hard lumps	1	6
11.	Grey hard nodular ragstone	2	6
10.	Greyish brown hassock	–	8
9.	Uneven hard grey ragstone band with oysters 4 in to –		8
8.	Greyish brown hassock	–	6
7.	Light grey ragstone, with fibrous structure in part ... ... 6 in to –		8
6.	Greyish brown hassock	–	6
5.	Slightly nodular fine light grey ragstone ...	–	6
4.	Greyish brown hassock with doggers of ragstone up to 9 in thick	1	6
3.	Light grey fine ragstone	–	9

<table>
<tr><td></td><td>Ft</td><td>in</td></tr>
</table>

	Ft	in
2. Greyish brown hassock with small (2 in × 6 in) nodules of ragstone	1	6
1. Light grey speckled ragstone ... ... seen for	–	6

The bottom of the above section is about 4 ft above the lowest ragstone band, which floors most of the quarry. The strata are folded locally about north-east-trending axes, into two synclines and a gentle anticline, which are probably of superficial origin. Casey (1961, p. 523) recognized the above two sections to be in the *deshayesi* Zone, with *Cheloniceras parinodum* Casey, *Deshayesites* of the *involutus* and *grandis* groups in the lower two layers of ragstone, and *Australiceras gigas* (J. de C. Sowerby), *Lithancylus grandis* (J. de C. Sowerby), and *Cheloniceras cornuelianum* (d'Orbigny), above.

Towards the base of the beds near Mersham the courses of ragstone are broken and there is much decalcification. This disintegration, which is probably the result of frost action, may have been accompanied by superficial movement (p. 66). A quarry (056391) 400 yd S.E. of Mersham church exhibits the following:

	Ft	in
Soil and decalcification residue (see below) ...	1 ft to 8	0
Broken weathered ragstone	2 ft 6 in to 4	6
Hassock with small ragstone nodules, slightly banded appearance resulting from partial decalcification	2	6
Rubbly ragstone formed of numerous hard small nodules	1	7
Hassock with numerous small light grey ragstone nodules and slight decalcification banding ...	2	0
Massive hard fine grey ragstone	7 in to –	9
Light grey hassock with large nodules of grey speckled ragstone 1½ ft above base and nodular fine-grained ragstone 0 to 7 in thick, 6 in below top	4	6
Massive hard light grey, speckled sandy ragstone...	1	3

Fossils collected include *Cheloniceras* (*Cheloniceras*) *crassum* Spath, *Ch. cornuelianum*, *Deshayesites* aff. *grandis* Spath, *Australiceras gigas* (J. de C. Sowerby) and *Tropaeum hillsi* (J. de C. Sowerby). Greenish clayey sand, 1½ ft thick, was formerly exposed below the lowest ragstone band which floors most of the quarry. As the escarpment edge is approached the beds become more rubbly, an irregular southerly dip appears and the decalcification products, which consist of contorted and piped well-banded light and dark brown clayey sand with calcareous geodes and patches of green remanié Sandgate Beds, assume their thickest development.

In the railway cutting (065385) just west of Smeeth Station Topley (1875, p. 117) saw 4 ft of grey sandy Atherfield Clay overlain by Hythe Beds composed of 8 ft of brown clayey sand beneath 1 ft of soft yellowish stone. This section probably showed the full local development of the basal sands of the formation. The following fossils from the railway cutting ½ mile west-north-west of Smeeth Station (about 062386) were found in the lowest few feet of the Hythe Beds (Casey 1961, p. 523): *Cheloniceras* aff. *parinodum*, *Deshayesites deshayesi* (d'Orbigny), *D. multicostatus* Swinnerton and *D. consobrinoides* (Sinzow).

The clayey sand at the base of the Hythe Beds also occurs ⅓ mile south-west of Sellindge church; the outlier 350 yd S. of Water Farm seems to be comprised wholly of these sands.

The long outlier extending from Bank Farm, Aldington, to Stonegreen has been much worked, but no sections remain. At Clap Hill (059372) a quarry is being worked and exhibits the following:

		Ft	in
12.	Soil	–	3
11.	Massive grey to bluish grey ragstone in three courses alternating with thin beds of hassock	3	4
10.	Hassock with numerous ragstone nodules	1	9
9.	Nodular fine-grained ragstone	1	0
8.	Hassock with nodular grey ragstone (0 to 9 in) near base and small ragstone nodules and ferruginous staining near top	3	6
7.	Massive light grey ragstone	–	10
6.	Hassock	1	0
5.	Nodular brown and grey ragstone	1	3
4.	Hassock	2	0
3.	Coarse grey nodular ragstone	–	9
2.	Hassock with ragstone nodules and patches of concentric lines of ferruginous staining, some of which have ironstone cores	3	0
1.	Nodular ragstone	3	6

This quarry has yielded *Australiceras gigas*, *Cheloniceras parinodum*, *Deshayesites grandis* and *Dufrenoyia sp.*, showing that both the *deshayesi* and *bowerbanki* zones are present.

In the disused quarry (063369) 300 yd N. 10° W. of Aldington Corner the following section was noted:

		Ft	in
15.	Soil	1½ ft to 3	0
14.	Rubbly and weathered beds. Probably lowest 2 ft consist of hassock with two bands of nodular ragstone and uppermost 2 ft broken bedded ragstone	4	0
13.	Light grey medium-grained ragstone, generally massive but nodular in places	1	0
12.	Hassock with large nodular light grey ragstone up to 1 ft near base and line of smaller nodules at top. Nodules nearly coalesce in places ...	1	6
11.	Light bluish grey ragstone	–	11
10.	Hassock	–	4
9.	Dark grey massive ragstone	–	9
8.	Hassock	–	6
7.	Hard massive light grey ragstone	1	6
6.	Hassock	6 in to –	7
5.	Light grey nodular ragstone	10 in to –	11
4.	Hassock	4 in to 1	6
3.	Light grey massive ragstone	1	6
2.	Hassock, greenish brown, with line of grey ragstone nodules in centre, which in places expand to reach bed 3 above	2	0
1.	Light grey ragstone	1	0

In the above quarry bed 7 is correlated with bed 11 in the Clap Hill quarry, so that together the quarries expose nearly 30 ft out of a total thickness of about 40 ft for the Aldington area. The quarry near Aldington Corner shows that the bands of ragstone, and the nodular bands in particular, are somewhat variable, both in thickness and form ; bed 7, for instance, becomes nodular in part of the face and bed 5 thins away and regenerates. There is a gentle syncline present in the centre of the face and some slight festooning of the top of the rubbly bed 14.

Topley (1875, p. 117) recorded " five courses of stone, each from 12 to 18 inches thick " in a quarry (074360), now obscured, near Aldington church. At Aldington Knoll (071353), where the ground is extensively landslipped, there is a tiny outlier of the basal Hythe Beds sands capping the hill; a trench there exposed 3 ft of hassock with nodules of ragstone. From Aldington east-wards the upper limit of the landslips bordering Romney Marsh roughly coincides with the base of the Hythe Beds, the beds themselves forming a steep cliff to the crest of the escarpment.

In the trackway (091353) 50 yd S. of Manor Farm the upward sequence exposed is ragstone (just visible), below hassock with large ragstone nodules, 2 ft; light grey nodular ragstone with many oysters in centre, 2 ft; hassock with many small ragstone nodules, 1 ft 5 in; fine light grey ragstone, 2 ft. The uppermost three beds are also seen at a locality (093352) 400 yd E. 35° S. of this farm. J.G.O.S.

Barrowhill—Otterpool—Lympne—Sandling—Hythe.

In a temporary section (108374) on the west side of the Folkestone road at Barrowhill the following strata were noted: fairly soft pale grey sandy limestone, weathered brown, seen for 3 in, overlain in order by hard light grey fine-grained sandy limestone, 1 ft 6 in to 1 ft 8 in, passing up into soft greyish brown to brown shelly calcareous sand, 1 ft 2 in to 1 ft 4 in, soft dark green glauconitic sand with roughly elliptical doggers of harder glauconitic sandstone up to 2 ft long and 9 in thick near the top, 1 ft 8 in, overlain by Sandgate Beds (p. 81). The Hythe Beds yielded *Anomia laevigata* J. de C. Sowerby, *Chlamys robinaldina* (d'Orbigny), *Gervillella sublanceolata* (d'Orbigny) and *Gryphaeostrea canaliculata* (J. Sowerby). A similar succession was recorded by Casey (1961, p. 526). G.B.

The section of the Hythe Beds at the Otterpool Quarry (112366), south of the main road 250 yd north-east of Otterpool Manor, could well be taken, as Casey (1961, p. 523) suggested, as a standard for East Kent. His bed numbers are used in the following abridged version of the section as measured in 1957 :

		Ft	in
Sandgate Beds : (p. 82)	seen for	2	6
Hythe Beds:			
31–32. Green hassock, in parts indurated to greenish grey sandy limestone		7	0
29–30. Grey sandy limestone with phosphatic nodules, underlain by sandy hassock		2	0
24–28. Hard light grey sandy limestone in 3 beds 8 in to 1 ft thick, with intervening roughly laminated hassock		3	2
23. Hassock, well marked in quarry face, with occasional doggers of fine-grained lime-stone		1	0
22. Hard light grey limestone, in places shelly, roughly in 3 beds separated by impersistent hassock partings		3	0
21. Hassock	2 in to	–	4
14–20. Bluish grey limestone in 4 beds 1 ft to 1 ft 4 in thick, with intervening hassocks up to 6 in		5	10

		Ft	in
13.	Laminated hassock	–	4
12.	Grey, fine-grained limestone, full of moulds of shells in places. Merges into hassock above and below	4 in to –	6
11.	Hassock, even-grained, not laminated ...	–	6
10.	Limestone with flat top and bottom surfaces	6 in to –	8
9.	Hassock parting	–	2
2–8.	Hard, bluish grey finely glauconitic limestone in 4 beds 9 in to 1 ft 6 in thick, with 3 intervening hassocks each about 1 ft ...	8	9
1.	Bluish grey sandy clay (proved for 8 ft according to quarry manager) ... ... seen for –		6

The strata are slightly buckled, showing northward dips up to 5°, and at the east end of the north face a fault with a downthrow of 5 ft to the west. Casey (1961) assigned beds 1 to 8 to the *deshayesi* Zone, beds 9 to 28 to the *transitoria* Subzone and beds 29 to 32 to the *meyendorffi* Subzone of the *bowerbanki* Zone. Additional to ammonites and other fossils listed by Casey (1961), the following fossils were collected from bed 2: *Sellithyris sella* (J. de C. Sowerby), *Sulcirhynchia hythensis* Owen; *Limatula tombeckiana* (d'Orbigny), *Venilicardia* cf. *protensa* (Woods); *Cymatoceras radiatum* (J. Sowerby): and the following from beds 8 to 30: *Sellithyris sella, Sulcirhynchia hythensis*; *Cucullaea* cf. *fittoni* (Pictet and Campiche), *Glycymeris sp., Linotrigonia (Oistotrigonia) ornata* (d'Orbigny), *Resatrix (Resatrix) hythensis* Casey, *Septifer lineatus* (J. de C. Sowerby), *Venilicardia sp. nov.*; and *Neohibolites sp.* Fossils found loose included *Serpula sp.; Camptonectes cottaldinus* (d'Orbigny), *Exogyra latissima, Gervillella sublanceolata* and *Pinna robinaldina.*

Casey (1961, p. 523) described bed 30 as an *Exogyra* bed, but though it is sandy and lies at about the same horizon as does the *Exogyra* Bed at the Goldwell Quarry (p. 69), bed 30 has not the fine-grained matrix that characterizes the *Exogyra* Bed of the Maidstone district. In the writer's opinion there is no evidence that the Maidstone *Exogyra* Bed extends farther east than the Goldwell Quarry, nor, in view of its behaviour as a diachronous shell-bank separating an upper and a lower group of fine-grained limestone beds (Worssam 1963, p. 31, fig. 10) would it be expected to continue eastward of where the upper of those groups, which is last seen at the Goldwell Quarry, dies out.

Dr. R. Dearnley reports: Specimens of ragstone from Otterpool Quarry (E 28171–7) are similar to the ragstone layers from the Great Chart outlier except that very minor amounts of iron-ore locally occur as scattered dusty aggregates (E 28171–4, beds 2–16) and as discrete patches of pyrite (E 28177, bed 30) often associated with glauconite. The presence of occasional phosphatic (organic) grains in E 28173 (bed 12), E 28176 (bed 26) and E 28177 (bed 30) contrasts with the apparent absence of such grains in the Great Chart outlier. One specimen (E 28177) has a much higher proportion of quartz grains to calcite matrix than any other rock examined: it is transitional to a calcareous sandstone. Detrital zircon grains were noted in one specimen (E 28176).

A disused quarry in the triangle of roads just west of Shipway Cross (123351) showed 10 ft of strata, comprising seven ragstone beds, 5 in to 2 ft 6 in thick, and intervening hassocks up to 9 in each. All the ragstones were fine grained and not notably sandy. They resembled beds 8 to 20 at the Otterpool Quarry, but the two sections could not be matched in detail. Casey (1961, p. 523) assigned the beds exposed here to the top of the *deshayesi* Zone and the lower part of the *bowerbanki* Zone. Some 9 ft of similar beds were noted in a section on the escarpment near Bellevue (109351), ¼ mile W.N.W. of Lympne church.

A borehole at Sandling Park, 300 yd S.W. of the house, proved 18 ft of interbedded " Green sandstone " and " Green sand and clay " from 109 ft 6 in downwards, the base not being reached (Whitaker 1908, p. 189). A section on the escarpment 800 yd S. 22° W. of Pedlinge Court (138345) shows some 22 ft 6 in of Hythe Beds overlain by Sandgate Beds. In the lowest 7 ft 6 in the limestones are hard, light grey and fine-grained ; in the next 9 ft 6 in they are light grey and sandy and develop a honeycombed surface under weathering ; and in the highest 5 ft 6 in the hassock is green and the limestones are greenish grey and very glauconitic. The sequence matches in general the upper part of that at Otterpool Quarry, above bed 14 or 16. At the top of the second group is a shelly limestone that corresponds in position and lithology to bed 30 at Otterpool. Beds equivalent to the lowest two groups are to be seen in another escarpment section, 250 yd to the west.

Of the quarries above Hythe, Drew (in Topley 1875, p. 116) remarked that fourteen courses of stone from 1 to 2 ft thick, with layers of hassock 6 in to 3 ft thick, made together 40 ft of strata. Simms (1843, p. 207) recorded the thickness of " quarry rock " at Hythe as being 48 ft, below which he proved in a shaft 14 ft of sand and stone above Atherfield Clay. From these records Topley (1875, p. 116) inferred that the thickness of the Hythe Beds at Hythe is about 60 ft, though Casey (1961, p. 520) considered this excessive. The presence of loose yellow medium-grained sand near the base of the formation was confirmed by temporary exposures in 1955, 350 yd W. 10° N. of Hythe parish church.

The quarries on the hill top (Tanner's Hill) above Hythe now show only the top part of the Hythe Beds. The largest of them (163352), 350 yd N. 10° E. of the parish church, showed:

	Ft	in
Greenish grey laminated hassock ... ... seen for	3	0
Soft, weathered sandy limestone with cherty patches ...	–	6
Green hassock	1	2
Dark greenish grey, highly glauconitic, hard sandy limestone. Small black phosphatic nodules scattered in the lower half	1	10
Dark green laminated hassock ... ... about	1	0
Dark green, glauconitic indurated hassock, almost a limestone	1	0
Dark green, very glauconitic hassock	2	0
Massive, dark greenish grey, very glauconitic limestone ; top surface flat. Exfoliates where weathered, the edges and corners of joint blocks becoming rounded ... ... 2 ft to	2	6
Dark green glauconitic hassock	–	5
Dark greenish grey, glauconitic shelly limestone, with layers crowded with Exogyra, Terebratulids and Trigoniids ... ... 9 in to	1	0
Dark green glauconitic hassock, with an 8-in layer of soft flaky greenish limestone 6 in from the base ...	2	3
Light grey sandy, massive limestone ; the top surface flat. A parting plane 11 in from the base	2	3
Hassock	–	2
Light grey nodular limestone	–	4
Fawn, laminated clayey hassock	–	4

	Ft	in
Light grey, hard, fine-grained limestone, slightly sandy, and full of shells	1	0
Hassock ... ... seen for –		1
Total about	20	0

Another old quarry (159352) 350 yd N. 22° W. of the parish church, showed about 13 ft of interbedded greenish grey hassock and massive ragstone. In a small disused quarry (164360) just north of the railway embankment 300 yd E. of Saltwood Castle the green rag and hassock beds are reduced to a total thickness of 3 ft 6 in. The section in 1955 showed 4 or 5 ft of grey sandy ragstone and hassock beds very similar to the six lowest beds of the Tanner's Hill section, overlain in order by dark green glauconitic hassock, 6 in ; massive greenish grey glauconitic limestone, 9 in to 1 ft thick ; and dark green glauconitic loam seen for 2 ft 6 in, with a line of weathered phosphatic nodules, taken as the base of the Sandgate Beds, 2 ft up. Similar sections are displayed in a small quarry 75 yd W. of here, in a cutting (170352) 250 yd S.E. of the disused Hythe Station, and in a disused small quarry (168363) 800 yd E. 28° N. of Saltwood Castle.

Old quarries along the former railway between Hythe Station and Seabrook are now overgrown. In a partly filled quarry (184356) ½ mile E. 10° S. of Scene Farm 7 ft 9 in of grey limestone and hassock were overlain by a 1-ft bed of greenish grey limestone, capped by 7 ft of brickearth. The highest of the grey limestone beds yielded echinoid radioles, Serpula sp., 'Cardium' sp., Chlamys robinaldina, Exogyra latissima, Plicatula carteroniana d'Orbigny and Pterotrigonia caudata (Agassiz). This is the old Horn Street Quarry of which Casey (1961, p. 523) recorded that the ammonites preserved in old collections are of transitoria age. A roadside section (187357) ¼ mile south-south-west of Horn Street exposed a bed of hard, light grey shelly limestone overlain by 2 ft 9 in of greenish grey, glauconitic limestone and hassock, capped by Sandgate Beds.   B.C.W.

Sandgate—Folkestone—borehole records.

Topley (1893, p. 43) recorded that the Hythe Beds were once quarried at Wellington Terrace, Sandgate. Slipped masses of steeply dipping Hythe Beds form prominent reefs at about the low tide-mark along most of the coast from opposite Shorncliffe Camp to ½ mile south-west of Folkestone Harbour.

At Mill Point (221351), 1650 yd. W. 25° S. of Folkestone Harbour Station, a broad ragstone platform is uncovered at low tide. The Hythe Beds here have been slightly displaced by landslipping and are intersected by marked joints trending north-east and north-west. Near the bottom of the beach slope 1700 yd W. 22° S. of the above station, and resting in hollows in a bed of ragstone, there is a nodule bed composed of phosphatic and ferruginous boxstones. The nodules in this bed range up to about 1 ft in length and 4 in. in thickness ; their partially mammillated ferruginous skins envelop cores containing grey sandy limestone pebbles among and squeezed into dark grey, phosphatic gritty siltstone. The coarse grains consist of dark siliceous pebbles, quartz pebbles, phosphatic nodules, green grains and, according to Casey (1961, p. 525), green calcareous sandstone pieces. The cores of the nodules are also fossiliferous. Overlying the nodules locally are about 3 in of brown to grey ragstone. Basal gritty clays of the Sandgate Beds are present in close proximity, but are not seen in contact with the Hythe Beds in this exposure. The nodule bed is exposed again some 150 yd westward, where it is repeated three times by slipping or perhaps faulting movements, which also repeat Sandgate Beds in part of the beach. The nodule bed misleadingly appears as three distinct layers within the Hythe Beds, but in fact the gritty basal clays of the Sandgate Beds have been seen directly overlying the nodules here, without any intervening ragstone. In this exposure the boxstones are less gritty, less fossiliferous and contain more of the ragstone pieces.

The boxstones were examined petrographically by Dr. Dearnley, who first found that dolomite was present in the nodule cores. He states:

A specimen from the foreshore at Mill Point, Folkestone (E 29104) shows a very irregular junction between two different rock types. One is a fine-grained calcite-mudstone with scattered quartz fragments, locally recrystallized aggregates of calcite grains probably replacing organic fragments, and occasional grains of glauconite and pyrite. The other rock consists almost entirely of dolomite (0·05 mm grain size) together with a few scattered phosphatic fragments, and grains of glauconite and quartz. The interfingering and 'wispy' nature of the contact suggests that there may have been some plastic flow before final consolidation.

Casey (1961, pp. 523–5) found that the ragstone underlying the nodule bed contained an ammonite fauna similar to the top of the *transitoria* Subzone and considered the nodule bed to be a conglomeratic base of the Sandgate Beds, pressed into the top of the ragstone. He regarded the overlying ragstone as a raft. Furthermore he found that the nodules contained three elements: firstly, debris from the *meyendorffi* Subzone, consisting of rolled greenish sandstone, rolled ragstone pebbles and black nodules; secondly, buff phosphatic nodules of *buxtorfi* age, yielding the zonal ammonite; thirdly, an indigenous fauna of *nutfieldiensis* age, of brachiopods and lamellibranchs including *Cyclothyris latissima* J. de C. Sowerby, *Cyrtothyris cyrta* (Walker), *C. uniplicata* (Walker); *Acesta longa* (Roemer), *Arca dupiniana* d'Orbigny, *Chlamys robinaldina, Entolium orbiculare* (J. Sowerby), *Exogyra tuberculifera, Gryphaeostrea canaliculata, Limopsis dolomitica* Casey, *Proveniella regularis* (d'Orbigny), *Pseudocardia sp. nov.* and *Thetironia minor* (J. de C. Sowerby). He recorded also *Myopholas* cf. *semicostata* (Agassiz), apparently bored into the ragstone. A specimen of this fossil in the Geological Survey collection is associated with the grey phosphatic material and partially encased by the ferruginous skin of the nodules. The nodule bed is thus a non-sequential deposit, spanning in time the *buxtorfi* Subzone and probably the major part of the *martinioides* Zone. The underlying *meyendorffi* Subzone was probably destroyed by erosion beforehand. The non-sequence may have extended into the *nutfieldiensis* Zone, as indicated by the indigenous fauna, but in that case a return to normal Hythe Beds deposition must have taken place, during which time the overlying ragstone was deposited. This is considered unlikely and the indigenous fauna is believed to be of facies origin.

The presence of dolomite in a condensed sequence is unusual, as is its squeezed interrelation with limestone. It may have formed during the non-sequence in a similar manner to that in fissures in concrete within Alexandria harbour (Hume 1914), and as a jelly enveloping limestone fragments which were partially in the same state.

There is more evidence of erosion of the Hythe Beds before the Sandgate Beds were laid down. The greenish ragstones of Hythe are missing and the 3-inch ragstone overlying the nodules is cut out in the western exposure where the nodules deceptively appear to be the basal bed of the Sandgate Beds.

Whitaker (1908, pp. 136–8) recorded 23 ft of Hythe Beds in the borehole at the Metropole Hotel, Folkestone (216354), as rock and light grey marl, and he gave two records of a borehole in Tontine Street (231362) in which the Hythe Beds were 28½ ft and possibly 34 ft thick respectively.   J.G.O.S.

Lamplugh, Kitchin and Pringle (1923, p. 37) were not able to assess the thickness of Hythe Beds at the Abbot's Cliff Borehole, although they considered that all the divisions of the Lower Greensand were present. In the Ottinge Borehole, about 3½ miles from the nearest outcrop of Hythe Beds, these authors (1923, p. 54) doubtfully allotted 63 ft 2 in of strata to this formation; their classification was a reinterpretation of the MS. notes by Boyd Dawkins, as published by Whitaker (1908, p. 224), and the beds termed Hythe Beds consisted of 11 ft 2 in of hard green sandstone, partly calcareous and likened by Dawkins to ragstone, with softer green sandstone, resting on 26 ft of dark brown sandy clay and 26 ft of dark brown clay. This abnormal thickness of Hythe Beds, coupled with the small thickness of beds allotted to the Sandgate Beds (p. 84), suggests that the revised classification may be wrong, and that the so-called Sandgate Beds and the uppermost 11 ft 2 in of the "Hythe Beds" should be termed Folkestone Beds, giving this formation a total thickness of 102 ft, while the underlying 52 ft of dark brown sandy clay and clay should be referred to the Sandgate Beds, the Hythe Beds being absent. It is noteworthy that Lamplugh, Kitchin and Pringle (1923, p. 48) found nothing in the log of the Elham Borehole, 1¼ miles N.N.E. of Ottinge, to suggest the presence there of Hythe Beds.

G.B.

Sandgate Beds

Great Chart—Westwell—Hinxhill—Smeeth—Brabourne Borehole.

At Goldwell Quarry (p. 69), the Sandgate Beds rest with a sharp junction on an apparently eroded surface of the ragstone bed forming the top of the Hythe Beds. They comprise dark green glauconitic sandy clay with some phosphatic nodules, 1 ft thick; overlain by doggers, 6 in to 1 ft thick, of dark green glauconitic limestone containing phosphatic nodules; and dark green highly glauconitic loam with wisps of pale greyish clay, seen for 10 ft. Some of the phosphatic nodules were fossiliferous, yielding 'Terebratula' depressa auctorum non Lamarck, Platithyris sp. and lamellibranchs (Worssam 1963, p. 50).

An old quarry (984410) ⅜ mile S. 22° E. of Great Chart church is now filled in, but in 1951 it showed 4 to 5 ft of dark green highly glauconitic clay or loam with a layer of phosphatic nodules at the bottom, taken as marking the base of the Sandgate Beds. It was probably from this quarry that the brachiopods and lamellibranchs recorded by Topley (1875, p. 129), Gregory (1895) and Kirkaldy (1937) were obtained. Casey (1961, p. 526) noted that the only ammonite recorded from here was too immature to be identified more closely than Cheloniceras [sensu lato], so the exact position of the nodule bed in the zonal sequence is unknown. In an adjacent quarry to the south the Sandgate Beds consist of 3 ft of glauconitic loam with phosphatic nodules in the lowest 6 in. There may be a slight unconformity at the Hythe Beds–Sandgate Beds junction between these quarries.

Specimens from a borehole at the Waterworks at Westwell (p. 66) indicate Sandgate Beds between 158 ft and 209 ft depth. The main part of the formation here comprises dark grey or black silty clays, some specimens being glauconitic, while the top 3 ft consist of dark greenish grey fine clayey sand overlain by dark grey clay mingled with some sand. The junction with the Folkestone Beds has the characters of a passage by alternation.

A trial borehole for the Henwood Pumping Station (p. 71) proved "clay and some sand", 13 ft, probably Folkestone Beds, overlying "dark sand and loam", 55 ft, probably the full thickness of the Sandgate Beds, on Hythe Beds.

A road cutting (975448) just east of Home Farm shows about 10 ft of black silty clay, with fine yellow sand of the Folkestone Beds a few feet above in a nearby section. The presence in the top part of the Sandgate Beds of grey silt or yellow-weathered glauconitic fine sand was indicated by temporary sections (018433) about ½ mile S.W. of Bybrook; by others (025425) some 300 yd south of Foxglove; and by exposures (042414) in a cutting on the Folkestone road just east of Willesborough Street.

B.C.W.

In one borehole made in 1899 at Hinxhill Pumping Station (055431) hard black clay, brown mottled clay and black clay struck between 180 and 202 ft are regarded as Sandgate Beds (Whitaker 1908, pp. 155–6). A later boring (1942) on approximately the same site passed successively downwards through the following strata below a depth of 170 ft: black sandy clay with green sand streaks, 20 ft; grey sandy clay, 5 ft: black sandy clay, 5 ft; dark green sandy clay, 15 ft; dark green sandy clay with some green sand, 15 ft; medium grey sandy clay, 10 ft. The thickness of 70 ft is not excessive for the region described in this memoir; it contrasts, however, with the 20 to 30 ft indicated by the outcrop north-west of Hatch Park. However, in view of the rapid changes in thickness of the Sandgate Beds along the outcrop and the presence of ragstone and hassock in the boreholes at Henwood (p. 71), it is concluded that the Hinxhill bores are not deep enough to reach Hythe Beds and that they therefore furnish no evidence of the dying out of the ragstone and hassock beneath the cover of Sandgate Beds.

According to Worrall (1954, p. 191) the base of the Sandgate Beds in the quarry at Mersham (056391) contained soft phosphatic concretions but no fossils had been found. Near Smeeth church the beds form a conspicuous rise above the Hythe Beds plateau which is lost, as a result of faulting, 600 yd N. of Water Farm and subdued near Stone Hill where there is a local southerly dip.

An outlier north-east of Court-at-Street consists entirely of sandy clays at the base of the formation, in a much-weathered condition; 4 ft of glauconitic sandy clay were noted in a temporary exposure (093358) at the roadside 700 yd N. 32° E. of Court-at-Street.

Lamplugh and Kitchin (1911, p. 36) classified 98 ft of strata in the Brabourne Borehole as Sandgate Beds, as follows, in downward order: dark green loamy clay, 20 ft; dark green loamy sand, 11 ft; brown sand, 4 ft; loamy sand, 4½ ft; loamy clay, 3½ ft; grey loamy sand, 10 ft, base at 237 ft; greenish grey fine-grained silt, loam and marl with occasional semi-indurated sandy bands and layers of pyritous clay: also some hard claystone nodules, soft brown phosphatic nodules, rod-like concretions or markings, and some veins of gypsum: slickensided surfaces plentiful throughout: no 'Kentish Rag' seen, and the whole series akin to the Sandgate Beds of Dover, 45 ft, base at 282 ft. The lowest 44 ft was compiled from cores and the remainder from the driller's report. Casey (1961, p. 519) has shown that most of the lowest 45 ft must now be regarded as Atherfield Clay (p. 67). The thickness remaining, 53 ft at maximum, is still considerably greater than that found at outcrop in Hatch Park.   J.G.O.S.

Sellindge—Otterpool—Saltwood—Seabrook.

The south side of a railway cutting (105377) 150 yd S.E. of Somerfield Court showed the following upward sequence, the sediment in each layer passing up into that of the bed above: dark grey to black micaceous loamy to clayey silt, roughly laminated, seen for 2 ft 6 in; grey glauconitic loamy fine sand, weathering to orange, about 1 ft 6 in; dark greyish green glauconitic loamy silt with some clay pellets, 8 to 10 in; pale grey to greenish grey, slightly glauconitic silty clay, seen for 2 ft 6 in.   B.C.W.

At the roadside at Barrowhill (p. 75) Hythe Beds were overlain by a bed of grey glauconitic phosphatic nodules 5 to 8 in thick, but increasing to as much as 1 ft 4 in in places where its base sagged into the Hythe Beds as a result possibly of frost-wedging. The nodules, averaging about ½ in diameter, were set in a matrix of brownish grey glauconitic sand, together with much fossil wood. The nodule bed was overlain by soft glauconitic sand, seen for 2 ft, coloured pale to dark green and brown in bands 1 to 2 in thick and containing scattered phosphatic nodules, and about 10 in above its base a fairly well-marked band of soft incipient phosphatic nodules. Fossils from the nodules include: wood; *Serpula sp.*; *Oblongarcula oblonga* (J. de C. Sowerby), *Praelongithyris* cf. *praelongiformis* Middlemiss, *Sellithyris* aff. *sella, Sulcirhynchia hythensis*; *Ampullina sp., Dicroloma sp., 'Turbo' sp.*; cf. *Astarte cantabrigiensis* Woods, *Chlamys robinaldina, Cucullaea glabra* Parkinson, *Exogyra sp., Gervillella sublanceolata, Grammatodon (Nanonavis) carinatus* (J. Sowerby), *Gryphaeostrea canaliculata, Isognomon ricordeana* (d'Orbigny), *Linotrigonia (Oistotrigonia) ornata, Mactromya vectensis* (Woods), *Neithea quinquecostata* (J. Sowerby), *Panopea gurgitis plicata, Plicatula carteroniana, P. placunea, Pseudaphrodina ricordeana* (d'Orbigny), *P. sp.* (=*Nucula albensis* Gregory *non* d'Orbigny), *Pseudocardia sp., Pseudolimea parallela* (J. de C. Sowerby), *Pterotrigonia mantelli* Casey, *Resatrix* cf. *hythensis, Thetironia minor, Venilicardia sowerbyi* (Woods); and *Eutrephoceras sp.* The exposure was mentioned by Worrall (1954, p. 191), and Casey (1961, p. 526) noted the presence of rare specimens of *Cheloniceras (Epicheloniceras) buxtorfi* (Jacob) and *Ch. (E.) sp. nov.*, and stated that the whole assemblage suggested a condensed deposit equivalent to the upper part of the Hythe Beds (*martinioides* Zone) of the Maidstone area.   G.B.

A similar section of Sandgate and Hythe Beds, including a 1-ft layer of nodules and bored wood, with fossils, at the junction, was recorded by Topley (1875, p. 129) from a small quarry, now filled in or overgrown, at Grove Bridge, the railway bridge 250 yd N. of the section described above. He added that some other sections in the immediate neighbourhood did not show the nodules.

At the Otterpool Quarry (p. 75) the base of the Sandgate Beds is marked by a 6-in bed of small, buff, fine-grained phosphatic nodules, some fossiliferous, scattered in a sandy matrix. This bed is overlain by dark green loamy sand, seen for 2 ft.

The strata in a 36-ft deep well (142367) at Sandling Park, 300 yd S.W. of the house, are not recorded, but the well was probably dug to the base of the Folkestone Beds. A later borehole sunk in the well, proved clays and sands to 109 ft 6 in below surface, on Hythe Beds (Whitaker 1908, p. 189). The thickness of the Sandgate Beds may therefore be about 75 ft.

On the north side of the railway cutting (159362) 250 yd N.E. of Saltwood church the following strata at the top of the Sandgate Beds were noted in 1955:

		Ft
Soil and hillwash		1
Pale grey waxy clay, slightly silty and glauconitic ...	1 ft 6 in to	2
Grey to fawn glauconitic loamy silt or fine sand passed down to	... about	2
Dark grey to fawn silty clay ... ... about passed down to		2
Black, laminated silty clay with wisps of greenish black glauconitic silt ... ... seen to		10

Sands of the Folkestone Beds crop out on the hill-slope just above the cutting. The waxy clay at the top of the section resembles that found in the lower part of the Folkestone Beds near Willesborough and near the top of the Sandgate Beds near Seabrook. X-ray analysis by Mr. B. R. Young (films X 2022, A) has shown that the clay contains abundant montmorillonite but there is a little quartz present.

At a section (133345) on the escarpment just over $\frac{5}{8}$ mile W. 35° S. of Pedlinge Court, small phosphatic nodules are sparsely scattered between 1 and 2 ft above the top ragstone bed of the Hythe Beds, in greyish green glauconitic loamy sand which totals 4 ft thick and is overlain with a sharp junction by dark green, very glauconitic clay or loam, seen for 10 ft. In a similar section 800 yd S. 22° W. of Pedlinge Court a line of small cream-coloured phosphatic nodules, forming the base of the Sandgate Beds, occurs 1 ft above the top ragstone bed of the Hythe Beds, in glauconitic loamy sand that is seen for 4 ft. A similar line of nodules was seen in a quarry 300 yd E. of Saltwood Castle (p. 78).

In a roadside section ¼ mile south-south-west of Horn Street the base of the Sandgate Beds is marked by a ½ to 1-in band of soft orange-brown ironstone, above which are seen 2 ft of glauconitic loam. Topley (1875, p. 129) noted that in some sections on the hill-side between Hythe and Shorncliffe, ferruginous layers and concretions replaced the phasphatic nodules at the base of the Sandgate Beds.

Small exposures (188352) in the banks of the lane (Sandy Lane) which leads from Seabrook to Shorncliffe indicate a total thickness of about 90 ft of Sandgate Beds. Near the base occurs grey to fawn plastic clay, and a little higher up, fine-grained greenish grey glauconitic sand. About mid-way in the Sandgate Beds a bed of fine-grained yellow sand is seen for 4 ft. The upper half of the formation appears to consist predominantly of clays. Interbedded greyish green glauconitic silts and black clays are seen. Just below the base of the Folkestone Beds, a trench 20 yd north of the lane showed 1 ft of green clayey silt, overlain by 1 ft of pale grey waxy clay, with grey silty clay seen for 6 in above it. X-ray analysis of the waxy clay by Mr. Young (films X 2023, A) shows that it contains abundant montmorillonite, with an appreciable amount of quartz ; a small amount of one of the kaolin group of minerals also appears to be present.   B.C.W.

Sandgate—Folkestone—borehole and shaft records.

Price (1874b, pp. 136–8) divided the Sandgate Beds of the Sandgate and Folkestone area into four ; no thicknesses of the divisions were quoted, but his overall thickness of 80 ft in Folkestone is an underestimate. At Sandgate the upper limit of the beds is buried beneath landslip and lies somewhat below the steep cliffs of the slip-scars in the western part of the town. Springs occur at or near the top of the beds in the town and between there and Cheriton. On the roadside (200351) 440 yd W. 16° S. of the church with spire in Sandgate 3 ft of brown and grey sandy clay are exposed. In a bank (205353), which is probably an old slip-scar, 50 yd W. 30° N. of this church, 2 to 4 ft of dark grey sandy clays with lenses up to 3 in thick of fine glauconitic ('pepper and salt') sand were noted ; a few yards away, 4 ft of grey fine glauconitic sand underlie 2 ft of rubbly grey and brown sandy mudstone. None of the exposures is of strata *in situ*. Two exposures (206355) 400 yd N. 30° E. of the church with spire show 3 ft of fine grey and brown micaceous sand, and at a slightly higher horizon, 5 to 6 ft of grey and brown somewhat clayey sand were noted (207355) 420 yd E. 38° N. of this church. These sands are near the top of the formation and are perhaps 10 to 15 ft thick. In a driveway (206354), 300 yd E. 8° N. of the above church, 1 ft of coarse dark green glauconitic sand is exposed beneath 3 ft of coarse brown sand. This section lies near the base of the Sandgate Beds.

The descending sequence of the 118 ft of Sandgate Beds in the borehole at the Metropole Hotel, Folkestone (216354) is: green loamy sand, 29 ft ; dark sand and clay, 8 ft ; green sand, 79 ft ; green sand and stone, 1½ ft ; green sand, 6 in (Whitaker 1908, p. 136).

The contact with the overlying Folkestone Beds (pp. 93–4) is exposed near the Toll-bar (220353), 1550 yd W. 18° S. of Folkestone Harbour Station, where 4 ft of fine grey and brown micaceous sands, with crumbling moulds of fossils and a slightly hardened and sparsely bored upper surface, were noted. The sands, which become slightly clayey at the bottom of the exposure, are similar to those occurring at the top of the beds in Sandgate. Fitton (1836, pp. 123–4) recorded a section " at a prominent point of the cliff ", probably near the Toll-bar, showing, below

Folkestone Beds, between 40 and 50 ft of greenish grey and yellow sand, with very large concretionary nodules, consisting of slightly consolidated greenish sand, about 5 or 6 ft below the top of the section. Fitton also noted a section in the undercliff which must have been within landslip. Topley (1875, p. 128) described a section " a little east of Mill Point " (221353), 1620 yd W. 20° S. of Folkestone Harbour Station, showing the following upward succession : clayey sand, slate-coloured at top, but greenish below, 2 ft ; clayey sand, slate-coloured in mass, but green when broken, interbedded below with thin layers of sand ; the lower part of this bed was often concreted into large lenticular masses 18 in thick, showing on the weathered surface curious markings, somewhat worm-like, but twisted about and intermixing, which weathered brown and stood out from the mass which weathered light green, 18 to 24 ft ; below greenish sandstone weathering brown, 1 ft. In 1956 one of the concretions described by Topley, wreathed in ironstone and measuring 5 ft long and 1 ft thick, was visible near the base of the beds. In green silts, which were also exposed, were numerous pyrite nodules, some of which enclosed pyritized wood, and also a lens, 2 in thick, of green and blue fuller's earth. The lowest beds are gritty in texture (p. 78). The strata lie within landslip, and presumably Topley's section was similarly placed.

In a borehole in Tontine Street (231362) Sandgate Beds were described by the drillers as 92 ft thick and by Topley (*in* Whitaker 1908, pp. 137–8) as 107½ ft, composed of sandy clay and sand, 101½ ft, on coarse sandstone, 9 in, on green sandy clay, 5 ft 3 in.

Casey (1961, p. 528) saw the junction with the Folkestone Beds in excavations for the promenade below the East Cliff. He described the beds as yellow-green silty sand. Price (1874b, p. 137) recorded " black sands " of his lowest division of the beds at low tide level east of the harbour. This division could not occur hereabouts without there being a structural disturbance (see also Casey 1961, p. 524). Topley (1875, pp. 138-9) described this exposure, which is now hidden by marine sand, as " dark, loose, somewhat clayey sand " and listed fossils, most of which were found in " dark coloured ferruginous masses ".

At the harbour, two trial borings on the British Railways Pier (235358), 150 yd E. 20° S. of the station, proved green sand and clay of the Sandgate Beds at 29 ft 3 in and 30 ft below O.D. They pierced the beds for 35 ft 9 in and 31 ft 9 in respectively without proving the base.   J.G.O.S.

In the Folkestone Borehole, Lamplugh, Kitchin and Pringle (1923, pp. 29–31) allotted 89 ft of strata to the Sandgate Beds, on the evidence of the driller's log. The Dover No. 1 Borehole showed 72 ft of dark grey and greenish grey glauconitic sandy clay and clayey sand. In the Ottinge Borehole (Lamplugh, Kitchin and Pringle 1923, p. 54) the thickness allotted to the Sandgate Beds was 15 ft 9 in, consisting of loose greensand and hard green sandstone ; however the correctness of this classification has been queried above (pp. 79–80).

In the Fredville Borehole the Sandgate Beds comprised : 5 ft of slightly glauconitic muddy sand, on 14 ft of mingled greensand and clay in varying proportion, on 14 ft of sand and clay, on 5 ft of dingy grey silty clay with small nodules, on 5 ft of greenish grey micaceous silt with glauconitic grit at the base (43 ft in all). Ritchie (1920, p. 178) recorded 36 ft of Sandgate Beds in the Snowdown Colliery No. 3 Shaft, but Brown (1923, pl. 1, fig. 7) showed only about 21 ft.

The Adisham Borehole revealed a reduction in thickness to 8 ft 5 in, consisting of brown glauconitic sand (6 ft 5 in) on dark grey clayey sand (2 ft), the Folkestone Beds being absent (Pringle 1928, p. 78) ; the Sandgate Beds were probably 10 ft thick at Walmestone (Lamplugh, Kitchin and Pringle 1923, pp. 181–2), while at Harmansole 23 ft 6 in of strata were apparently of Sandgate Beds aspect, although Lamplugh, Kitchin and Pringle (1923, pp. 112, 114) did not specifically separate them from the thin development of Folkestone Beds above. G.B.

FOLKESTONE BEDS

Westwell—Ashford—Willesborough Lees.

Cornes (*in* Dewey and others 1925, p. 262) mentioned that the railway cutting near The Warren (000443) showed a well-defined pebble band full of small pebbles of quartz and lydite. This was not at the base of the Folkestone Beds, as he supposed, however, but must have been at least 20 ft above it.

In a borehole at the Waterworks at Westwell (p. 66) the Folkestone Beds are assigned a thickness of 118 ft. Specimens show that at about 40 ft depth glauconitic clay at the base of the Gault rests on pale green loamy sand with phosphatic nodules. The bed recorded as " ironstone " at 46 ft 6 in to 48 ft 9 in in the driller's log (Buchan and others 1940, p. 45) is in fact a phosphatic nodule band. Green sands occur down to about 50 ft, below which depth the bulk of the formation down to 131 ft consists of fine to medium-grained greenish grey sand with scattered glauconitic grains and some mica. Below 131 ft occurs medium to coarse-grained or pebbly sand, interbedded with seams of dark grey, glauconitic sandy clay. The base of the Folkestone Beds is taken at 158 ft, at the base of the lowest bed of pebbly sand. An unusual rock type is a dark brown coarse-grained sideritic glauconitic sandstone from 138 ft, which in thin section (E 24962, described by Mr. R. W. Elliot) shows rounded and subangular grains (from about 0·2 to 1·28 mm diameter) of quartz, granulitized quartz, orthoclase, microcline, glauconite, chert, and glauconitic mudstone and sandstone, with shell fragments, in a matrix composed of small rhombs and grains (generally 0·02 to 0·04 mm diameter) of siderite ($\omega = 1·80$), with interstitial limonite. The thickness of the rock is not known ; specimens labelled 137–138 ft and 144 ft were of medium to coarse-grained greyish green sand, so the brown sideritic sandstone was evidently less than 6 ft and may have been only a foot or two thick. It resembles iron ore from the Lower Greensand of Seend, Wiltshire (E 12658–60), described by Hallimond (1925, pp. 88–9, and fig. 36, plate vii), except that the latter contains more grains of limonite-stained clay than of quartz, and includes limonite ooliths. Hallimond considered that, being a ferric mineral, glauconite could only originate in an oxidized facies, and not in a ferrous ore, and that therefore cementation by siderite of the Seend ore was a late phase. However, his suggestion that it might even have been so late as to have been connected with the present land-surface, the reduction of ferric ores by vegetable products being not uncommon, could not be applied to the rock from the depth of 138 ft in the borehole at Westwell.

A pit on the north-east side of Sandyhurst Lane (005457), $\frac{3}{8}$ mile W. by S. of Lenacre Hall, exposes some 28 ft of sands. The lowest 20 ft are strongly current-bedded to the south-east, with glauconitic streaks and lenticles of pale grey chert or cherty sandstone up to 6 in thick developed along the foreset beds ; the upper 8 ft comprise 2 to 3 ft of yellowish brown sand with unevenly scattered tiny quartz pebbles, overlain by medium-grained sand with faint horizontal banding. The chert contains occasional shells, including *Neithea sp.* cf. *morrisi* (Pictet and Renevier). Casey (1961, p. 541) recorded that just below the soil at the top of this section he found the basal nodule-bed of the *mammillatum* Zone with *Sonneratia kitchini*. He suggested that the sands with seams of cherty sandstone, from their resemblance to those in the East Cliff at Folkestone, were in the *regularis* Subzone. A pit 200 yd to the south, on the opposite side of the lane, exposes 30 to 35 ft of slightly glauconitic sand, strongly current-bedded in their lower part but less so at higher levels, and containing occasional concretionary masses of dark brown ferruginous sandstone but not chert. Above are seen 8 to

10 ft of poorly graded, fine-grained to pebbly sand with a general resemblance to the pebbly sand in the pit north of the lane.

A section (027427) on the north side of the Ashford by-pass road 950 yd S. 22° W. of Conningbrook showed 9 ft of pale grey, fine-grained sand including, 6 ft above its base and an estimated 10 to 20 ft above the base of the Folkestone Beds, a 9-in seam of pale grey waxy clay resembling a fuller's earth. An X-ray analysis by Mr. B. R. Young reveals that the clay consists mainly of mont-morillonite, but there is a very small amount of quartz present as impurity (X-ray film X 2113). Similar beds at about the same horizon were noted in a temporary section on the south side of the Ashford road 500 yd S. 22° W. of the chapel at Willesborough Lees, where, resting with a sharp base on fine grey sand seen for 4 ft, the clay was 1 ft 9 in thick and passed up to fine white sand seen for about 10 ft. Further traces of the clay were seen in a low cutting on the main road at Willesborough Street (039417).

A sand-pit (039433) 800 yd S. 20° W. of Blackwell Farm showed:

	Ft
Gault:	
Glauconitic clay　...　...　...　...　...　...　...	1
Folkestone Beds:	
Glauconitic loam packed with pale grey coarsely sandy phosphatic nodules with *Douvilleiceras sp.*　...　... about	2
Gritty coarse greenish brown sand ...　...　... seen for	1
Unexposed (probably coarse-grained sands)　about 10 ft to	15
Loamy or clayey coarse glauconitic sand　...　...　...	2½
Greenish grey clay with scattered glauconite and sand grains (in places very glauconitic)　...　...　...　...　...	2½
Greenish grey to buff loamy coarse glauconitic sand with scattered clay pellets, and quartz and siltstone pebbles, especially in lower part. Base sharp and uneven ...　...	5
Fine-grained, white slightly micaceous sand with buff slightly loamy layers, and small ferruginous concretions　...　... seen for about	8

The *mammillatum* nodule bed forms a prominent break of slope around outliers of Gault clay which cap hill tops near Blackwall Farm.

Clay seams in the lower part of the Folkestone Beds are the probable cause of springs in the wooded valleys between Willesborough Lees and Hinxhill. B.C.W.

Hinxhill—Brabourne Borehole—Brabourne Lees—Stone Hill.

At Wye College, a boring for water (055469), 150 yd E. 15° S. of Wye church, encountered Folkestone Beds at a depth of 250 ft and proved " Green sand ", 6 ft, resting upon " Sand ? ", 10 ft (Whitaker 1908, p. 219). Nodules of the *mammillatum* Zone yielding *Douvilleiceras sp.* were noted in a ditch (046433) 100 yd W. of Goodcheap Farm, where they were underlain by brown sand and overlain by Gault clay.

At Hinxhill Pumping Station the 1942 borehole (p. 81) struck the Folkestone Beds at 55 ft and proved the following descending sequence to 170 ft: soft grey sandstone, 2 ft; dark sandy clay, 7 ft; greensand and nodules, 8 ft; medium grey sand, 9 ft; hard bed, compact sand, 1 ft; black clay with green sandy streaks, 5 ft; fine grey sand, 18 ft; dirty sand with black clay, 4 ft; fine grey sand, 11 ft; hard dirty sand with clay, 50 ft. In the older borehole (1899) the record shows " Rock and shells " (thought to be the *mammillatum* nodule bed) 1½ ft, with base at 57 ft depth. The succession below is as follows: green sand and pyrite, 23½ ft; brown clay, 6½ ft; hard sand, 8½ ft; rotten brown clay, 7½ ft; overlying 77 ft of sand (Whitaker 1908, pp. 155–6).

Grey clays, with which small springs are associated, crop out 300 yd S.W. of Goodcheap Farm. Other dark grey clays are seen as short streaks in brown gritty sand for 4 ft overlying 1 ft of grey and brown medium to coarse-grained sand 90 yd E. 42° N. of Ouseley (050423). The coarse sand at the top of the Folkestone Beds (p. 54) is perhaps only 10 ft in thickness where it forms a plateau near Hinxhill. East of the woodland north of Quarrington this upper-most division of the Folkestone Beds thickens and apart from the uppermost few feet is generally finer in grain. It rests directly on the fine sands of the lowest division and this arrangement of fine sands below and coarse sands above continues eastwards in varying proportions.

A small pit (049421) in the middle division 200 yd S. 10° W. of Ouseley shows 3 ft of medium-grained well-bedded dark brown sand with light grey seams. A section at Fallon (073420), 750 yd E. 24° N. of Seeley, includes the *mammillatum* Zone: 4 in of hard greenish ferruginous sand, iridescent in places, with streaks of phosphate, numerous gritty phosphatic nodules, very small quartz pebbles and green siltstone pebbles, are overlain by 2½ ft of brown coarse speckled sand with occasional phosphatic nodules, below 1 ft of very dark brown ferruginous sand with phosphatic nodules, in turn overlain by 6 in of Gault. The following fossils were found: *Cucullaea glabra, Beudanticeras newtoni* Casey and *Douvilleiceras sp.*

No cores of the Folkestone Beds were taken in the Brabourne Borehole. The beds were struck at 66 ft and proved as follows: dark grey sand with pieces of black sandstone (? *mammillatum* Bed), 6½ ft; with 111½ ft of grey sand, apparently all incoherent and without stone bands, below (Lamplugh and Kitchin 1911, p. 36).

An old sand-pit (074408) 100 yd S. of Brockham shows very well the general south-easterly dip of the current bedding in the 12 ft of buff-brown ferruginous coarse sand exposed. Grey clays, sandy clays and glauconitic clays crop out on the north and east sides of a hill 650 yd W. 40° S. of Brockham. The clays fall northwards with a local increase in dip and appear to be impersistent. Some 2 ft of brown and grey clay with 'race' nodules, at a horizon very near the top of the Folkestone Beds, are exposed (075404) 550 yd S. 8° E. of Brockham. At Brabourne Lees, 250 yd W. 10° S. of the church, a partially degraded pit (080402) shows 1 to 4 ft of drift overlying 12 ft of buff to brown coarse current-bedded sand, with some iron-pan near the surface in parts. Towards the eastern end of the pit a line of light grey gritty, dark-speckled phosphatic nodules of the *mammillatum* Bed is seen and is overlain by 3 ft of gritty brown coarse sand. South of the roadway (084401) 400 yd E. 40° S. of the church at Brabourne Lees, sand at a somewhat lower horizon is exposed below 0 to 9 ft of drift (p. 233). The sands, seen for 24 ft, are current-bedded, buff in colour, with irregular bands and streaks of ferruginous brown staining. At the top of part of the section is a good example of a woodland podsol, consisting of a hard dark brown iron-pan of irregular thickness, overlain by 2 ft of grey leached sand with flints imbedded.

Fine sand crops out south of Brabourne Lees and is faulted against Sandgate Beds in part. Some 450 yd S. of Pound House is a fine section (090401) in the Granary Court Sand-pit (Plate IIIB) showing up to 7½ ft of drift (p. 266) and 7½ ft of Gault (p. 99), overlying Folkestone Beds as follows:

		Ft
4. Coarse brown ferruginous sand with phosphatic nodules ...		2¼
3. Coarse green clayey sands with fibrous fossil wood; phosphatic nodules at base		4

G

Ft

2. Coarse current-bedded buff and brown speckled sand with
   occasional ferruginous lumps   50
1. Striped, coarse gritty sand and coarse buff sand, current-bedded
   with grey clay seams and streaks. Very gritty in uppermost
   6 in to 1 ft, with quartz and fine sandstone pebbles up to
   ½ in long and a few ferruginous tubes infilled with leached
   sand and solid ferruginous rods   15

Numerous small slump structures were noted among the current-bedding. The
following fossils were collected from beds 3 and 4 (*mammillatum* Zone):
Perissoptera sp., Inoceramus salomoni d'Orbigny, *Linotrigonia (Oistotrigonia)
fittoni* (Deshayes), *Resatrix (Dosiniopsella) vibrayeana* (d'Orbigny); *Beudanticeras
newtoni* and *Douvilleiceras mammillatum* (Schlotheim); *Cyclothyris sp.* was
found in the underlying sands. "Arborescent polyzoa" in sparsely distributed
nodules immediately below bed 3 have been found by Casey (1961, p. 535). He
equated these nodules with similar nodules at Swan Lane (below) and Sandling
Junction (p. 89) and used this horizon as evidence of an unconformity at the
base of the *puzosianus* Subzone.

The lower very coarse sands are exposed again (092399) 580 yd S. 17° E. of
Pound House. A lower face shows 4 ft of dark and light grey striped current-
bedded sand in which the darker beds are a very coarse grit, while a 1-in
clayey seam truncates the foresets about 1 ft below the top of the bed; these are
overlain by coarse dark grey gritty sand with streaks and seams of dark grey
clay. An upper face shows 12 ft of coarse brown speckled sand with irregular
ferruginous-stained layers, the uppermost 6 ft being current-bedded.

Between Lilyhole and Stone Hill fine sands crop out on the upper slopes of
the valley, while springs mark the top of the underlying Sandgate Beds. The
coarse sands occupy the higher ground below the river gravels, and south of Park
Farm form a marked dip-slope inclined 1½° to the north-east.   J.G.O.S.

Stanford—Sandling Junction—Newington—Horn Street.

At a sand-pit (File's Pit) (119392) on Swan Lane, 450 yd N. 15° W. of Hope
Farm, the section in 1955 was:

		Ft	in
5–9. Gault : (p. 99) ... ... seen for		7	0
Folkestone Beds:			
4. Closely packed phosphatic nodules in two layers, in a partly cemented yellow and red weathered glauconitic loam matrix. Fossils are few but include *Inoceramus salomoni* and *Douvilleiceras mammillatum*		1	0
3. Green, slightly loamy, fairly coarse-grained glauconitic sand, weathered brown		–	10
2. Bed of close-packed well-rounded pale grey phosphatic nodules, with many fossils, including *Anchura sp., Teredo* in wood, *Beudanticeras newtoni* and *Douvilleiceras mammillatum*		–	6
1. Greenish grey coarse and gritty glauconitic sand, current-bedded to south and east ... ... about		15	0

The north–south fault shown through here on the geological map appears
on the face of the pit as a set of three or four small faults close together, with
a total downthrow west of about 6 ft. Casey (1961, p. 535) recorded that 4 to
10 ft below the top of the lowest bed are sparsely distributed phosphatic nodules
with arborescent polyzoa, exactly like those found in bed 5 of the *anglicus*
Subzone of Sandling Junction. He noted the top 12 to 16 in of this bed to be

patchily indurated into a yellowish sandrock and riddled with the dark-coloured vertical pipings as seen below the *mammillatum* Zone at Sandling Junction, and the bed to be overlain with a sharp junction by bed 2. He placed the three nodule layers (in beds 2 and 4 of the above section) in the *puzosianus* Subzone.

An outcrop 200 to 300 yd wide of coarse sands at the top of the Folkestone Beds has been traced south-eastwards from the Swan Lane pit as far as Perry Wood. In the underlying part of the Folkestone Beds between these places occur much fine-grained yellow sand and, particularly near the base of the formation, loamy greenish grey sand. Springs issuing from these fine-grained sands give rise to the marshes of Gibbins' Brook, west of Brook Farm (p. 281), and of adjacent valleys.

The Sandling Junction Sand-pit (147371) is 200 yd N. of the railway station. The following section of Folkestone Beds is abridged from Casey (1961, p. 533):

			Ft	in
17–19.	Gault: (p. 99)	seen for	5	0
	Folkestone Beds (total about 62 ft):			
16.	Band of phosphatic nodules in a matrix of green sandy clay, weathering reddish brown ; in places two lines of nodules may be made out		1	0
10–15.	Interbedded glauconitic sand and sandy limestone...		12	5
9.	Clusters of small black phosphatic nodules and pebbles disposed in a gently undulating line. Occasional doggers of sandy limestone ; matrix coarse yellowish green sand ... ...1 in to		–	3
8.	Sharp yellow sand with lines of iron-staining ...		3	0
7.	Reddish brown sandstone (Red Bed)		1	0
6.	Yellowish sand with abundant small pebbles, partially indurated		1	0
5.	Very coarse sand with glauconitic and clayey laminae, steeply current-bedded. Phosphatized and semi-phosphatized nodules at the base... 1 ft 6 in to		2	9
4.	Coarse sand with clayey streaks ... ... 3 ft to		3	10
3.	Sand as above but steeply current-bedded		12	0
1–2.	Pale sands with wisps and pockets of bluish clay. Rotted ironstone concretions, mainly in top 3 ft, absent from lowest 10 ft		25	0

Beds 1 to 9 represent the *anglicus* Subzone, beds 10 to 15 the *milletioides* Subzone and bed 16 the *puzosianus* Subzone. In 1955 the quarry face showed only 36 ft of beds below the Gault, down to 5 ft below the top of bed 2. The stone bands, in beds 7 and 15, comprised chert layers up to 6 in thick and massive doggers of hard pale grey sandy limestone 6 in to 2 ft thick and up to 4 or 5 ft long. Casey (1939 ; 1960b ; 1961, pp. 533–5) listed fossils from this pit and discussed their mode of preservation.

A disused roadside sand-pit (159365) 750 yd S. by W. of Stone Farm showed 2 ft of fine greyish sand overlain by about 12 ft of greenish grey medium to coarse-grained sand with gritty and clayey streaks and, about 3 ft up, a 1 ft thick irregular ironstone bed.

Fine-grained sands in the lower part of the Folkestone Beds crop out in Sandling Park and are exposed in the eastern entrance-cutting to the Saltwood Tunnel (163370) for 20 to 30 ft up from its floor. Within them, at 17 ft up, is an 8-inch bed of soft orange-brown fine-grained ferruginous sandstone, containing tiny quartz and siltstone pebbles and fossils including *Corbula* cf. *truncata* J. de C. Sowerby, ' *Lucina* ' *sp.*, *Pterotrigonia sp.*, *Resatrix* (*Dosiniopsella*) *sp.* ; and an indeterminable ammonite, probably *Hypacanthoplites*. At about this level ferruginous concretions with fossils were found in making the tunnel (Simms 1843, p. 208), which, according to Casey (1961, pp. 531, 533), are on the same horizon as the ironstone concretions in bed 2 of the Sandling Junction Pit ; the fossils include the type of *Hypacanthoplites simmsi* (Forbes). In the top 50 ft of the cutting are occasional exposures of medium to coarse-grained sands. The presence of a spring 40 yd west of the western tunnel mouth (153369) indicates that the Sandgate Beds are there not far beneath the cutting floor. The total thickness of the Folkestone Beds here must be at least 100 ft and is possibly 120 feet. Simms (1860, pp. 70, 93, 142) described the ground excavated for a pilot heading along the line of the tunnel as dark-coloured sand and clay, nearly black, which became lighter as it dried. The water it held made it of the consistency of soft mud, but when the heading was drained the ground was a dry porous sand. Topley (1875, p. 128) followed Simms (1843 ; 1860) in describing the Saltwood Tunnel as being partly driven through the Sandgate Beds, though Drew, in the Old Series Geological Map (Sheet 3), showed it wholly in Folkestone Beds.

A quarry (166371) north of the railway and ¼ mile east of the tunnel (the Beachborough Hill pit of Casey 1939, p. 371) showed in 1955 the following beds:

		Ft
5.	Greenish grey, coarse glauconitic sand, with oval doggers of hard sandy limestone 6 in to 2 ft thick and some cherty seams 2 to 3 in thick. A nearly continuous line of limestone doggers occurs at the base ... ...seen for about	12
4.	Greenish grey, coarse and pebbly current-bedded sand, with occasional black phosphatic nodules midway ... ... 2½ ft to	3
3.	Coarse glauconitic sand, strongly current-bedded in several layers. Iron-staining occurs in streaks along current-bedding planes, and makes irregular concentric patterns crossing the bedding	28
2.	Pebbly sand with pebbles of quartz, quartzite and siltstone, mostly of ¼ to ½-inch diameter ... ... .up to	2
1.	Greyish white glauconitic sand, very even-grained ... seen for	4

A lane cutting 200 yd W.S.W. of Frogholt (175375) exposes 10 ft of coarse sand (overgrown) overlain by a 1-ft bed of phosphatic nodules in a clay matrix, overlain in turn by 3 feet of weathered glauconitic Gault clay. A large disused sand-pit (180369) 600 yd S.S.W. of Newington church showed (in 1955) 4 ft of glauconite-speckled pale grey current-bedded medium-grained sand, with chert seams along the bedding planes ; overlain by 2 ft of greenish grey coarse-grained compact sand with scattered phosphatic nodules ; overlain in turn by greenish grey glauconitic sand with massive doggers of limestone and bands of chert, seen for 15 ft. These beds appear to correspond with beds 3, 4 and 5 of the Beachborough Hill pit. Casey (1961, p. 532), giving the locality as just south of the railway bridge at Newington, recorded a much more complete section, seen in 1939 (see Fig. 1).

Near the crest of the escarpment ¼ mile S. 35° W. of Scene Farm an old sand-pit (172355), reopened in 1955, showed 20 ft of current-bedded glauconitic sand comprising alternate beds each 5 to 10 ft thick of, respectively, pale grey medium-grained sand and grey ill-graded fine to coarse-grained loamy sand, weathering brown. Some concretionary ironstone was present. About 5 ft of brown sandy loam hillwash with angular flints and weathered sandy limestone fragments capped the section.

A cutting on the disused Canterbury railway, where it is crossed by the main Folkestone road (190370), shows 4 ft of greenish grey to orange compact sand, current-bedded to the south, with occasional pale grey, coarsely sandy phosphatic nodules scattered in the top 1 ft. This is the lower of the two nodule beds formerly seen here (Topley 1875, p. 147; Casey 1961, p. 532), and Casey regarded it as equivalent to the 'main *mammillatum* Bed of Copt Point; the top nodule bed, 2 ft above it, he assigned to the *puzosianus* Subzone. He noted that the nodule beds could be followed on the north side of the cutting on the main Dover–London line (187369) for about 150 yd, due south of the Star Inn, Newington. In the most westerly exposure the lower nodule bed was missing. West of this point as far as Brabourne all exposures of the *mammillatum* Zone show the *puzosianus* Subzone only.

In the valley at Horn Street the base of the Folkestone Beds is marked by numerous springs. A sand-pit (190352) south of Sandy Lane (p. 83) and on the 200-ft contour exposes 12 ft of fine-grained 'pepper-and-salt' glauconitic sand including, near the top of the face, a 2 to 3-ft bed of grey glauconitic silty clay. This clay must be at least 40 ft above the base of the Folkestone Beds.

B.C.W.

Cheriton—Shorncliffe Station—Sandgate.

On the north side of the railway cutting (194368), 1 mile W. of Shorncliffe Station, beds of hard, greyish white to white cherty sandstone, up to 21 inches thick, are intermittently exposed. Dr. Dearnley has described a thin section from one of these beds (E 28880) as a medium-grained sandstone consisting of subangular to rounded quartz grains, organic fragments partly replaced by microcrystalline silica, and occasional rounded phosphate and glauconitic grains, in a cementing matrix composed almost entirely of chert.

An old pit around gardens at Wood Cottages (197372), 650 yd E. 24° S. of Danton Farm, displays 7 ft of coarse sand, brown near the surface, and white, brown and glauconitic below.

In a trench (209372) 1450 yd W. 4° N. of Park Farm the Folkestone Beds comprised 6½ ft of fairly coarse glauconitic white sand enclosing scattered rounded 'doggers' of coarse glauconitic sandstone up to 2 ft by 4 ft, some of which were surrounded by a thin shell of ferruginous sand, with, in addition, near-horizontal layers of brown iron-stained sand 1 to 2 in thick and 1½ ft below the top a layer of similar sand some 6 in thick which passed laterally into an irregular lens 3¼ ft thick; overlain by the *mammillatum* Bed, consisting of a persistent bed of yellow, iron-rich sand, with many phosphatic nodules, brown, iron-cemented patches and phosphatic patches and containing ammonite fragments and other fossils, 1 ft 1 in; below light yellowish brown coarse sand with fragments of fossil wood, 2 ft 7 in; below Gault (p. 100). Dr. Dearnley described a thin section from a comparatively firmly cemented part of the *mammillatum* Bed (E 28881) as a glauconitic sandstone poorly cemented by haematite, and consisting of irregularly sized subangular fragments of quartz, occasional feldspar and vein quartz and rare microquartzite, with rounded grains of glauconite quite common. He also contributed the following description of a dogger in the sand below (E 28882): A glauconitic calcareous quartzitic sandstone consisting of relatively well sorted and rounded grains of quartz, some vein quartz and occasional microquartzite and chert, together with numerous rounded grains of glauconite and a few detrital shell fragments in a matrix of calcite plates.

Dr. Casey recorded (in MS.) that the *mammillatum* Bed in the Cheriton area, and in particular in an old pit, now obliterated, in the grounds of Harcourt School, consists of several feet of loosely compacted clayey sand with scattered phosphatic nodules.

A sand-pit (215370), on Folkestone Golf Course, 950 yd W. of Park Farm, shows 7 ft of coarse white to brown glauconitic quartz sand, with occasional hard iron-stained masses.

G.B.

An old sand-pit (221369) 350 yd S. 37° W. of Park Farm and a nearby trench showed 10 to 15 ft of unbedded, coarse, loose, glauconitic sand, with a band of large doggers of exceedingly hard, coarse sandstone, tabular or tending to be spherical or of the form of coalesced spheres 2 ft in diameter. Just west of the junction of Park Farm Road and Park Road (224367), a sewer trench, running east–west, displayed 20 ft of sand and current-bedded glauconitic cherty sandstone and 160 yd to the east-south-east a 19-ft-deep shaft in Park Road (226367) and a tunnel leading from the bottom of it for 50 yd to the junction of Park Road and St. John's Church Road showed coarse-grained greyish green glauconitic sand with horizontally lying impersistent beds of massive sandstone, generally about 2 ft thick.   B.C.W.

In Shorncliffe Camp, where the beds are about 100 ft in thickness, temporary sections (192358) 2030 yd W. 22° S. of Shorncliffe Station showed 6 ft of coarse grey calcareous glauconitic sand with bands of coarse gritty siliceous sandstone up to 2 in thick. A small pit (191357) 2200 yd W. 28° S. of this station shows 9 ft of light grey micaceous medium to fine-grained sand with thin (up to 1 ft) bands of harder ferruginous sand and occasional reddish brown or red ironstone nodules. Near the old redoubt (193353) 2240 yd W. 33° S. of Shorncliffe Station 4 ft of hard light grey calcareous sand with two bands of dark grey-hearted siliceous sandstone containing moulds of fossils were exposed and 100 yd to the westward a similar thickness of grey micaceous medium-grained sand, somewhat lower in the beds, was seen in a lane side. In the grounds of Encombe (198352), 1800 yd S. 43° W. of Shorncliffe Station, Fitton (1836, p. 122), in describing what he considered to be his middle division (Sandgate Beds) of the Lower Greensand, recorded the presence of a line of ferruginous nodules. The exposure was extremely poor in 1956 but one fossiliferous ironstone nodule was observed *in situ* about 35 ft below the top of the cliff and well within the Folkestone Beds. Fossils from this nodule are *Lamellaerhynchia caseyi* Owen; '*Natica*' sp.; *Cardita*' sp., *Gervillella sublanceolata, Ensigervillia forbesiana, Lucina* sp., *Pterotrigonia mantelli* and *Thetironia minor*.

Fitton (1836, pp. 119–20) also described a section " about midway between Risborough and the east of Sandgate ", probably at a place (201357) about 1300 yd W. 40° S. of Shorncliffe Station and now in landslip. He noted about 16 ft of Folkestone Beds with concretionary calcareous and siliceous stone bands, chert bands and a 2 to 8-in pebble bed of quartz, " flinty slate, and dark green specks ". In a landslip scar (201364) 910 yd W. 2° S. of Shorncliffe Station, 9 ft of grey sand are exposed, with three irregular but well-defined chert seams up to 6 in thick, one of which is enclosed by a grey gritty sandstone dogger $2\frac{1}{2}$ ft thick. This reveals the relative ages of the two forms of cementation in the Folkestone Beds at least at this locality.

There is coarse glauconitic sand with seams of grey and white chert up to 6 in thick containing *Oxytoma pectinatum* (J. de C. Sowerby) and some enclosing green siltstone pebbles, exposed for 7 to 10 ft in the cutting on the Sandgate–Folkestone road (212355) 1050 yd S. 11° E. of Shorncliffe Station. Fitton (1836, p. 120) figured the cherts following the current-bedding in this locality. Casey (1961, p. 531) recorded the bottom stone band of the *regularis* Subzone with large *Douvilleiceras* and *Leymeriella* hereabouts. Near the western Gault outlier in Folkestone, temporary sections (213357) 860 yd S. 12° E. of Shorncliffe Station entered brown to green clayey sand of the *mammillatum* Zone to $3\frac{1}{2}$ ft; weathered gritty phosphatic nodules were strewn at the top of the section. *Beudanticeras newtoni, Douvilleiceras* cf. *monile* (J. Sowerby) and *D.* aff. *orbignyi* Hyatt were collected.

Mackie (1860, p. 125) stated that his "junction bed" (Sulphur Band) at the top of the Folkestone Beds could be observed along the cliff top for half a mile west of the Leas. The promenade now conceals this horizon. Topley (1875, p. 147), however, recorded a section at the "west end of the Lees" showing a different development of the *mammillatum* Zone from that at Copt Point, as follows:

	Ft.	In.
" *a*. Clay, with green grains	2	0
b. Phosphatic band, with bored wood	0	6
c. Clayey sand, with green grains	1	3
d. Phosphatic band, with bored wood	0	1
e. Sand, with fewer green grains	1	0
f. Phosphatic band	0	1
g. Sand, as next above, with a nest of phosphatic nodules in one place	3	0

In such a section as this the exact junction is not obvious; in fact there appears to be a gradual passage between the true Gault Clay (*a*) and the sand at the bottom, which is no doubt part of the Folkestone Beds."

Between Sandgate and Folkestone Harbour the Folkestone Beds form the steep main cliff above the undercliff. The sandstone and chert beds crop out at many localities. Near the martello tower (211353) 1220 yd S. 7° E. of Shorncliffe Station 20 ft of coarse greyish white sand with chert seams and doggers of sandstone up to 2 ft thick are exposed, while 3 to 4 ft above the sandstone doggers there is a band of small (up to ½-in) black phosphatic nodules and quartz pebbles.

Folkestone—borehole and shaft records.

The borehole at the Metropole Hotel, Folkestone (216354), proved 51 ft of Folkestone Beds, from a horizon near to the top of the formation down to the lowest stone band (Whitaker 1908, p. 136). Immediately west of the Toll-bar 1550 yd W. 18° S. of Folkestone Harbour Station is a section described by Fitton (1836, p. 123). The section exposed here in 1956 was as follows:

	Ft
Sand with chert seams	seen for 2
Coarse grey sandstone with green siltstone pebbles and grains	1
Coarse green and grey soft sand with ragged chert seams	10
Coarse grey cherty sandstone with phosphatic nodules (to ½ in) and green siltstone pebbles in lowest 6 in ...	1½ ft to 2½
Coarse light grey hard sand with scattered phosphatic nodules (½ in), large oysters and ragged chert seams (to 6 in in thickness)	7½
Green glauconitic soft sand with harder lumps ...	seen for 3
Unexposed strata	about 5
Hard grey speckled sandstone, nodular, cherty at base...	2
Hard grey and brown speckled sand, nodular at base, current-bedded in part with occasional ½-in quartz pebbles and brown phosphatic patches. There are scattered bands of phosphatic nodules (usually ½ in in size) 6 in, 4½ ft and 9 ft below top of bed. Chert is common and occurs either as incompletely cemented impersistent beds up to 6 in. in thickness with ragged appearance where weathered, or as larger doggers, more completely cemented, but with honeycomb weathering at top and bottom	19

Ft

Interbedded brown and coarse brown-speckled sand
with occasional black phosphatic nodules at base ...   1½
Sandgate Beds–bored in places at top: (p. 83) ...   seen for 4

The section at this locality noted by Casey (1961, p. 531) included higher
Folkestone Beds for 13 ft and must have represented almost the complete
thickness of the beds. Immediately east of the Leas Cliff Hall (225356), 1000 yd
W. 15° S. of Folkestone Harbour Station, a section at the horizon of Folkestone
stone is as follows, in upward order: coarse grey sandstone with cherts, 3 ft;
below coarse glauconitic sand with doggers of cherty sandstone 1 ft thick and
scattered ragged cherts, 7 ft; unexposed strata, 2½ ft; grey cherty sandstone with
quartz pebbles and ½-in long phosphatic nodules, 1½ to 2 ft; unexposed strata,
4 ft; sands with chert seams, about 4 ft.

Casey (1961, p. 531) noted black phosphatic nodules with an *anglicus* Subzone
fauna near the bottom of Remembrance Road (231358), 300 yd W. 20° N. of
Folkestone Harbour Station, and Mackie (1851, p. 262) mentioned a section
(232364) in London Street, 600 yd N. 10° W. of this station, showing Folkestone
Beds with Gault overlying.

In the East Cliff there is an almost continuous section from east of the harbour
to Copt Point showing Folkestone Beds beneath the lower beds of the Gault.
The lowest 15 ft approximately of the Folkestone Beds are normally obscured,
but a full description of the beds and their fossils was given by Casey (1961, pp.
527–31). The basement bed, which is the First Division of Price (1874b), is 1 ft
thick, and consists of firm glauconitic sand with pockets of buff siliceous rock and
black phosphatic nodules, containing a fauna characteristic of the *anglicus*
Subzone of the *jacobi* Zone, white and green-veined quartz, black chert, and
green sandstone fragments up to 1 in long. Locally in the base or centre of
the bed are ferruginous phosphatic concretions 6 to 8 in long, containing a
fauna of the *rubricosus* Subzone. The nodules and concretions are thickly
encrusted with oysters and polyzoa. The overlying 2 ft (Price's Second Division)
consist of clayey greensand with shell fragments and very small black chert
pebbles, overlain by a bed, 0 to 2 in thick, of small phosphatic nodules and
black chert pebbles. This is overlain by greyish green, glauconitic calcareous
sandstone 1 ft 9 in thick, which forms the base of Price's Third Division. The
remainder of the obscured strata comprise clayey sands with one course of
spicular sandstone. The section below the Gault (p. 100) near the eastern
end of the promenade (236363) 620 yd N. 33° E. of Folkestone Harbour Station
is:

	Ft	in
15. Sulphur Band: phosphatic nodules set in a glauconitic grey to yellow clayey matrix	1	4
14. Brown clayey sand	–	7
13. Coarse grey to green speckled sand, very gritty in uppermost 1 ft. Main *mammillatum* Bed apparently reduced to a gritty seam with greyish brown phosphatic nodules 1½ ft below Sulphur Band	5	0
12. Hummocky band of hard coarse grey grit ... ...10 in to 1		4
11. Coarse grey gritty sand	1	3
10. Gritty grey chert ... ... up to –		9
9. Coarse green to brown gritty speckled sand	–	6
8. *Sonneratia kitchini* Bed: brown gritty phosphatic nodules scattered in coarse gritty sand ... ... 4 in to –		8
7 Coarse green to brown gritty speckled sand, very green and gritty in uppermost 6 to 8 ft	16	5

Ft in

6. Ragged grey gritty sand with round coarse gritty sandstone
doggers and ironstone nodules up to 2 in at top   ...   2   3
5. Coarse khaki to brown sand   4   0
4. Ragged grey chert seam   ...   .   –   8
3. Coarse grey to brown gritty sand with chert lenses and a
chert seam near base   9   3
2. Sandstone with cherty patches   1   9
1. Coarse grey to brown sand with chert seams ...   ...   seen for 4   6

The chert seams are particularly variable, some die out entirely, others become
less carious and better defined. Westward of the measured section the sandstone
of bed 2 becomes more prominent and well defined at its upper and lower
surfaces and the distinctive sandstone nodules (bed 6) die out. To the east
a doggery sandstone appears in bed 1 and two seams (up to 3 in thick) of hard
lumpy ironstone develop about 5 ft below the top of bed 9.   J.G.O.S.

The Folkestone Beds section at Baker's Gap, about 30 yd short of the eastern
end of the promenade, was described in detail by Casey (1961, p. 528). It is now
largely obscured, but some 120 yd farther east the following sequence below the
Gault is exposed in the cliff (239364):

Ft in

11. Sulphur Band: pyritous phosphatic nodules embedded in
glauconitic sand   ...   .1 ft to 1   3
10. Greenish brown glauconitic clayey sand   ...   .about –   9
9. Greenish white glauconitic sand. Main *mammillatum* Bed,
consisting of brown phosphatic nodules, fragments of
ammonites and lamellibranchs, and quartz pebbles up to
½ in. in diameter, occurs between 1 ft 4 in and 2 ft below
the base of bed 11   ...   . about 6   0
8. Hummocky bed of hard glauconitic sandstone   ...   ... up to 1   3
7. Greenish white glauconitic gritty sand; a particularly gritty
layer with quartz pebbles, rounded phosphate fragments
and angular phosphatic nodules about the middle repre-
sents the *Sonneratia kitchini* Bed   4   0
6. Sandstone, white and cherty in the basal 3 in   ...   ... up to 1   6
5. Greenish white glauconitic sand, with small brown phosphatic
nodules and shell fragments dispersed between 9 in and
1 ft below the top. (Farther east nodules are only rarely
to be found at this horizon)   1   6
4. Impersistent bed of brown ferruginous sandstone   ...   –   3
3. Greenish white glauconitic sand   ...   .   –   9
2. White cherty sandstone, with soft patches of sand. Occa-
sional large lamellibranchs   ...   . up to 1   0
1. Sand, with lenticular beds of white cherty sandstone. (To
the east a persistent sandstone bed 6 in thick is developed
about 3 ft below the top)   8   0

The *mammillatum* Zone thickens eastwards. About midway between Baker's
Gap and the sewer outfall at Copt Point the Main *mammillatum* Bed is locally
represented by brown phosphatic nodules set in sandstone some 10 in thick, with
its top 3 ft below the Sulphur Band, but 10 yd west of the sewer it consists of
isolated masses of sandstone up to 9 in thick, enclosing phosphatic nodules up
to 3 in. in diameter, quartz pebbles up to ½ in in diameter, and fossils; in places
it is underlain by up to 6 in of coarse glauconitic sandstone. East of Copt Point
the main nodule bed is more generally indurated, forming a reef on the foreshore.

Dr. P. A. Sabine has described a specimen of the Sulphur Band (E 28888), collected at Copt Point (242365), 1200 yd E. 34° N. of the Folkestone Harbour Station and 75 yd east of the sewer outfall, as follows: An ochreous-weathering dark grey nodular glauconitic phosphate-rock with narrow irregular pyrite veins. The rock is seen in thin section to consist of abundant rounded to subangular quartz grains mainly 0·2–0·7 mm across, and glauconite grains of similar size, in a turbid pale brown extremely fine-grained groundmass of high refringence (presumably phosphatic), containing minute flakes of mica. A pebble of impure fine-grained sandstone is present. There are occasional brown patches perhaps of purer phosphate. The rock is traversed by narrow branching veinlets of pyrite (with a trace of a colourless fibrous mineral present). The Sulphur Band, here about 5 in thick, rests upon the following beds:

		Ft	in
10.	Dark brownish grey clayey sand, with streaks and patches of dark grey clay	3	0
9.	Loose greenish white glauconitic sand, brown-stained in places; rare pyritic and phosphatic concretions up to 3 in long at the top ... ... 9 in to	1	0
8.	Main *mammillatum* Bed: rounded masses of loosely cemented glauconitic sand, enclosing brown phosphatic nodules and fossils ... ... up to	–	10
7.	Loose greenish grey glauconitic sand, brown-stained in places	3	0
6.	A more or less persistent bed of coarse glauconitic quartz sandstone, in doggers up to 9 in, but normally less than 5 in, thick. Comminuted white shell fragments are common ... say	–	5
5.	Coarse brown glauconitic quartz sand ... ... 9 in to	–	10
4.	Rounded masses of glauconitic quartz sandstone up to 7 in thick, but usually less than 4 in, with a thin (½-in) grey muddy or phosphatic layer containing white comminuted shell fragments near the base in places ... ... say	–	4
3.	Coarse greenish brown glauconitic quartz sand, with a few grey clay streaks; the whole loosely and variably cemented, so that it weathers to produce a honeycombed appearance ...	3	6
2.	Greenish brown to grey gritty sandstone	2	7
1.	Greenish brown gritty sand	–	6

A concretion from the top of bed 9 (E 28887) was described by Dr. Dearnley as follows: A medium-grained glauconitic sandstone, consisting of subangular to rounded grains of quartz and glauconite. Some glauconite replaces organic (polyzoan) fragments. The matrix consists of pyrite and a light brown nearly isotropic fine-grained aggregate of clay minerals with scattered pyrite crystals altering to haematite. Local patches of a fibrous colourless mineral occur, probably gypsum, forming a cement between the quartz grains.

The sandstones in beds 6 and 8 were similar in thin section (E 28885–6) to that in bed 4 (E 28884), which Dr. Dearnley described as a glauconitic calcareous quartzitic sandstone, similar to E 28882 (p. 91) but well cemented and containing relatively abundant fragments of a glauconite-cemented siltstone together with occasional fragments of a phosphatic chert. Some intergranular phosphatic cement occurs locally.

The succession of interbedded unconsolidated sand and hard sandstone is eroded by the sea, the sand being removed first, leaving the sandstone beds undercut; these eventually collapse and leave a jumble of sandstone blocks on the foreshore. Some of the fallen masses are large and show a sequence of hard and soft beds dipping seawards. The doggers are commonly very elongated along one axis, trending about E. 35° N. Major joints in the sandstones strike between E. 12° N. and E. 2° N., while minor joints strike S. 42° E. and occasionally N. 43° E.

On the foreshore (242371) east of the Roman Villa near Copt Point, the Sulphur Band can be traced where it protrudes through thin Marine Beach Deposits, its sinuous outcrop being due to minor structural undulations. It is here both underlain and overlain by about 1 ft of coarse glauconitic clayey sands, which blend with the Beach Deposits.

In the Ottinge Borehole the thickness of beds thought by Lamplugh, Kitchin and Pringle (1923, p. 54) to represent the Folkestone Beds was 75 ft 1 in, consisting of greensand with hard cherty and calcareous beds, but the full thickness may be 102 ft (pp. 79–80). These authors (1923, p. 29) put the thickness of the formation in the Folkestone Borehole at 95 ft, based on the borer's log. In the Dover No. 1 Borehole 8 ft of dark green coarse glauconitic sandstone and unconsolidated sand represented the division, while in the Shakespeare Colliery Shafts the Sulphur Band is underlain by about 1½ ft of dark green sandy clay of the *puzosianus* Subzone, resting on 3 to 4 ft of irregular concretionary calcareous sandstone (Lamplugh and Kitchin 1911, pp. 8–9 ; Casey 1961, p. 531).

At Fredville 2 ft of "Clayey glauconitic grit, with brown phosphatic nodules and small pebbles of quartz, lydite, etc., up to ½ inch diameter, some worn and polished, some subangular" were ascribed to "Base of Gault and Folkestone Beds" by Lamplugh, Kitchin and Pringle (1923, pp. 90–2). Ritchie (1920, p. 178) recorded 1 ft of Folkestone Beds in the Snowdown Colliery No. 3 Shaft, while Brown (1923, pl. 1, fig. 7) also showed a small thickness of "greensand rock". About 2 ft of Folkestone Beds were thought to be present at Walmestone (Lamplugh, Kitchin and Pringle 1923, p. 181). The Harmansole Borehole showed 1½ ft of fine-grained quartzose sands, which probably represented the Folkestone Beds (Lamplugh, Kitchin and Pringle 1923, pp. 112–4).   G.B.

GAULT

Kennington—Wye—Hinxhill—Brabourne Lees.

An old clay-pit at Goat Lees, Kennington (015455), now overgrown, was described by Jukes-Browne (1900, p. 84) as showing, in 1896, 6 to 8 ft of dark grey clay overlain by 14 to 16 ft of grey and bluish grey clay with phosphatic nodules. Gault fossils from Kennington in the Geological Survey collection, presumably from the Goat Lees pit, have been redetermined by Dr. Casey as follows: *Inoceramus concentricus* Parkinson, *I. anglicus* Woods, *Nucula (Pectinucula) pectinata* J. Sowerby ; *Dentalium (Fissidentalium) decussatum* J. Sowerby ; *Dimorphoplites doris* Spath, *D. parkinsoni* (Spath), *D. tethydis* (Bayle), *D. chloris* Spath, *Anahoplites planus* (Mantell), *A. daviesi* Spath, *Euhoplites lautus* (J. Sowerby), *E. opalinus* Spath, *E. truncatus* Spath, *E. proboscideus* (J. Sowerby) *E. armatus* Spath, *E. trapezoidalis* Spath, *E. sublautus* Spath, *E. inornatus* Spath, *E.sp.*, *Dipoloceras cristatum* (Brongniart), *Hysteroceras orbignyi* (Spath), *Mortoniceras* cf. *pricei* Spath, *Hamites attenuatus* J. Sowerby, *H. maximus* J. Sowerby, *H. compressus* J. Sowerby, *H. intermedius* J. Sowerby, *H. incurvatus* Brown ; *Eutrephoceras clementinum* (d'Orbigny). Dr. Casey reports that this assemblage indicates a range of strata corresponding to Beds IV to IX of the Gault at Folkestone.

The presence of Gault clay at 10 to 15 ft depth beneath peat and alluvium in the valley north of Hinxhill was proved by augering 500 yd N. by E. and 800 yd N.N.W. of Blackwall Farm. On the slope north of the valley, at the site of an old clay pit (041448) 800 yd N. of the farm, ploughing in 1955 turned up light and dark grey clay with phosphatic nodules yielding *Acila (Truncacila) bivirgata* (J. de C. Sowerby), *Inoceramus concentricus, I. sulcatus* Parkinson ; *Euhoplites inornatus, Hamites gibbosus* J. Sowerby, *H. intermedius* J. Sowerby *distinctus* Spath, *H. intermedius opalinus* Spath, *Hysteroceras orbignyi* and *Neohibolites minimus* (Miller). These indicate the *inflatum* Zone, *orbignyi* Subzone.　　B.C.W.

A borehole at Wye College (p. 86) passed down through Head (p. 227) and Lower Chalk　(53 ft) into Gault at a depth of 65 ft (Whitaker 1908, p. 219). The total Gault thickness was 185 ft, recorded in descending sequence as follows: " Blue clay, with carbonate of lime ", 16 ft ; " Bed full of green sand ", 5 ft ; " Blue clay ", 74 ft ; " Black clay ", 51 ft ; " Sandy rock ", 1 ft ; " Black clay with flints [? nodules] in top 10 feet ", 38 ft.

At the brickworks (049445) 750 yd S. 15° E. of Spider's Castle 22 to 23 ft of beds are exposed, consisting of very dark grey marl which lightens in colour after it has been dug and becomes brown mottled. Small dark grey phosphatic nodules and grey calcareous nodules were noted. Other nodules present have a core of crystalline gypsum surrounded by a red ferruginous and sandy layer and an outer shell of gypsum. There are also some cementstones up to 5 inches in length. Fossils indicate that the section is opened in Upper Gault, *inflatum* Zone, *varicosum* Subzone, at a horizon between Beds IX and X at Folkestone. Cornes (*in* Dewey and others 1925, pp. 263–4) recorded that the upper beds of the Lower Gault were once exposed and he listed fossils from this section. Fossils recently collected by the Geological Survey include: *Nielsenicrinus cretaceus* (Leymerie) ; ' *Aporrhais* ' sp., ' *Solarium* ' sp., *Dentalium (Fissidentalium) decussatum ; Nucula (Leionucula) ovata* Mantell ; *Anahoplites* cf. *planus, Beudanticeras beudanti* (Brongniart), *Epihoplites deluci* (Brongniart), *Euhoplites alphalautus* Spath, *E. vulgaris* Spath, *Hysteroceras* aff. *bucklandi* (Spath), *H. orbignyi, H.* aff. *simplicicosta* Spath, *Idiohamites spiniger* (J. Sowerby), *I. spinulosus* (J. Sowerby), *I. subspiniger* Spath, *I. tuberculatus* (J. Sowerby), *Mortoniceras* cf. *pricei, M. (Deiradoceras) sp.* [juv.] cf. *cunningtoni* Spath, *Puzosia spathi* Breistroffer, *Scaphites (Eoscaphites) subcircularis* Spath and *Neohibolites minimus.*

A borehole (062441), which is not precisely located but is approximately 500 yd W. 14° S. of Brook church, at about the centre of the broad Gault outcrop, reached the underlying Folkestone Beds at 102 ft depth. A temporary section here showed numerous small calcareous nodules overlying 5½ ft of brown mottled clay and below 1½ ft of soil.

At Hinxhill Pumping Station (p. 81) the base of the Gault was reached at 55½ ft depth in the borehole sunk in 1899 and at 55 ft in that sunk in 1942. The Gault was not cored in the Brabourne Borehole. The following descending sequence is extracted from the section given by Lamplugh and Kitchin (1911, p. 35), which is based almost entirely upon the driller's records: drift, 4 ft ; dark blue clay, 57 ft ; sandy clay with green veins, 3 ft ; on black sand (? phosphatic and pyritous base), 2 ft. A total thickness of 62 ft was proved.

The clays are dug intermittently at two localities (088409) close together some 550 yd N. 30° W. of Pound House and degraded sections up to 6 ft deep have been noted. In the north-western digging near the road fork, the following were collected from the *cristatum* Subzone of the *lautus* Zone: *Inoceramus concentricus, I. sulcatus, Anahoplites planus, Neohibolites minimus.* Fossils from this subzone were also found in the surface debris of the south-eastern digging, including *Nucula (Pectinucula) pectinata* and *Dimorphoplites? silenus* Spath. The clays *in situ* here yielded fossils of the lower *subdelaruei* Subzone, as follows: *Inoceramus concentricus, Nucula (Leionucula) albensis* d'Orbigny, *Nucula (Pectinucula) pectinata, Anahoplites mantelli* Spath, *Dimorphoplites* aff. *niobe* Spath, cf. *D. niobe, Euhoplites loricatus* Spath (late form), *E.* aff. *meandrinus* Spath, trans. to *E. lautus bilobus* Spath, *Hamites attenuatus, H. subrotundus* Spath, and *Notopocorystes stokesi* (Mantell).

At a locality (090401) immediately east of Brabourne Lees, 450 yd S. of Pound House, the Folkestone Beds (pp. 87–8) are overlain by 2½ ft of green glauconitic clay, below 5 ft of mottled clay, underlying up to 7½ ft of drift (p. 226). The fossils, most of which were picked off the graded clay slope, indicate the presence of the *dentatus–spathi* Subzone. They include *Hoplites* cf. *canavarii* Parona and Bonarelli, *H. dentatus* (J. Sowerby), *H.* aff. *dentatus densicostatus* Spath, *H.* aff. *escragnollensis* Spath, *H. paronai* Spath, *H. persulcatus* Spath and *H.* aff. *spathi* Breistroffer. J.G.O.S.

Stanford—Sandling Junction—Folkestone—borehole and shaft records.

At a sand-pit (File's Pit) (119392) 450 yd N. 15° W. of Hope Farm, the section of Gault in 1955 was:

	Ft
Head: (p. 229) ... ... up to	4
Gault:	
9. Greyish brown, weathered clay ... ... seen for	4
8. A single line of fine-grained phosphatic nodules, which weather whitish. Many fossils, including *Inoceramus concentricus, Nucula (Pectinucula) pectinata, Hoplites* aff. *dentatus, H. dentatus densicostatus* and *H. spathi.*	
7. Dark greyish brown, slightly glauconitic stiff clay, in places appearing to pass down to the clay below	1
6. A line of phosphatic nodules spaced at 2 to 3-ft intervals. *Hoplites?* collected.	
5. Dark green, highly glauconitic clay	2
1–4. Folkestone Beds: (p. 88) ... ... seen for	17

Topley (1875, p. 147) mentioned a pit ¼ mile north of Stanford church showing 15 ft of clay, with two layers of phosphatic nodules 1½ ft apart. This, and a nearby pit mentioned in the Directory of British Fossiliferous Localities (1954, p. 60) as showing Lower Gault of the *intermedius* to *niobe* subzones, are now overgrown.

The Sandling Junction Sand-pit in 1955 showed, above the Folkestone Beds (p. 89), 2 ft of dark green glauconitic clay with a line of phosphatic nodules along its top, overlain by grey to fawn, weathered stiff clay, seen for 3 ft. A waterworks borehole (148377) made in 1935 near Postling Wents proved 46 ft of Gault on Lower Greensand. The record of another hole at Bluehouse (158379) (Whitaker, 1908, pp. 187–8) suggests that some of the Gault may there be cut out by faulting (p. 231). It proved Gault to 77 ft only, on calcareous sandstone forming the top of the Folkestone Beds.

Near Summerhouse Hill the Gault outcrop is much cut up by faulting. At each end of the down-faulted outlier west of Bargrove, respectively 600 yd N.W. and 300 yd S.W. of the farm, glauconitic clays at the base of the Gault were noted near the boundary with Folkestone Beds sands. The outlier of Gault between Bargrove and Scene Farm has not hitherto been known to exist. It occupies a hollow on the Folkestone Beds dip-slope. There is an old overgrown clay-pit in the wood 250 yd N.E. of Scene Farm, and 300 yd E. of the farm a temporary exposure of greenish grey glauconitic clay and a few coarsely sandy phosphatic nodules indicated the base of the Gault. B.C.W.

Topley (1875, p. 147) recorded that the Folkestone Beds–Gault junction was formerly exposed in the railway cutting ½ mile north-east of Cheriton church, where the section was:

"

	Ft.	In.
a. Sandy clay with phosphatic nodules	2	0
b. Yellowish brown sand	2	0
c. Nodules in brown sand	0	6
d. White and buff sand with stone in places, false-bedded ...	6	0 "

In the active pit of Folkestone Brickworks (205376), 1100 yd W. 15° S. of Castle Hill, the following section was seen:

	Ft	in
Light grey slightly micaceous clay	6	0
Light grey clay, discoloured yellow along cracks and joints; darker grey and blocky, with less yellow staining, between 3½ ft and 6 ft down. Occasional lenticular masses of reddish brown ferruginous siltstone at the top, 3 inches thick and up to 2 ft long, and intricately bored ... ... about	7	0
Clay, dark grey at base, becoming lighter grey upwards; slightly micaceous	5	6
Layer of phosphatic nodules	–	2
Dark grey slightly micaceous clay, with occasional phosphatic nodules	–	11
Layer of dark brown phosphatic nodules and lenses of reddish brown siltstone with intricate light brown borings ...	–	3
Dark grey slightly micaceous blocky clay with occasional phosphatic nodules	2	6

These beds span the Lower Gault–Upper Gault junction. The lowest 3 ft 10 in represent the *lautus* Zone, *cristatum* Subzone, and yielded the following fossils: *Pseudocardia tenuicosta* (J. de C. Sowerby), *Inoceramus concentricus, I. concentricus subsulcatus* Wiltshire, *I. sulcatus, Pinna sp.* ; *Anahoplites* cf. *daviesi ornatus* Spath, *A. planus, A.* aff. *planus, Dimorphoplites chloris, D.? silenus, D.* cf. *tethydis, D. sp., Dipoloceras* aff. *cristatum, D. sp.* [juv.] cf. *bouchardianum* (d'Orbigny), *Metaclavites compressus* (Parona & Bonarelli), *M.* aff. *compressus, M. trifidus* (Spath), *E. armatus, E. inornatus, E. proboscideus, E. sublautus, E. trapezoidalis* Spath, *E. trapezoidalis formosus* Spath, *Hamites compressus, H. gibbosus, Hysteroceras* cf. *symmetricum* (J. de C. Sowerby), *H.* [*Dipoloceras?*] *symmetricum, H. sp.* [juv.] cf. *subbinum* Spath, *H. sp., Neohibolites minimus* and fish debris. The remainder of the section represents the *inflatum* Zone, *orbignyi* Subzone; fossils collected from these beds were as follows: *Serpula sp., Perissoptera sp. Dentalium* (*Fissidentalium*) *decussatum, Pseudocardia tenuicosta, Inoceramus anglicus, I. sulcatus* ; *Anahoplites planus, Dimorphoplites? silenus, D.?* aff. *silenus, Metaclavites trifidus, Euhoplites ochetonotus* (Seeley), *E. Sublautus, Hamites incurvatus, H. intermedius, Hysteroceras orbignyi, H. orbignyi* trans. to *H. subbinum,* and *Neohibolites minimus.*

A deteriorated clay-pit (203374) 1300 yd E. 2° N. of Danton Farm yielded loose fossils of the *orbignyi* Subzone, including *Serpula sp.* ; *Pseudocardia tenuicosta, Inoceramus concentricus, I. sulcatus* ; *Euhoplites* cf. *opalinus, Hamites intermedius* and *Neohibolites minimus.*

In a trench 1450 yd W. 4° N. of Park Farm the Folkestone Beds were overlain by 1 ft 2 in of grey, brown and yellow clay with phosphatic nodules, which was succeeded in turn by 5 ft of greenish grey and brown clay with occasional phosphatic nodules.   G.B.

A tiny outlier of Gault caps high ground in the western part of Folkestone town, 800 yd S. 15° E. of Shorncliffe Station. A larger outlier in the town centre is partly concealed by brickearth. In the East Cliff (236363) 620 yd N. 33° E. of Folkestone Harbour Station 8 ft of grey shaly and gypsiferous marl, with the 'dentatus band' of phosphatic nodules 2 ft above the base, were exposed resting upon the Sulphur Band (p. 94).   J.G.O.S.

On the foreshore south-east of Folkestone Warren, slipped masses of Gault and Chalk are mingled. The extrusion of Gault on the foreshore, which formerly provided excellent opportunities of collecting fossils, has been checked by measures taken to stabilize the Warren landslips (pp. 295–6).

In the Ottinge Borehole, Dawkins (1905, p. 30) put the thickness of the Gault at 127 ft; at Elham the borer's log suggested 116 ft, and at Ellinge 164 ft, while at Lower Standen and Abbot's Cliff estimates of 146 ft and 143 ft respectively are close to the figure obtained by Muir Wood and Casey in The Warren (p. 57). Farther east, at Farthingloe, the Gault is 127 ft thick, and at Shakespeare Colliery Lamplugh and Kitchin (1911, p. 6) gave it 135 ft, but at Dover No. 1 Borehole, between the two, the thickness is 145 ft 3 in, made up as follows:

	Thickness Ft in		Depth Ft in	
Chalk: ... ... to			403	1
Gault:				
Firm grey glauconitic marl; highly glauconitic with occasional ½-inch brown phosphatic nodules below 411 ft 3 in	10	6	413	7
Grey glauconitic silty marl; phosphatic nodules at 416 ft; glauconitic content decreases below 417 ft 7 in ...	5	5	419	0
Grey silty mudstone	5	0	424	0
Grey fossiliferous mudstone with scattered phosphatic nodules; colour becomes progressively darker downwards	90	9	514	9
Dark grey silty mudstone with phosphatic nodules; glauconitic at intervals below 520 ft 5 in	8	9	523	6
Dark grey fossiliferous shaly mudstone; traces of finely-divided pyrite; occasional phosphatic nodules; glauconitic below 535 ft 6 in	16	4	539	10
Dark grey shaly silty mudstone with glauconite, finely-divided pyrite and phosphatic nodules; becoming sandy and highly glauconitic below 546 ft	8	6	548	4
Folkestone Beds: (p. 97).				

The base of the Gault was readily recognized but the top was less certainly ascertained: the base of the Chalk was taken at 403 ft 1 in, beneath a bed of highly glauconitic marl; a bed of similar lithology 10 ft below yielded Albian fossils, and was taken to represent Bed XII of the Gault, while Bed XIII was presumably represented by the intervening marl, although palaeontological evidence was lacking.

At Ropersole, Dawkins (1905, p. 30) gave the Gault a thickness of 119 ft, while Ritchie's figure of 191 ft in the Snowdown Colliery No. 3 Shaft (1920, p. 176) differs widely from the 148 ft given by Lamplugh, Kitchin and Pringle (1923, pp. 89, 90) for the Fredville Borehole. At Walmestone 153 ft were recorded, at Adisham 168 ft, at Littlebourne 158 ft, and at Harmansole 180 ft. At Chilham, Dawkins (1913, p. 362) gave a thickness of 224¼ ft, but Lamplugh, Kitchin and Pringle (1923, p. 122) found this unacceptable, and suggested 180 ft.   G.B.

Chapter V

PALAEONTOLOGY OF THE GAULT

INTRODUCTION

THE FOSSILS of the Gault provide a glimpse of life in a fairly shallow muddy-bottomed sea which covered much of southern and eastern England towards the end of the Lower Cretaceous epoch. The formation is best known for its molluscs, fishes and crustacea and for its well-preserved microzoa. Compared with that of the Lower Greensand the Gault fauna is marked by a great increase in prominence of the Ammonoidea, the rarity of clear-water organisms such as brachiopods, sponges and polyzoa, and the absence of ponderous, thick-shelled bivalves like *Exogyra latissima* (Lamarck) and *Gervillella sublanceolata* (d'Orbigny), which were presumably adapted to living in shallow, current-swept waters. The Gault is poorer in echinoids, brachiopods and sponges than the Lower Chalk, but richer in gastropods and bivalves. It tells us practically nothing about life on land at the time of its deposition, since its yield of terrestrial organisms comprises only some pine-cones, drift wood and the remains of a few pterodactyls which had been carried out to sea. Evidence from other sources makes it certain, however, that during the Gault period the dinosaurs still dominated the land fauna (although mammals had been in existence since the Trias), that flowering plants had established themselves among the conifers and cycadophytes and that birds co-existed with air-borne reptiles.

Attempts have been made to deduce the depth and temperature of the Gault sea on the assumption that the habitat of fossil assemblages was similar to that of their analogues among the Recent faunas. Thus, Price (1875), working on the mollusca, deduced that the Lower Gault sea-bed did not exceed 100 fathoms in depth, though he believed that the Upper Gault had been deposited in deeper water. Jukes-Browne (1900, pp. 413–5), by using only the bivalves, estimated a depth of 150 to 180 fathoms for the Lower Gault sea and 200 fathoms for that of the Upper Gault. Chapman (1898), in studying the foraminifera, applied bathymetric data of Recent species to arrive at figures of 830 fathoms and 860 fathoms for the depths of the Lower Gault and Upper Gault seas respectively. Khan (1950b), however, has pointed out that Chapman's method of calculation is statistically unsound and that his figures are excessive; he has himself expressed the belief that the foraminifera indicate that the Gault of south-east England was laid down in a temperate and relatively shallow marine environment.

Concerning the temperature of the Gault sea, Price (1879) took up the suggestion of earlier authors that the Cambridge phosphate bed was produced by cold northerly currents. He asserted that decapod crustacea were characteristic of the Gault phosphatic nodule-beds and that in the present-day North Atlantic Ocean this group of organisms is most numerous in cold-water areas. On the other hand, the recently introduced method of palaeotemperature measurement by assaying oxygen isotopes in fossil organic carbonates suggests that the Gault was deposited during a period of climatic maximum. Data derived from analyses of guards of the belemnite *Neohibolites minimus* (Miller) from the Gault of Folkestone give temperatures of 20° to 23·3°C (Bowen 1961), compared with 6° to 17·5°C which obtain in the Strait of Dover at the present time.

Some of the oldest and most familiar names of Gault fossils originate in the works of J. Parkinson (1811, 1819), J. Sowerby (1812–22) and J. de C. Sowerby (1823–46), Mantell (1822) and W. Fitton (1836). Many were first named by the great French palaeontologist d'Orbigny, who obtained much of his material from Wissant, on the north coast of France, opposite Folkestone. Gault fossils, mostly from Folkestone, figure prominently in the monographs of the Palaeontographical Society, such as those on the Cretaceous Lamellibranchia (Woods 1899–1913), Cretaceous Entomostraca (Jones 1850; Jones and Hinde 1890), Crustacea (Bell 1858–62; Woods 1924–31) and Lower Cretaceous belemnites (Swinnerton 1955). The Ammonoidea of the Gault were made the subject of a special monograph by Spath (1923–43). Charles Darwin himself dealt with the Cirripedes (Darwin 1851–5).

Catalogues issued by the British Museum (Natural History) are another source of information on Gault fossils, notably for Reptilia and Amphibia (Lydekker 1888–90), Fishes (Woodward 1889–1901), Nautiloid Cephalopods (Foord 1891) and Cirripedes (Withers 1935). Among the numerous papers devoted to special aspects of Gault palaeontology, mention may be made of those by Starkie Gardner (1873–80) on Gastropoda and Scaphopoda, Chapman and Sherborn (1893) on Ostracoda, Carruthers (1866–71) on pine-cones, Chapman (1891–8) and Khan (1950a, 1952) on Foraminifera, and Casey (1936–60) on Ammonoidea.

SYSTEMATIC ACCOUNT

Foraminifera

Chapman's work on the foraminifera of the Gault is now out of date and a complete revision has yet to be attempted. Khan has given some attention to the fauna of the Lower Gault, primarily that of Copt Point, Folkestone. He has found that while some species like *Hormosina folkestoniensis* Khan, *Vaginulina neocomiana* Chapman and *Spiroplectinata annectens* (Parker and Jones) have a short vertical distribution, others such as *Bolivina textilarioides* Reuss, *Pleurostomella alternans* Schwager, *Siphogenerina asperula* (Chapman), *Epistomina spinulifera* (Reuss) and *Arenobulimina macfadyeni* Cushman, though having a longer vertical range, occur in greatest abundance only within restricted limits. Using the frequency-occurrence of certain species, Khan established the following three zones in the Lower Gault:

3. *Epistomina spinulifera–Arenobulimina macfadyeni* Zone (Beds VII & VIII), characterized by the index species and *Pleurostomella alternans* Schwager, *Spiroplectinata annectens* (Parker and Jones), *Bifarina tenuilissa* Tappan and *Tritaxia pyramidata* Reuss.

2. *Siphogenerina asperula* Zone (Beds III to VI), characterized by the index species and *Haplophragmoides latidorsatum* (Bornemann) and *Lagena sulcata* (Walker and Jacob).

H

1. *Bolivina textilarioides–Pleurostomella alternans* Zone (Beds I & II), characterized by the index species and *Hormosina folkestoniensis* Khan, *Vaginulina neocomiana* Chapman, *Dorothia gradata* (Berthelin), *Haplophragmoides latidorsatum* (Bornemann), *Verneuilina chapmani* Ten Dam, and *Marssonella oxycona* (Reuss).

The adherent foraminifer *Bullopora laevis* (Sollas) has been found on shells and nodules.

Porifera, Anthozoa, Annelida, Echinodermata, Brachiopoda

Among the phosphatic nodules at the base of the greensand Bed XII are a few rolled Lithistid sponges, such as *Jerea pyriformis* Lamouroux and *Siphonia tulipa* Zittel, the occurrence being a feeble reflection of the rich Lithistid faunas of the Blackdown and Warminster Greensands.

Serpula antiquata J. Sowerby, *Hamulus curvatus* (Gardner), *Rotularia polygonalis* (J. de C. Sowerby), *Sarcinella socialis* (Goldfuss) and *Glomerula gordialis* (Schlotheim) are the chief representatives of the Annelida. The last forms winding threads and loops on the outsides of nodules and larger fossils. Burrows lined with fish-scales have been attributed to a Terebelloid worm '*Terebella*' *lutensis* Bather. The structure known as *Granularia,* consisting of tiny ovoid pellets arranged in a tube, was also thought to be of annelid origin by Bather (1911).

Echinoids are not common and are invariably crushed. *Hemiaster asterias* Forbes and *H. bailyi* Forbes occur in most of the beds, especially Bed III. *Polydiadema wiltshirei* (Wright) is a Lower Gault species ; *Typocidaris gaultinus* (Forbes) is characteristic of Bed X. Columnals of the crinoid *Nielsenicrinus cretaceus* (Leymerie) (=*Pentacrinus fittoni* of authors) are also found in Bed X. The mobile Comatulids *Glenotremites aequimarginatus* (Carpenter) and *Palaeocomaster loveni* (Carpenter) and the micro-crinoid *Styracocrinus peracutus* (Peck) have all been found in the Gault of Folkestone.

EXPLANATION OF PLATE IV

Fig. 1. ' Button-coral ', *Discocyathus fittoni* (Edwards and Haime), Lower Gault. (GSM. 108208).

Fig. 2. Gastropod, *Anchura carinata* (Mantell), Lower Gault. (GSM. 1739).

Fig. 3. Uncoiled ammonite, *Heteroclinus nodosus* (J. Sowerby), Lower Gault. (GSM. 1785).

Fig. 4. Bivalve, *Inoceramus concentricus* Parkinson, Lower Gault. (GSM. 21145).

Fig. 5. Bivalve, *Inoceramus subsulcatus* Wiltshire, Lower Gault. (GSM. 21156).

Fig. 6. Bivalve, *Inoceramus sulcatus* Parkinson, Upper Gault. (GSM. 21161).

Fig. 7. Belemnite, *Neohibolites minimus* (Miller), Lower Gault. (GSM. 92911).

Fig. 8. Ammonite, *Hysteroceras orbignyi* (Spath), Upper Gault. (GSM. 108470).

Fig. 9. Ammonite, *Euhoplites truncatus* Spath, Lower Gault. (GSM. 31024).

Fig. 10. Crab, *Notopocorystes stokesi* (Mantell), Lower Gault. (GSM. 1854).

Fig. 11. Bivalve, *Nucula (Pectinucula) pectinata* J. Sowerby, Lower Gault. (GSM. 1641).

Fig. 12. Uncoiled ammonite, *Hamites gibbosus* J. Sowerby, Lower Gault. (GSM. 60788).

All figures natural size.

FOSSILS FROM THE GAULT OF FOLKESTONE

Gault Anthozoa are mostly small button or cup corals and are met with more frequently in the lower beds, particularly in Bed V, which Price termed the 'coral bed'. The commonest are *Bathycyathus sowerbyi* Edwards and Haime, *Discocyathus fittoni* Edwards and *Trochocyathus harveyanus* Edwards.

The muddy waters of the Gault did not suit the brachiopods, though *Kingena lima* (Defrance) and '*Terebratula*' *dutempleana* (d'Orbigny) (= '*T.*' *biplicata* of authors) may be successfully sought for in Bed X.

Mollusca (Bivalvia)

All the beds yield *Inoceramus*, of which the best known is *Inoceramus concentricus* Parkinson (Beds I to XI). *Inoceramus anglicus* Woods (Beds VIII to XIII), *I. tenuis* Mantell (Beds XII and XIII) and *I. crippsi* Mantell (Bed XIII) also occur, the last becoming more common in the Chalk above. The radially fluted *I. sulcatus* Parkinson (sometimes placed in a separate genus or subgenus, *Actinoceramus*) is confined to Beds VIII and IX and is an easily recognizable guide to the junction of the Lower and Upper Gault. It is accompanied by a curious hybrid form, *I. subsulcatus* Wiltshire, specimens of which show every stage of transition from *I. concentricus* to *I. sulcatus;* seemingly conspecific forms are found as far afield as Mexico and Russia. *Inoceramus* is the most likely source of the fossil pearls that are found as occasional curiosities in the Gault. Next in abundance are the 'nut' shells, especially *Nucula* (*Pectinucula*) *pectinata* J. Sowerby and *N.* (*Leionucula*) *ovata* Mantell. These two occur all through the Gault, but *N.* (*L.*) *albensis* d'Orbigny, *Acila* (*Truncacila*) *bivirgata* (J. de C. Sowerby) and *Mesosaccella mariae* (d'Orbigny) are more typical of the Lower Gault. Other long-ranging bivalves are the scallop *Entolium orbiculare* (J. Sowerby), the Spondylid *Plicatula gurgitis* Pictet and Roux, *Pseudocardia tenuicosta* (J. de C. Sowerby), a member of the Carditidae, and the Arcid *Nanonavis carinata* (J. Sowerby). Oysters (e.g. *Gryphaeostrea canaliculata* J. Sowerby sp. and *Pycnodonte vesicularis* Lamarck sp.) are more commonly found in the Upper Gault. Species confined to or especially characteristic of one particular bed include *Linotrigonia fittoni* (Deshayes) (Bed I), *Corbula gaultina* Pictet and Campiche (Bed II), the Arcid *Cucullaea* (*Idonearca*) *glabra* Parkinson (Bed VIII), the Arcticid *Proveniella quadrata* (d'Orbigny) (Bed VIII), the scallops *Chlamys elongata* (Lamarck) (base of Bed XI) and *Aequipecten beaveri* (J. Sowerby) (Bed X), and the Aucellid *Aucellina coquandiana* (d'Orbigny) (Bed XII). Drift wood is commonly riddled with the crypts of bivalves whose timber-boring mode of life was comparable with that of the modern shipworm, the principal forms being *Opertochasma constrictum* (Phillips), *Terebrimya gaultina* (Woods) and *Xylophagella zonata* Casey. *Astarte, Lucina, Anomia, Pinna, Isognomon, Pseudolimea, Thracia* and *Eopecten* are some of the many other genera represented.

Mollusca (Gastropoda and Scaphopoda)

Gault gastropods are rich in variety and numbers. *Nummocalcar fittoni* (Roemer) (=*Solarium ornatum* of authors) ranges through almost the whole of the formation and is a survivor of the ancient family that is typified by the Palaeozoic *Euomphalus*. Another ancient family (but with living representatives), the Pleurotomariidae, is exemplified in the Gault by *Leptomaria pricei* Cox, *Conotomaria folkestonensis* Cox and *Pleurotomaria plicata*

(J. Sowerby). The last is fairly common in Bed VIII and passes up into Bed IX. The Aporrhaidae, a family which reached its maximum in the Cretaceous and is characterized by finger-like extensions of the aperture, is well represented. This family includes the long-ranging *Anchura carinata* (Mantell) and *Tessarolax retusum* (J. de C. Sowerby) ; *A. carinella* (d'Orbigny) and *A. marginata* (J. de C. Sowerby) are found in the higher beds of the Lower Gault ; *Dimorphosoma calcaratum* (J. de C. Sowerby) is confined to Bed II. The present-day whelks have their counterparts in *Buccinofusus clementinus* (d'Orbigny) and *Sipho gaultinus* (d'Orbigny), the winkle in *Gyrodes genti* (J. Sowerby), all of which occur sporadically through the Gault. Other typical Gault gastropods are the Cerithioids *Mesalia* (*Bathraspira*) *tecta* (d'Orbigny) and *Metacerithium trimonile* (Michelin), the Scalariids *Confusiscala dupiniana* (d'Orbigny) and *Claviscala clementina* (d'Orbigny), the Turritellid *Torquesia vibrayeana* (d'Orbigny), the limpet '*Acmaea*' *tenuicosta* (d'Orbigny), the Solariids *Semisolarium moniliferum* (Michelin) and '*Trochus*' *conoideus* J. Sowerby, and the Opisthobranchs *Acteon affinis* (d'Orbigny) and *Ringinella inflata* (d'Orbigny). The little *Bellerophina minuta* (J. Sowerby) (Beds V to VII), once thought to be a cephalopod, is now believed to be the first of a group of pelagic prosobranchs, the Heteropoda.

The two families of scaphopoda, the Dentaliidae and the Siphodentaliidae, are both present in the Gault. The former is represented chiefly by the ubiquitous *Dentalium* (*Fissidentalium*) *decussatum* (J. Sowerby), resembling an elephant's tusk in miniature ; *Cadulus* (*Gadila*) *gaultinus* Gardner belongs to the latter family. Small hooked or curved tubes formerly ascribed to *Siphodentalium* belong properly to the annelid *Hamulus*.

Mollusca (Cephalopoda)

The richness of the ammonite fauna of the Gault is not due entirely to favourable conditions of preservation but reflects a world-wide burst of evolutionary activity that took place in the group towards the end of the Lower Cretaceous. With few exceptions Gault ammonites belong to the ' Trachyostraca ', the highly ornamented types that are believed to have flourished in the shelf seas, and nearly all of these belong to the three superfamilies Hoplitaceae, Acanthocerataceae and Turrilitaceae. The first takes in the ventrally grooved ammonites such as *Hoplites*, *Euhoplites*, *Anahoplites* and *Dimorphoplites*, which dominate the Lower Gault, and *Epihoplites*, *Callihoplites* and the rare *Discohoplites*, which characterize the Upper Gault. Also included among the Hoplitids are *Lepthoplites*, *Pleurohoplites* and *Arrhaphoceras*. These occur sparsely in the highest bed of the Gault and have a raised siphonal line foreshadowing the Cenomanian keeled family Schloenbachiidae. Gault ammonites with a keel at some stage of growth or with a median row of nodes on the venter belong to the Acanthocerataceae. *Hysteroceras*, *Prohysteroceras*, *Mortoniceras* and *Dipoloceras* are typical keeled genera ; *Neophlycticeras*, with cockscomb venter, is rather scarce. Although represented in the *dentatus* Zone by *Mojsisovicsia*, and rare *Eubrancoceras* and *Oxytropidoceras*, this superfamily did not become important until the top of the Lower Gault, after which it gained supremacy over the Hoplitaceae. The Turrilitaceae comprises heteromorphs or abnormally coiled types such as *Hamites*, *Anisoceras*, *Heteroclinus*, *Pseudhelicoceras*

Turrilitoides and *Mariella* (but not the Scaphitids, which are judged as constituting a separate superfamily). Except for the Desmoceratid *Beudanticeras,* which is not uncommon at the junction of the Lower and Upper Gault, the 'Leiostraca', or smooth ammonites, are known from the Gault only by a few chance finds of *Hypophylloceras* (Phyllocerataceae), *Tetragonites, Pictetia* (Lytocerataceae), *Desmoceras, Puzosia* and *Uhligella* (Desmocerataceae). These ammonites are thought to have preferred open waters and to have had their European centre of dispersal in the Mediterranean region. The Placenticerataceae are another exotic element in the Gault ; this superfamily is represented by isolated finds of the Engonoceratid *Engonoceras iris* Spath in the *dentatus* Zone of the Lower Gault and by the Placenticeratid *Hengestites applanatus* Casey in the *inflatum* Zone of the Upper Gault. The Engonoceratidae, with their curiously modified (' pseudoceratitic') suture-lines, characterize Albian deposits of equatorial regions and seldom strayed into the northern waters where Hoplitids abounded. A short-term penetration into the Gault Hoplitid province by the Haplocerataceae is.indicated by the occurrence of the primitive Binneyitid *Falciferella milbournei* Casey in Bed II. That the Gault sea was open to immigrants from the north was proved by the discovery in Bed VIII at Folkestone of the Canadian genus *Gastroplites,* a member of an Arctic branch of the Hoplitidae.

In contrast to the Ammonoidea, Nautiloid cephalopods are relatively infrequent and belong to a few long-ranging genera of world-wide distribution. *Eutrephoceras clementinum* (d'Orbigny), *E. bouchardianum* (d'Orbigny) and *Cymatoceras albense* (d'Orbigny) occur in the Lower Gault and the earliest beds of the Upper Gault. Calcified beaks of Nautiloids, known as *Rhynchoteuthis,* have also been found.

Guards of the belemnite *Neohibolites minimus* (Miller) and varieties are ubiquitous in the Gault of this region and because of their resistant nature make up a large part of the weathered-out fossil debris.

A list of Ammonoidea from the Gault of Folkestone is given in Table 2.

TABLE 2

List of Ammonoidea from the Gault of Folkestone

	BED
Hypophylloceras subalpinum (d'Orbigny) Spath	VIII
Tetragonites kitchini (Krenkel)	II
Pictetia astieriana (d'Orbigny)	I
Desmoceras latidorsatum (Michelin)	II
Beudanticeras beudanti (Brongniart)	VIII–X
„ „ *ibiciforme* Spath	IX
„ *sphaerotum* (Seeley)	IX
„ *subparandieri* Spath	VIII–IX
Puzosia (Puzosia) spathi Breistroffer	X
Puzosia (Anapuzosia) provincialis Parona & Bonarelli ...	II
Uhligella derancei Casey ...	II
„ „ *erugata* Casey	II
Hoplites (Isohoplites) eodentatus Casey ...	I
Hoplites (Hoplites) dentatus (J. Sowerby)	I
„ „ „ *robustus* Spath	I

BED

Hoplites (*Hoplites*) *spathi* Breistroffer	... ...	I
„	„	*paronai* Spath	... ...	I
„	„	*persulcatus* Spath	... ...	I
„	„	*rudis* Parona & Bonarelli	... ...	I
„	„	*similis* Spath ...	... ...	I
„	„	*canavarii* Parona & Bonarelli	... ...	I
„	„	*benettianus* (J. de C. Sowerby)	... ...	I
„	„	*escragnollensis* Spath	... ...	I
„	„	*dentatiformis* Spath ...	... ...	II
„	„	*spp. nov.*	... ...	I–II
Anahoplites praecox (Spath)	... ...	I–II
„	*intermedius* Spath	... ...	I–II
„	*mantelli* Spath	... ...	I–II
„	*planus* (Mantell)	... ...	I–XI
„	„	*gracilis* Spath	... ...	... I–VIII
„	„	*discoideus* Spath	... ...	... I–VIII
„	*splendens* (J. Sowerby)	... ...	II–VII
„	*pleurophorus* Spath ...	... ...	IV
„	*picteti* Spath ...	... ...	... IX–X
„	*daviesi* Spath ...	... ...	VII
„	„	*ornatus* Spath	... ...	VII
„	*sp. nov.*	... ...	XI
Dimorphoplites niobe Spath	... ...	... II–IV
„	*doris* Spath	... ...	IV
„	*pinax* Spath	... ...	... II–V
„	*glaber* Spath	... ...	II–VIII
„	*chloris* Spath	... ...	II–VIII
„	*tethydis* (Bayle) ...	... ...	III–VIII
„	*biplicatus* (Mantell)	... ...	IV–VII
„	*parkinsoni* (Spath)	... ...	V–VIII
„	*? silenus* Spath	... ...	VIII
Metaclavites metamorphicus (Spath)	... ...	VIII
„	*trifidus* (Spath)	... ...	... IX–X
„	*compressus* (Parona & Bonarelli) ...	... ...	VIII–IX
Epihoplites denarius (J. de C. Sowerby) ...	... ...	X
„	*deluci* (Brongniart)	... ...	X
„	*glyptus* Spath ...	... ...	X
„	*gibbosus* Spath	... ...	... X–XI
Semenovites gracilis (Spath)	... ...	X
„	*iphitus* (Spath)	... ...	X
Callihoplites catillus (J. de C. Sowerby) ...	... ...	XI
„	*patella* Spath	... ...	XI
„	*strigosus* Spath	... ...	XI–XII
„	*auritus* (J. Sowerby)	... ...	XI
„	*seeleyi* Spath	... ...	XIII
„	*tetragonus* (Seeley) ...	... ...	... XIII
„	*glossonotus* (Seeley)	... ...	XII
„	*leptus* Spath ...	... ...	XII
„	*pulcher* Spath	... ...	XIII
Lepthoplites falcoides Spath	... ...	XIII
Pleurohoplites subvarians Spath ...	... ...	XIII
Arrhaphoceras substuderi Spath ...	... ...	XIII
Euhoplites truncatus Spath	... ...	... V–IX
„	*lautus* (Parkinson)	... ...	V–VIII
„	„	*bilobus* Spath	... ...	V

	BED
Euhoplites alphalautus Spath	X–XI
„ *microceras* Spath	II–IV
„ *subtabulatus* Spath	II
„ *aspasia* Spath	II–V
„ „ *cantianus* Spath	V
„ *loricatus* Spath	II–V
„ *meandrinus* Spath	II–V
„ *pricei* Spath	II–IV
„ *subtuberculatus* Spath	II–IV
„ *opalinus* Spath	V–IX
„ *bucklandi* Spath	V–VIII
„ *nitidus* Spath	V–VIII
„ *proboscideus* Spath	V–X
„ „ *intermedius* Spath	VIII
„ „ *ultimus* Spath	X
„ *armatus* Spath	VIII–X
„ *trapezoidalis* Spath	VIII–IX
„ „ *formosus* Spath	VIII
„ *serotinus* Spath	VIII–IX
„ *solenotus* (Seeley)	VIII–IX
„ *ochetonotus* (Seeley)	VIII–IX
„ *sublautus* Spath	VIII–IX
„ „ *monacanthus* Spath	VIII–IX
„ *subcrenatus* Spath	IX
„ *inornatus* Spath	IX
„ *vulgaris* Spath	IX–XI
„ *boloniensis* Spath	IX–X
Discohoplites coelonotus (Seeley)	XII
Gastroplites cantianus Spath	VIII
Engonoceras iris Spath	III
Neophlycticeras (*Neophlycticeras*) *brottianum* (d'Orbigny) ...	VIII–IX
„ „ *gibbosum* Spath	VIII–IX
Neophlycticeras (*Eotropitoides*) *jayeti* Breistroffer	VIII–IX
Neophlycticeras (*Protissotia*) *itierianum* (d'Orbigny)	VIII
„ „ *orion* Casey	VIII
Stoliczkaia sp.	XIII
Oxytropidoceras roissyanum (d'Orbigny)	I
„ *mirapelianum* (d'Orbigny)	I
„ *sp. nov.*	I
„ *cantianum* Spath	VIII
„ „ *excentricum* Spath	VIII
Mojsisovicsia (*Dipoloceroides*) *subdelaruei* (Spath)	IV
„ „ *remota* (Spath)	IV
„ „ *spinulosa* (Spath)	IV
„ „ *cornuta* (Pictet)	V
„ „ *equicostata* (Spath)	V
Dipoloceras cristatum (Brongniart)	VIII
„ *corbulatum* Spath	VIII
„ *multispinosum* Spath	VIII
„ *fredericksburgense* Scott *britannicum* Breistroffer	VIII
„ *pseudaon* Spath	VIII–IX
„ „ *moniliforme* Spath	VIII–IX
„ *bouchardianum* (d'Orbigny)	VIII–IX
„ „ *alticarinatum* Spath	VIII
Mortoniceras (*Mortoniceras*) *inflatum* (J. Sowerby)	XI–XII

BED

Mortoniceras (*Mortoniceras*) *inflatum picteti* Spath	...	...	...	XI
,,	,,	,, *gibbosum* Spath	...	... XI
,,	,,	*pricei* Spath ...	... ...	... IX–XI
,,	,,	,, *intermedius* Spath ...	...	... IX
,,	,,	*geometricum* Spath	... ...	... X
,,	,,	*fissicostatum* Spath	...	... XI
,,	,,	,, *ascendens* Spath	...	... XI
,,	,,	*commune* Spath ...	... ...	... XI
,,	,,	*potternense* Spath	... ...	... XI
,,	,,	*evolutum* Spath ...	... ...	... XI
,,	,,	*rostratum* (J. Sowerby) ...	...	... XIII
,,	,,	*kiliani* (Lasswitz) *altonense* Breistroffer	...	... XII
,,	,,	*rigidum* Spath ...	... ...	... VIII
Mortoniceras (*Deiradoceras*) *cunningtoni* Spath...	...	...	... IX–XI
,,	,,	,, *flexuosum* Spath	...	... IX–XI
,,	,,	*bipunctatum* Spath	... ...	... IX–X
,,	,,	*devonense* Spath ...	... ...	... IX–X
,,	,,	*albense* Spath ...	... ...	... IX–X
,,	,,	,, *transitorium* Spath	...	... IX–X
Cantabrigites cantabrigensis (Spath)	... ...	...	... XII
Elobiceras pseudelobiense (Spath)...	... ...	...	... XI
Prohysteroceras (*Goodhallites*) *goodhalli* (J. Sowerby)	...	... IX–XI
,,	,,	,, *shenleyensis* Spath ...	... IX–XI
,,	,,	,, *tuberculatum* Spath ...	... IX–XI
,,	,,	,, *aplanatum* Spath	... ... IX–XI
,,	,,	*candollianum* Spath *non* Pictet	... XII
,,	,,	*delabechei* Spath	... ...	... X
,,	,,	,, *robustum* Spath	... ...	... X
Prohysteroceras (*Neoharpoceras*) *coptense* Spath	...	... XI
Eubrancoceras cricki (Spath)	... ...	... ...	... II–III
,,	*spp.*	... ...	... ...	... I–III
Hysteroceras varicosum (J. de C. Sowerby)	...	...	... IX–X
,,	,, *binodosum* (Stieler)	... ...	...	... IX–X
,,	*binum* (J. Sowerby)	... ...	...	... IX–X
,,	*subbinum* Spath	... ...	...	... IX–XI
,,	*carinatum* Spath	... ...	...	IX–XII
,,	,, *ascendens* Spath	... ...	...	IX–XII
,,	*orbignyi* (Spath)	... ...	...	VIII–XI
,,	aff. *choffati* Spath ...	... ...	...	... IX
,,	*bucklandi* (Spath) ...	... ...	...	... XI
,,	*capricornu* Spath	... ...	...	... VIII
,,	*pseudocornutum* Spath	... ...	...	... VIII
,,	[*Dipoloceras?*] *symmetricum* (J. de C. Sowerby)	... VIII
,,	*simplicicosta* Spath	... ...	...	... VIII
,,	*serpentinum* Spath ...	... ...	...	... VIII
Hengestites applanatus Casey	... ...	...	... XI
Falciferella milbournei Casey	... ...	...	... II
Eoscaphites circularis (J. de C. Sowerby)	...	...	... X
,,	,, *depressus* Spath ...	... ...	...	... X
,,	,, *rugosus* Spath	... ...	...	... X
,,	*subcircularis* Spath ...	... ...	...	... X–XI
Scaphites simplex Jukes-Browne ...	... ...	...	... X–XII
Mariella cf. *escheriana* (Pictet) ...	... ...	...	... XI
Proturrilitoides densicostatus (Passendorfer)	... ...	...	V
Pseudhelicoceras robertianum (d'Orbigny)	... ...	...	... IX–XI

					BED
Pseudhelicoceras robertianum ornatum Spath	...	...	...	...	IX–XI
„ *pseudoelegans* Spath	...	...	...	...	VIII
„ *subcatenatum* Spath	...	...	...	...	II
„ *? gaultinum* Spath	...	...	...	...	X
Anisoceras armatum (J. Sowerby)		...	...	...	XII
„ *perarmatum* (Pictet & Campiche)	...	...	...	XII–XIII
„ *saussureanum* (Pictet) ...	...	...	...	XI–XII
„ *subarcuatum* Spath	...	...	...	...	X–XI
Protanisoceras moreanum (Buvignier)	...	...	...	...	I
„ cf. *halleri* (Pictet & Campiche) ...	...	...	...	II
Heteroclinus nodosus (J. Sowerby)	...	...	...	...	I–II
„ „ *splendens* Spath	...	...	...	...	I–II
„ *flexuosus* (d'Orbigny)	...	...	...	...	I–II
Metahamites sp. nov.	...	...	...	...	I
Idiohamites tuberculatus (J. Sowerby)	...	...	...	...	IX–XI
„ *spiniger* (J. Sowerby)	...	...	...	...	IX–XI
„ *subspiniger* Spath	...	...	...	...	IX–X
„ *turgidus* (J. Sowerby)	...	...	...	...	X–XI
„ „ *robustus* Spath	...	...	...	...	X–XI
„ „ *subannulatus* Spath	...	...	...	...	X–XI
„ *spinulosus* (J. Sowerby)	...	...	...	...	X–XI
„ *favrinus* (Pictet)	...	...	...	...	X–XI
„ *ellipticoides* Spath	...	...	...	...	X
„ *? incertus* Spath	...	...	...	...	X–XI
„ „ *costatus* Spath	...	...	...	...	X–XI
Hamitoides? rusticus Spath	...	...	...	...	VIII
Hamites attenuatus J. Sowerby	...	...	...	...	I–IV
„ *tenuicostatus* Spath	...	...	...	...	II–IV
„ *rotundus* J. Sowerby	...	...	...	...	I–V
„ *subrotundus* Spath	...	...	...	...	IV–VII
„ *compressus* J. Sowerby	...	...	...	...	II–VIII
„ „ *gracilis* Spath	...	...	...	...	II–VIII
„ *incurvatus* Brown	...	...	...	...	VIII–IX
„ *maximus* J. Sowerby	...	...	...	...	V–IX
„ „ *rectus* Brown	...	...	...	...	V–IX
„ *gibbosus* J. Sowerby	...	...	...	...	II–VIII
„ *tenuis* J. Sowerby	...	...	...	...	V–VIII
„ *intermedius* J. Sowerby	...	...	...	...	IX
„ „ *opalinus* Spath	...	...	...	...	IX
„ „ *distinctus* Spath	...	...	...	...	IX
„ (*Stomohamites*) *sp.* cf. *parkinsoni* (Fleming)	...	...	...	XI
„ (*Psilohamites*) *bouchardianus* d'Orbigny	...	...	...	X
Ptychoceras (*Mastigoceras*) *adpressum* (J. Sowerby)	...	...	...	X
Lechites gaudini Pictet and Campiche	...	...	...	...	XI

Crustacea

Decapod crustaceans (crabs, shrimps and lobsters) are fairly common on some horizons, especially the nodule beds. The crab *Notopocorystes stokesi* (Mantell) is found in practically all the beds but reaches its maximum in the line of nodules at the top of Bed III, named the ' crab bed ' by the early collectors. Other crabs are *Eucorystes broderipi* (Mantell), *Necrocarcinus bechei* (Deslongchamps), *Homolopsis edwardsii* Bell and *Etyus martini* Mantell. The lobster *Homarus longimanus* G. B. Sowerby ranges up from the Lower Greensand. Valves of the pedunculate cirripedes (stalked barnacles) *Cretiscalpellum unguis* (J. de C. Sowerby) and *Pycnolepas rigida* (J. de C. Sowerby) and, less commonly, *Scalpellum (Arcoscalpellum) arcuatum* Darwin and *Zeugmatolepas mockleri* Withers, occur principally in the lower part of the Upper Gault.

Though ostracods are limited to a few species in the lower part of the Gault, individuals are locally very abundant. By the incoming of new species the fauna becomes more varied with time and in the upper half of the Gault at least twenty-two species are recognizable. *Cytherella ovata* (Roemer), *C. muensteri* (Roemer), *Cythereis reticulata* Jones and Hinde, *C. lurmannae* Triebel and *Protocythere auriculata* (Cornuel) are the commonest species at Folkestone, with *Cythereis bonnemai* Triebel, *Cytherelloidea stricta* (Jones and Hinde), *Schuleridea jonesiana* (Bosquet), *Platycythereis gaultina* (Jones) and *Protocythere lineata* (Chapman and Sherborn). *Cytherelloidea chapmani* (Jones & Hinde), *Haplocytheridea nana* Triebel and *Eucythere trigonalis* (Jones) are fairly common, whilst those of rare occurrence are *Paracypris gracilis* (Bosquet), *Bairdia subdeltoidea* (Münster), Auctt., *Bythocypris harrisiana* (Jones), *Dolerocytheridea bosquetiana* (Jones & Hinde), *Haplocytheridea rara* Triebel, *Eucytherura ansata* (Weingeist), *Monoceratina umbonata* (Williamson), *Isocythereis fortinodis* Triebel and *I. fissicostis* Triebel. As yet insufficient detailed collecting has been done to determine whether or not the distribution of ostracods in the Gault has any zonal significance.

Pisces

The Gault yields a good fish fauna, especially from its upper beds. Isolated teeth and vertebral discs of elasmobranchs (sharks and rays) are common. Most belong to the shark *Lamna appendiculata* Agassiz, though *Acrodus laevis* Woodward, *Lamna macrorhiza* Cope, *Isurus mantelli* Agassiz, *Cestracion canaliculatus* Egerton, *Corax pristodontus* Agassiz, *Notidanus lanceolatus* Woodward, *Scapanorhynchus subulatus* Agassiz and species of *Hybodus*, *Synechodus* and *Ptychodus* also represent this class. Dental plates of Chimaeroids, a group which includes the living rabbit-fish, have been found and are identified as belonging to *Ischyodus thurmanni* Pictet and Campiche, *Edaphodon sedgwickii* (Agassiz) and *E. laminosus* Newton. The Teleostomi, or bony fishes, are represented by *Portheus gaultinus* Newton, *Thrissopater salmoneus* Günther, *Apateodus glyphodus* (Blake), *Syllaemus anglicus* (Dixon) and *Protosphyraena ferox* Leidy. Fish remains also include ear-bones (otoliths) and coprolites. Burrows attributed to Terebelloid worms are commonly lined with fish-scales.

Reptilia

Teeth and bones of the large marine reptiles *Ichthyosaurus campylodon* Carter, *Polyptychodon interruptus* Owen and *Cimoliasaurus constrictus* Owen (= *Mauisaurus gardneri* Seeley) have been collected. An incomplete skeleton of the last species, estimated to be 12 ft in length, was found in Bed II at Copt Point by John Griffith, the professional fossil-collector of Folkestone in the last century. Associated with the bones were about a quart of rolled and smooth-surfaced quartz pebbles, evidently ' gastroliths ' or stomach-stones (Price 1879, p. 15). Remains of the turtles *Chelone benstedi* Owen,

Ch. jessoni Lydekker, and *Rhinochelys elegans* Lydekker, and of the flying reptile *Ornithocheirus compressirostris* Owen, have been recorded from the Upper Gault. From Bed X of the Upper Gault Price (ibid., p. 21) reported the discovery by Griffith of crushed bodies resembling crocodile eggs ; one of these is now in the Geological Survey Museum.

Plantae

Pine and *Sequoia*-like cones from the Gault, mostly from Beds IX and X, were described by Carruthers under the names *Pinites gracilis, P. hexagonus, P. pricei, Sequoites gardneri* and *S. ovalis.* Pieces of coniferous drift wood are commonplace and the cycadophyte *Benettites* has also been identified. Resin has been recorded from Bed V. Some of the beds contain branching filaments described by early writers as ' fucoid markings '. Lenses of clay ironstone in Bed III and Bed V are full of *Chondrites,* a structure once thought to be of vegetable origin but now believed to be the work of some burrowing organism.   R.C.

Chapter VI

UPPER CRETACEOUS: CHALK

GENERAL ACCOUNT

CHALK IS, for the most part, a soft, remarkably pure limestone. Some investigators have regarded is as comparable with the *Globigerina* ooze, some as a chemical precipitate, while others have considered it to be primarily of organic origin. M. Black (1953) found that ordinary white chalk was composed of two powders: the finer consisted of coccoliths and their disintegration products, that formed a matrix in which were embedded molluscan debris, foraminifera and 'spheres' or *Oligostegina*, comprising the coarser fraction. In soft chalk the coarser powder had particles between 100 microns and 10 microns in diameter, the finer between 4 microns and one-half micron. Colloidal clay material finer than one-half micron was present in appreciable quantities in chalk marl and grey chalk, but in only very small amounts in white chalk. Coccoliths, the minute calcareous bodies produced by planktonic algae, were present in all stages of disintegration down to the individual component crystals, and Black found the size and shape of the latter to vary from one horizon to another, according to the prevalent type of coccolith. Of the larger particles, predominance of molluscan debris gave a gritty but friable rock ; chalk with abundant foraminifera or spheres was apt to be rather hard or nodular, while preponderance of coccolith material gave common soft chalk. In contrast, modern precipitated oozes contained relatively little shell material and hardly a trace of coccoliths, but an abundance of minute aragonite crystals.

The description by William Phillips in 1818 of the cliffs between Folkestone and Walmer constituted one of the earliest attempts to divide the English Chalk. The following palaeontological subdivisions are now recognized within the area of the one-inch sheets here described:

Formations	Stages	Zones	Subzone
Upper Chalk	Senonian	*Marsupites testudinarius*	
		Uintacrinus socialis	
		Micraster coranguinum	
		Micraster cortestudinarium	
Middle Chalk	Turonian	*Holaster planus*	
		Terebratulina lata	
		Inoceramus labiatus	
			Actinocamax plenus
Lower Chalk	Cenomanian	*Holaster subglobosus*	
		Schloenbachia varians	

The distribution of some of these subdivisions is shown in the sketch-map (Fig. 2).

In the present district the full thickness of the Chalk has been penetrated in several boreholes, but the records of only a few are reliable: at the Littlebourne Borehole, which started in Thanet Beds, 750 ft of Chalk are

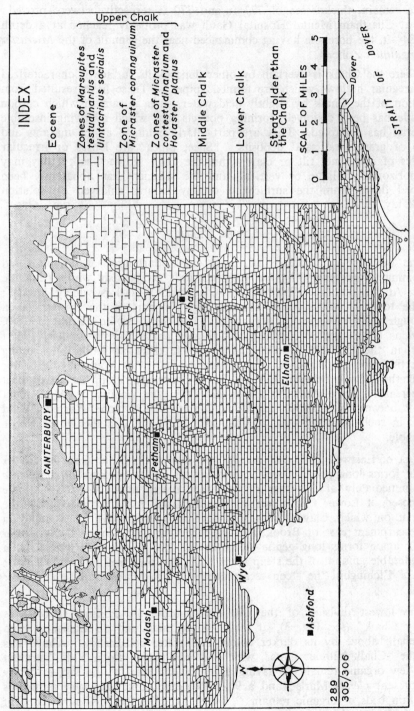

FIG. 2. *Sketch-map showing the distribution of some subdivisions of the Chalk*

present; at Adisham and West Court Farm boreholes, where no Eocene strata remain, thicknesses of 777 ft and 770 ft respectively were proved; and at Chartham Mental Hospital Gault was probably reached at a depth of 734⅔ ft, the borehole having commenced near the summit of the *Micraster coranguinum* Zone.

Where the Chalk is overlain by other deposits its surface is characterized by irregular hollows which are termed 'pipes'. These have resulted from solution of the Chalk by slightly acid water, such as rain in which carbon dioxide has been dissolved, forming pockets into which overlying insoluble material has collapsed. They are particularly common at the margins and base of gravel and sand deposits. 'Sheet pipes' are lenses or irregular masses of fine sand, silt or clay within the body of the Chalk; they may be horizontal, oblique or vertical, and the sediment has apparently been washed down from the surface to occupy fissures enlarged by solution (Kirkaldy 1950).   G.B., J.G.O.S.

LOWER CHALK

The Lower Chalk consists of marly chalk, dark grey at the base and becoming progressively lighter grey upwards to greyish white near the top. At the base the Chloritic (or Glauconitic) Marl is widely present; at the top the *plenus* Marls are lithologically part of the Lower Chalk, although palaeontologically they are allied to the *I. labiatus* Zone (p. 164). Dawkins (1913, pp. 362, 372) recorded grey chalk and chalk marl 196 ft thick in the Chilham Borehole and 200 ft thick at Ropersole, resting respectively on 24 ft and 16 ft of glauconite marl. At Westwell the Lower Chalk thickness is between 160 and 200 ft; eastwards along the escarpment it increases progressively, to about 225 ft in the vicinity of Wye, about 240 ft between Etchinghill and Pean, 258 ft at Folkestone Warren (Wood 1955, in reply to discussion, p. 463), and 262 ft 5 in in the Dover No. 1 Borehole.

West of Eastwell Park and on the east side of the Stour Gap the Lower Chalk forms long gentle slopes; north-east of Brook the base of the Chalk lies immediately at the base of the steep scarp, most of which is here composed of Lower Chalk, with Melbourn Rock forming the edge of the plateau on which Clay-with-flints is developed; south-east of a point on the escarpment east of Brook, and in particular near Stowting, the Lower Chalk again forms long gentle slopes at the foot of the escarpment, but a considerable amount of the steep scarp itself is also formed in Lower Chalk; east of Etchinghill the steep escarpment is again almost wholly in Lower Chalk.

The lowest division of the Chalk recognized by Phillips (1818) was 'Grey Chalk' (see Fig. 3), at least 200 ft thick and distinguished from the chalk above by its darker colour and softer consistency. Above this lay the 'Chalk without flints', which he subdivided into a 50-ft "stratum with few organic remains", having a bed of soft marl at the top, evidently the present *plenus* Marls; and a 90-ft "stratum containing very numerous and thin beds of organic remains", which included the Melbourn Rock of the Middle Chalk.

Hébert (1874), Potier and de Lapparent (1875) and Barrois (1876) wrote of the correlation of the English Chalk with that of France, where a scheme of palaeontological classification was being developed ; Barrois concluded that the ' Grey Chalk ' and the ' Chalk without flints and with few fossils ' of Phillips (quoted by Whitaker) represented the *Holaster subglobusus* Zone and the *Belemnites plenus* Zone respectively of the French classification.

The work of Price (1877) was based largely on his measurement with C. E. De Rance of the cliffs at Lydden Spout. He put the thickness of the Lower Chalk at about 198 ft : 24 ft were allotted to the ' Chalk Marl ', which he divided into two beds : Bed I, 14 ft thick, was the *Stauronema carteri* Zone, which had hitherto been identified as Upper Greensand, but which he distinguished on palaeontological grounds. Bed II, *Plocoscyphia maeandrina* Zone, was 10 ft thick (p. 120). Above this his Beds III to VII made up the ' Grey Chalk ', with a combined thickness of about 174 ft : Beds III, IV and V made up the *Ammonites rhotomagensis* and *A. varians* Zone (22¼ ft), and Bed V was known as the ' Cast-Bed ', 2¾ ft thick, from which the Lydden Spout spring emerged. Bed VI, 148 ft thick, represented the *Holaster subglobosus* Zone and Bed VII (4 ft) the *Belemnites plenus* Zone. Price reserved the term ' Lower Chalk ' for chalk above the *Belemnites plenus* Zone, and it was left to Penning and Jukes-Browne (1881), to give to the terms Lower, Middle and Upper Chalk the meanings that they now bear.

Jukes-Browne (1903) divided the Lower Chalk into nine beds, of which his Beds 1, 2, 4 and 9 corresponded to Price's Beds I, II, V and VII. He combined Beds III and IV of Price to form his Bed 3, but he subdivided Price's Bed VI to form Bed 5, grey fossiliferous marly chalk (19 ft), Bed 6, grey chalk with few fossils (60 ft), Bed 7, greyish chalk in thin beds with stony lumps and large ammonites (7 ft), and Bed 8, massive white chalk with *Holaster trecensis* and *Discoidea cylindrica* (53 ft).

Osman (1917, pp. 76–9) found the Lower Chalk to be 256 ft thick near Shakespeare Colliery, using the known elevation of the top of the *plenus* Marls in the cliff (135 ft above O.D.), and that of the Chloritic Marl in the Shakespeare Colliery No. 3 Shaft (121 ft below O.D.). At Lydden Spout his Bed 1, *Stauronema carteri* Subzone, Bed 2, *Plocoscyphia labrosa* Subzone and Bed 3, " Grey varying ", were 12, 72 and 22 ft thick respectively, and together made up the *Schloenbachia varians* Zone ; Bed 4, grey marly chalk, Bed 5, grey chalk, Bed 6, whitish chalk, and Bed 7, *A. plenus* Subzone, were 20, 60, 60 and 6 ft thick respectively, and constituted the *Holaster subglobosus* Zone.

The following lithological description is based upon that of the cliff-section by Jukes-Browne (1903, pp. 32–42).

The Chloritic Marl is a highly glauconitic marl, consisting largely of grains of dark green glauconite, with some quartz, in a varying amount of buff-coloured marly matrix. The underlying Gault is penetrated to a depth of 3 or 4 inches by pockets containing Chloritic Marl material. The proportion of glauconite decreases upwards, the transition to the overlying bed being poorly defined. In the Walmestone Borehole, Burr (1913, p. 732) noted 5 ft of glauconitic marl, and a similar thickness was present at Ottinge

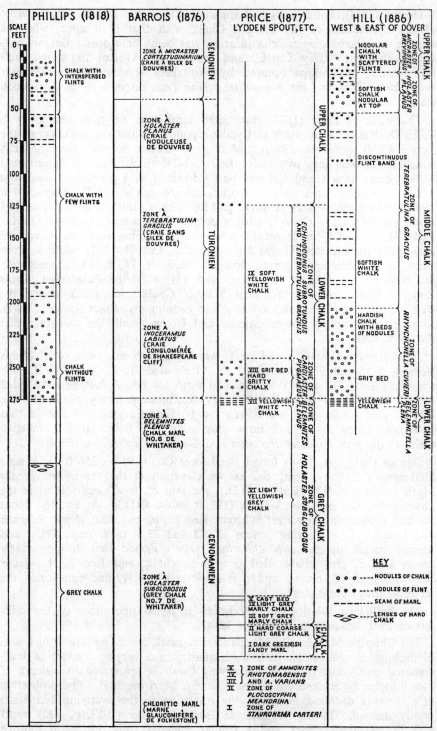

FIG. 3.   *Classification of chalk exposed*

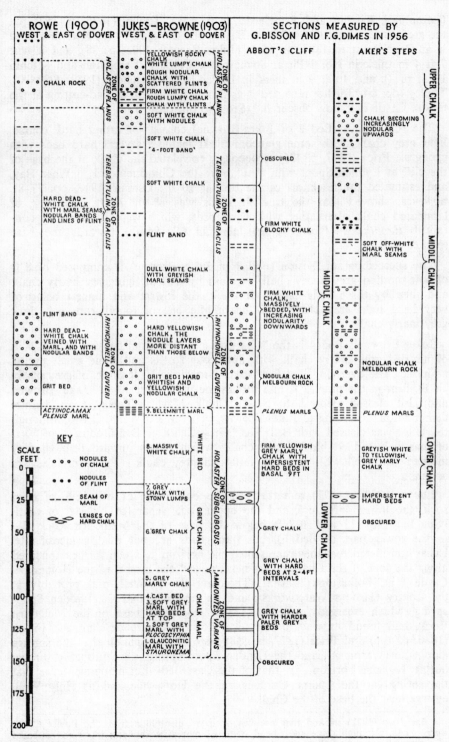

in cliffs between Folkestone and Dover

(Lamplugh, Kitchin and Pringle 1923, p. 54). Topley (1875, p. 152) put the greatest thickness in East Wear Bay at not less than 20 ft ; Kerr measured it at Copt Point in 1887, as 16 ft (Jukes-Browne 1903, p. 38), and it was 7 ft 4 in thick in No. 1 Pit at Shakespeare Colliery (Lamplugh and Kitchin 1911, p. 7), and 10 ft 9 in thick in Dover No. 2 Borehole (1958) (330400). Topley (op. cit.) further recorded it as 3 or 4 ft thick in a road-cutting at the foot of Castle Hill, now obscured. (See also pp. 126–7.)

The rock above (Bed 2 of Jukes-Browne) consists of rather hard, coarse, light grey chalk, with some glauconite. It includes several hard beds with sponges. Price (1877, p. 435) erroneously correlated the chalk at the base of the cliff at Lydden Spout with that above the Chloritic Marl in Wear Bay, and estimated the thickness of the bed at 10 ft ; Osman (1917, pp. 77–8), however, showed that the true thickness was about 72 ft, consisting of laminated chalk alternating with hard beds, which were thicker and contained *Exanthesis* [*Plocoscyphia*] *labrosus* (T. Smith) towards the base (p. 128–9).

The succeeding subdivision (Bed 3 of Jukes-Browne) is composed of 8 ft of soft mottled grey marly chalk overlain by 8 ft of light grey marly chalk and then by three beds each 1 ft thick, the lowest and highest being of very hard light grey chalk and the middle one of fossiliferous soft dark grey marly chalk.

Jukes-Browne's Bed 4 is the ' Cast-Bed ' of Price and consists of 2 ft 9 in of mottled grey marly chalk with numerous fossils and some small pyrite[1] nodules. The overlying Bed 5 of Jukes-Browne is 19 ft thick, of grey marly chalk, firm in the basal 2 ft and soft in the uppermost 10 ft. Jukes-Browne and Hill classed Beds 1 to 5 as ' Chalk Marl '. There is no definite demarcation, according to Jukes-Browne, between Bed 5 and Bed 6, although the latter was classed with Bed 7 as ' Grey Chalk '. Bed 6 consists of 60 ft of massively-bedded blocky grey chalk, which passes upward through grey mottled with yellowish grey, to yellowish grey chalk. Pyrite nodules are common.

Lucas (1908, p. 471) remarked on the occurrence of two beds of marl rock, the lower 4 ft thick and the upper 3 ft, separated by 3 ft of marl, lying between 117 and 127 ft below the base of the Melbourn Rock (i.e. in the lower part of Bed 6), but these beds are not readily identifiable. Lucas considered that they represented the horizon at which springs appeared along the foot of the Chalk escarpment, as at " the targets in the Horseshoe, Castle Hill, Folkestone ", at St. Thomas's (Holy) Well, and probably at the Cherry Gardens Waterworks ; and he supposed that the Lydden Spout spring, which emerged in Bed 4, 148 ft below the base of the Melbourn Rock, had found its way to this level down a fissure or channel. Fitton (1836, p. 106) and Price (1877, p. 438) were of the opinion that the springs along the escarpment, and the Lydden Spout spring, were probably thrown out at the same horizon—i.e. that of the Cast Bed—but mapping shows that the springs at the Cherry Gardens, at the Horseshoe and at Holy Well, emerge near the base of the Chalk.

---

[1] Bannister (1932) showed that nodules of iron disulphide from the Lower Chalk were pyrite, although some spear-head twins of marcasite showing warty excrescenses of pyrite did occur at Folkestone.

Bed 7, 7 to 10 ft thick, is massively bedded, firm, rather rough, yellowish grey marly chalk, with occasional nodules and rods of pyrite up to ½ inch in diameter : this bed is characterized by thin impersistent beds of hard stony chalk, which stand out in the face of the cliff. A thin section (E 28889) of one of these stony lenses was described by Dr. R. Dearnley as follows : Fine-grained limestone with abundant organic material. Local recrystallization of the matrix occurs producing large plates of calcite. A considerable proportion of the rock consists of small spherical organic bodies.

Bed 8, the ' White Bed ', is the equivalent of Phillips's ' Chalk without flints and with few organic remains ', and is 53 ft thick according to Jukes-Browne, but is about 57½ ft thick at Aker's Steps (p. 131) ; it consists of fairly soft yellowish grey chalk, passing up into greyish white chalk, with few fossils and some pyrite nodules. The *plenus* Marls (Bed 9) range in thickness from 4 ft 3 in to 10 ft. They comprise beds of soft yellowish grey or buff-coloured marl, interbedded with firm light yellowish grey marly chalk (Plate VA). Jefferies (1963) found that in the middle of the Anglo-Paris Basin the *plenus* Subzone conformed more or less to a standard succession of eight beds which were distinguishable by fauna and lithology ; erosion surfaces occurred at the bottom and top and at four intermediate horizons, and after each interval of erosion marl was deposited in response to fast erosion on land due to fall in sea level. He estimated (1962) that the temperature of deposition of the lowest bed (his bed 1) was probably below 5 to 10° C, that of bed 2 considerably above 16·9° C, and that of beds 4 to 6 about 16·9° C.

MIDDLE CHALK

In this memoir the Middle Chalk is taken to extend from the base of the Melbourn Rock to the base of a 4-ft bed of chalk with many flints, which Jukes-Browne (1903 ; 1904) adopted as the base of the *Holaster planus* Zone. Phillips's (1818) ' Chalk without flints ' comprised the uppermost 50 ft of the Lower Chalk and the lowest 90 ft of the Middle Chalk, the top being marked by a marl seam. The chalk above he called ' Chalk with few flints ' and about the junction between this and the overlying ' Chalk with interspersed flints ' he described a succession of marl and flint bands which suggests that the upper limit of the former was only slightly below the horizon now taken as the top of the Middle Chalk. Barrois (1876) assigned the ' Craie conglomérée de Shakespeare Cliff ' (including the Melbourn Rock) and the ' Craie sans silex de Douvres ' to his ' Zone à *Inoceramus labiatus* ' and ' Zone à *Terebratulina gracilis* ' respectively, with thicknesses of 25 and 30 metres, based on the measurement by Hébert.

Price (1877, p. 440) allotted 32 ft of " exceedingly hard gritty chalk " to his Bed VIII or ' Grit Bed ' (' Zone of *Cardiaster pygmaeus* ') ; his Bed IX was 118 ft thick, of soft yellowish white chalk (' Zone of *Echinoconus subrotundus* and *Terebratulina gracilis* ') and was measured from the top of the Grit Bed to the first line of flints. These two beds comprised his ' Lower Chalk '.

In 1886 Hill came nearer to the classification now accepted, when he described the Middle Chalk seen in the cliff above the western entrance of the Shakespeare Tunnel; he allotted 70 ft of nodular chalk, including the 32-ft Grit Bed, to the 'Zone of *Rhynchonella cuvieri*', which he considered agreed closely with the *Inoceramus labiatus* Zone of Barrois. His 'Zone of *Terebratulina gracilis*' consisted of 150 ft of softish white chalk with marl bands, with the lowest line of flints 60 ft above the base, and terminating in a marl seam which was closely underlain by two flint bands. These zones, together with his 'Zone of *Holaster planus*', made up his Middle Chalk; the last-named zone was 22 ft thick, with its top at a flint band some $13\frac{1}{2}$ ft above the top of the Middle Chalk as now defined. Rowe (1900) placed the top of the 'Zone of *R. cuvieri*' at a flint band 70 ft above the base of the Grit Bed.[1] Above this, 161 ft of chalk " as exposed " made up the 'Zone of *T. gracilis*', terminating in an open marl band between two pairs of flint lines. Jukes-Browne (1903 ; 1904) assigned thicknesses of 70, 160 and $43\frac{1}{2}$ ft to the zones of '*R. cuvieri*', *T. lata* and *H. planus* respectively, and believed that he had adopted the same plane for the top of the last zone as had Rowe ; for the junction between the *T. lata* and *H. planus* zones he took the base of a 4-ft bed of chalk with many flints which was $7\frac{1}{2}$ ft below the marl band used by Rowe. A comparison between the detailed measurements of Rowe (1900, p. 316) and Jukes-Browne (1903, p. 377) suggests that Rowe had omitted a thickness of 7 ft 6 in of chalk, and that the thickness of his 'Zone of *T. gracilis*" should be 168 ft 6 in.

At the Belmont and Throwley water boreholes the Middle Chalk is 233 ft and 225 ft thick respectively. West of Wye it forms a steep escarpment which is crowned by nodular chalk of the *H. planus* Zone ; its thickness is at least 200 ft and may be as much as 240 or 260 ft, depending on the dip. Within the Stour Gap mapping indicates a thickness of 200 ft. Eastward from Wye, the base of the Upper Chalk recedes from the crest of the escarpment and the Melbourn Rock rises until it forms the brow of the escarpment south of Arpinge and remains as outliers capping hills north of Cheriton. Farther east, in the cliffs, the base of the Melbourn Rock declines more or less evenly until it reaches shore level at the east end of Shakespeare Cliff ; the Middle Chalk forms the whole of the cliffs above, with the exception of a thin capping of Upper Chalk east of Lydden Spout, at Round Down and at Shakespeare Cliff. At Lydden Spout the Middle Chalk thickness is about 240 ft, while at Aker's Steps it is about 230 ft.

The Melbourn Rock (Plate VA) consists of hard, nodular, gritty, yellowish white chalk, with irregular horizontal streaks and partings of grey marl and many *Inoceramus* fragments, echinoids and other fossils. The thickness of the Grit Bed in the cliffs has frequently been given as 32 ft, but this would seem to be an arbitrary figure ; between Folkestone and Dover the chalk is intensely nodular for 38 to 43 ft above the *plenus* Marls, usually with a sponge bed near the top ; the chalk above is white and contains progressively fewer nodule beds, many of which are spongiferous ; partings and beds of marl up to 3 inches thick are common.

The first continuous flint band seen in the cliffs occurs 146 to 168 ft above the *plenus* Marls, although scattered flint nodules and impersistent

---

[1] The writer has been unable to locate this flint band, which is lower than any recorded by other authors.

(*A* 9666)

A.　The Lower Chalk–Middle Chalk boundary exposed at Shakespeare Cliff

B.　Thanet Beds and Woolwich Beds in Trenleypark Wood Sand-Pit

(*A* 9139)

bands occur up to 7 or 8 ft lower down. The succeeding chalk is massively bedded and towards the top becomes increasingly nodular, with several well-developed marl seams.

UPPER CHALK

In the Boughton-under-Blean Borehole some 283 ft of Upper Chalk are present, the top of the Chalk being here about 30 ft above Whitaker's Three-inch flint band (p. 125). In the Littlebourne Borehole the thickness of the Upper Chalk is around 300 to 350 ft. At Ropersole, 834 ft of Chalk were recorded, of which Dawkins (1900, p. 735) classified 480 ft as Upper Chalk; this allotment was probably excessive, but here, as in some other areas, the Upper Chalk thickness may approach 400 ft.

The Upper Chalk succession was worked out in the cliffs between Dover and Birchington, on the Ramsgate (274) and Dover (290) sheets; it is therefore not intended to set out here the stages in the development of the accepted classification. The following brief lithological description of the beds which, in upward order, comprise the Upper Chalk, is based upon the work of Rowe (1900) and Jukes-Browne (1904).

Holaster planus Zone. The chalk of the *Holaster planus* Zone is in general greyish white and harsh to the touch, with bands of hard cream-coloured nodules; these commonly exhibit a fibrous structure and are set in a grey marly chalk matrix. Flint nodules occur both dispersed and in bands, and vary greatly in size and in thickness of patina.

The Chalk Rock, which is found at the base of this zone in some parts of England, is not developed in the present district. The basal beds of the zone at Fan Hole, east of Dover, have been described by Jukes-Browne (1904, p. 138) :

	Ft	in
7. Rough chalk consisting of hard lumps in a softer matrix, some scattered flints and many fossils ...	2	6
Layer of flints	–	4
6. Smooth firm white chalk	1	4
A thin seam of marl.		
5. Smooth firm white chalk	1	4
Layer of flints	–	4
4. Rough lumpy chalk with *Holaster planus*	2	0
3. Seam of soft marl, forming a marked plane	–	2
2. Rough lumpy chalk with *Holaster planus*	3	6
1. Two layers of flints forming top and base of lumpy chalk with many scattered flints	4	0

Rowe took the base of the zone at the marl seam (bed 3), and his example was followed by White (1928a). In this memoir, however, the base is taken at the bottom of bed 1, following Jukes-Browne, because this horizon approximates to that used in the Maidstone and Chatham districts (Worssam 1963 ; Dines and others 1954). The beds are rarely seen in sections, and in practice the drawing of the line at the base of the Upper Chalk has been controlled palaeontologically, by the stage of evolution of the species of *Micraster* present and by the lowest occurrence in strength of a '*reussianum* fauna'. [The 'reussianum fauna' takes its name from the uncoiled ammonite Hyphantoceras reussianum (d'Orbigny). The term is usually applied to the assemblage of sponges and molluscs preserved as moulds in hard chalk in the lower part of the Holaster planus Zone, but a fauna of this type may be found from the upper part of the lata Zone to the bottom of the cortestudinarium Zone.]

In places, particularly around Chilham, hard nodular cream-coloured chalk is developed near the top of the H. planus Zone, in the position of the Top Rock of the Chiltern Hills area (p. 143).

In the eastern part of the Maidstone district the thickness of the planus Zone is estimated to be 45 to 50 ft (Worssam 1963). This compares with the thickness of 43½ ft arrived at by Jukes-Browne (1904, pp. 138–40) on the coast near Dover.

Micraster cortestudinarium Zone. The lithology of the Micraster cortestudinarium Zone was briefly described by Rowe (1900, p. 291) as " a hard nodular chalk, with marly veins and pockets, and bands of hard, yellow chalk-nodules at intervals. Flints in irregular nodular bands, with occasional thin tabular bands. Marly bands rare." Both Rowe and Jukes-Browne selected the uppermost band of yellow chalk nodules seen in the cliff-section as the top of the zone; the fact that they put the zonal boundary 15 ft and 8½ ft respectively beneath the 'Micraster coranguinum tabular' (see below) suggests that the horizon cannot be reliably located even in the cliff-section.[1] In addition Rowe recorded a tabular flint band 26 ft below the top of the zone, which he called the 'Micraster cortestudinarium tabular', and two strong flint lines 15 ft and 30 ft respectively below it, but these also have not been located in inland sections. The lower of these two flint lines he selected as the base of the zone, which was thus 56 ft thick. Jukes-Browne (1904, p. 141), using Hill's measurements, found the zone to be 75 ft thick, and considered that the discrepancy was due, in part at least, to variation in the thickness of the zone.

Micraster coranguinum Zone. Rowe (1900, p. 303) put the thickness of the Micraster coranguinum Zone at 280 ft, a figure that was repeated by White (1928, p. 22). Within the present district little direct evidence as to the thickness of the zone is available, but it is probably reduced to about 200 ft in the vicinity of Chartham Downs.

The chalk of the zone is white, soft and homogenous, except towards the base : on the coast White remarked that the lowest 40 ft were harsh to the touch, with obscurely lumpy structure in some bands, and in the west of the district here described nodular beds are found in places in the lowest part of the zone. Rowe found that the lowest fourth part of the zone was poorly fossiliferous : this is not everywhere the case within this district, although many chalk-pits considered to be near the base of the zone are not definitely dated because of the paucity of the faunal evidence.

The flint nodules of the zone are black, usually with thick white patinas, and occur both scattered and in regular bands spaced generally about 3 ft apart, except near the top of the zone where the separation is wider. Tabular flint bands occur both parallel and transverse to the stratification. The 'Micraster coranguinum tabular' is 15 ft and 8½ ft above the base of

---

[1] Mr. C. J. Wood (Ann. Rep. Inst. Geol. Sci. for 1966, 1967, p.91) has proposed to take the base of the Micraster coranguinum Zone in the cliffs of the Dover (290) Sheet at the upper of two persistent marl seams, which coincides with the first appearance of Micraster of coranguinum Zone type.

the zone as defined by Rowe and Jukes-Browne respectively at different parts of the cliffs. In addition the 'columnar band' of Bedwell (1874, p. 19), characterized by perpendicular columns of flint that rise from it at intervals, is 31 ft below a strong band of tabular flint known as 'Whitaker's Three-inch Band'; the latter consists of nodular flint masses, usually joined laterally by tabular flint, and it lies some 21 ft below the base of the overlying *Uintacrinus socialis* Zone.   G.B.

Uintacrinus socialis Zone and Marsupites testudinarius Zone. Rowe (1900, p. 296) divided his *Marsupites testudinarius* Zone into a lower *Uintacrinus* Band and an upper *Marsupites* Band. More recently the two bands have been raised to the status of zones. Lithologically these zones consist of even soft fine white chalk, usually in well-developed beds up to 2 to 3 ft thick. Flints are rare or absent but where present are black, with thin white cortex, and can attain large dimensions. No extensive sections similar to those on the Thanet coast are present within the district here described and accurate measurements of thickness are therefore impossible; however mapping and fossil collecting indicate that the two zones together probably do not exceed 100 ft, of which about 25 to 30 ft is in the *Marsupites testudinarius* Zone. This compares with a combined exposed thickness in the cliffs of 116 ft, according to Rowe, of which the uppermost 48 ft represented the *Marsupites* Band.

In the Thanet coast-sections (Rowe 1900, pp. 294–6 ; White 1928a, pp. 31–5) the base of the *Uintacrinus* Zone is defined by a layer of lumpy iron-stained chalk, termed 'Barrois's Sponge Bed'. Immediately overlying it there is a bed where *Conulus* is commonly found (the 'Conulus Band'). A combination of these two marker horizons with Whitaker's Three-inch Band enables the base of the zone to be traced with ease in the cliffs. Furthermore, an impersistent flint course (the 'Bedwell Line') conveniently marks the upper limit of the *Uintacrinus* Zone ; 2 to 3 ft below it there is a sponge bed, at the top of which a thin band of *Echinocorys scutata elevata* is present ; ammonites are common in the chalk between 4 and 28 ft below the Bedwell Line, these strata being known as the 'Ammonite Beds'. At several widely-spaced localities within this district *Conulus* has been proved to be very common at a level about 20 ft above Whitaker's Three-inch Band. Iron-stained chalk is also present, but is local and not developed at any one particular horizon. As is also the case in Thanet (Rowe 1900, p. 298), *Uintacrinus* is extremely rare in the lowest 20 ft or so of the zone ; furthermore the chalk for about 20 ft above Whitaker's Three-inch Band has a texture and occurrence of flint similar to that of the lower parts of the *Uintacrinus* Zone, and is usually as devoid of diagnostic fossils, so that, if Whitaker's Three-inch Band is not seen, the determination of horizon is generally uncertain.

Within the Canterbury (289) Sheet the base of the *Uintacrinus* Zone is taken at an horizon where *Conulus* is commonly found about 20 ft above a prominent flint band. This flint band is present over a wide area ; on the other hand *Uintacrinus* itself has been found at only one locality west of the Little Stour River (p. 155), but the lowest 20 ft or so of the zone, in which this index fossil is rare or absent, are believed to have a much wider extent westward. Such chalk is believed to be present west of the Blean, near Selling, Boughton church and Brogdale. Collins and Gill (1923, pp. 304–5) were the first to suggest that chalk of this zone was present west of the Blean, but they did not mention the exact locality referred to nor did they attempt to describe the zones of the chalk thereabouts. Chalk low in the *Uintacrinus* Zone is present on the higher ground of Chartham Downs and is believed to extend eastwards from there to Bridge and northwards to Canterbury. Usually in these localities only a few feet of this zone are seen at the top of sections opened in the *M. coranguinum* Zone, and these sections are described in this account under the heading of that zone.

Topographically the chalk of the *Uintacrinus* Zone does not give rise to any distinctive feature west of the Little Stour River, largely because the pre-Eocene bench cuts into successive levels of the Chalk and is the dominant feature of the Chalk landscape; however, to the east of the river, the even soft chalk of the *Uintacrinus* and *Marsupites* zones is thicker and forms distinctive fine open rolling country.

J.G.O.S.

DETAILS

LOWER CHALK

Sections showing Chloritic Marl. No trace of the Chloritic Marl was found west of the Great Stour valley, although Topley (1875, p. 152, footnote) recorded, on second-hand evidence, a thin bed of dark green sand between the Gault and the Chalk "near Eastwell'. Some small springs occur along the Gault–Chalk boundary eastward of Westwell. Other springs issue above the base of the Lower Chalk, among them the large spring at Westwell, which has eroded a deep gully east of the church, and that which rises ¼ mile west of Eastwell church and feeds the lake in Eastwell Park. Between Kennington Hall and Wye the boundary between the Gault and the Chalk is concealed by brickearth and its course as shown on the geological map is largely conjectural.

B.C.W.

At and south-east of Wye the base of the chalk is associated with several small springs: the Chloritic Marl has been proved immediately east of Silks Farm; at the mouth of the deep coombe (Devil's Kneadingtrough) on Broad Downs; south-east of Brabourne; and in the roadside south-east of Horton Park.

J.G.O.S.

Traces of glauconitic marl were noted near the foot of Tolsford Hill about 200 yd north-east of Bluehouse, and again on the southern slope of Summerhouse Hill; thence northward and eastward to Pean the bed is probably continuous and at least 2 ft thick.

B.C.W.

MS. notes by W. Whitaker on a 6-in map in the Geological Survey files record that Glauconitic Marl was found on the north side of the most northerly reservoir at the Cherry Garden Waterworks (211381). In the adjacent Terlingham Tunnel, driven some 1620 ft in a generally northerly direction into the chalk downs, beginning at 224 ft above O.D., Mr. S. C. A. Holmes noted interbedded hard and soft marl, with dip varying from about 10° N.N.E. to a small inclination in the opposite direction.

In the railway cutting at the south-west end of the Martello Tunnel (239372), 900 yd N. 13° W. of Copt Point, the Gault–Lower Chalk boundary was formerly exposed in a trench 30 yd from the tunnel entrance: the section was described by Casey (1949a), who commented that, although the 'pipings' of dark green Chloritic Marl into the uppermost 3 or 4 in of the Gault had been interpreted by some authors as indicating an interval of erosion, the transition was actually not abrupt, and that there was "no positive evidence of a non-sequence at this level."

In the low cliff (242372) 190 yd N.N.E. of the Roman Villa east of Folkestone some 10 ft of highly glauconitic marl, with wisps, tubes and lenses of buff-coloured clay, can be seen. The bed was formerly exposed in a pit (241367), now over-grown, on the east side of the hillock on which the most southerly martello tower stands ; in this area the thickness was estimated by Topley to be 15 ft (1875, p. 152).

The Chloritic Marl can also be seen on the shore south-east of Folkestone Warren, forming part of slipped masses of Chalk and Gault.   G.B.

Sections between Chloritic Marl and plenus Marls. A small quarry at the road junction (000469) 350 yd N.E. of Shoddington showed 8 ft of grey, soft, much-shattered chalk. Joints were inclined, and some could be traced for 2 to 3 ft through the broken chalk. Some surfaces showed slickensides.   B.C.W.

About 30 ft of firm grey chalk with some darker grey streaks and 'rust-balls' were seen in a small quarry (084437) by the roadside 650 yd E. 17° N. of Beddlestone Farm. Jukes-Browne (1903, p. 47) recorded " hard beds " and noted fossils collected from the road cutting north of Bulltown.

Some 14 ft of firm grey chalk with harder bands and pyrite nodules were noted (105419) immediately south-west of the cross-roads 400 yd N. 40° E. of Brabourne church, and 6 ft of similar chalk were seen immediately north-east of these cross-roads. An old quarry (113418) on the north side of the road 1050 yd E. 8° N. of this church showed 16 ft of shattered and disturbed firm grey chalk below 1 to 2 ft of chalky soil. In the road-bank of Stone Street (133400) 620 yd E. 6° N. of Horton Park, firm massive grey chalk is faulted against soft grey blocky chalk, 6 ft of each being exposed. Against the fault plane, which is sharply defined on the north side of the fault, is a mass, 6 ft broad, of soft small rather angular grey chalk fragments in a matrix of fine clayey sand, with desilicified flints, ironstones and quartz pebbles. The whole appears to represent an early stage in the development of a solution depression along the fault. The fault itself cannot be traced away from the section. Only 5 ft of rubbly grey chalk were noted in the extensive old pits on the east side of Stone Street at this place.
                                                                     J.G.O.S.

In the Ottinge Well of the Folkestone Waterworks Co. (172425) the *varians* Zone was shown to be at least between 125 and 135 ft in thickness, while in the railway cutting (167396) north of the Etchinghill–Arpinge road interbedded hard and soft grey marly chalk of this zone can be seen.   G.B.

The north side of the railway cutting (171393) east of Etchinghill shows a clean section of some 60 ft of laminated grey marly chalk, the lowest 15 ft being darker grey in colour and including, at 4-ft intervals, 4 to 6-in beds of hard marly limestone. Fossils including ammonites collected from a bed 15 ft above the floor of the cutting indicate a horizon in the *S. varians* Zone.

An extensive exposure of some 70 ft of chalk is provided by the Folkestone Limeworks Quarry (190381) on the escarpment ¼ mile E. 25° N. of Pean. The chalk is greyish white and blocky, with some hard, massive beds. A loose block yielded *varians* Zone ammonites, but the greater part of the section must be in the *subglobosus* Zone, for Melbourn Rock was exposed at the scarp crest.   B.C.W.

In a chalk-pit (191381) 700 yd N. 5° E. of Danton Farm about 10 ft of firm, brownish grey marly chalk, soft and greenish in colour when wet, can be seen, with scattered nodules of weathered pyrite up to 2 in in diameter ; in another pit (200381) 1250 yd E. 37° N. of Danton Farm, firm to hard grey marly chalk with a bed of greenish grey marl is seen 10 ft above hard grey marly chalk with ammonites including *Schloenbachia subtuberculata* (Sharpe) and *S. subvarians* Spath. The springs at Holy Well (223382) and up to 700 yd to the west, emerge from bedded firm fawn-coloured marl which is not far above the Chloritic Marl. Some 30 ft of greyish white massively bedded chalk can be examined 1100 yd N. of Copt Point, beside the minor road which is crossed by a foot-bridge at this place (241374); this is probably in Bed 6 of Jukes-Browne's classification.   G.B.

Osman (1917, p. 78) described a succession of sixteen bands of alternately harder and softer chalk within 43 ft of Bed 2 (p. 120) of the Lower Chalk, exposed in a channel cut in the chalk above the western end of the Martello Tunnel (239372), 270 yd N. 30° W. of the Roman Villa, Copt Point; the base of the lowest band was 21 ft above the top of the Chloritic Marl (Bed 1) and 40 ft above the top of the Gault. Allen (*in* Osman 1917, pp. 82–4) listed the fossils collected from these bands by Osman and the following list gives revised determinations for those specimens presented to the Geological Survey by Osman in 1930. Band 1 is at the top, and Band 16 at the base of the section.

		Height of base above top of Gault Ft
Band No.		
1.	*Serpula umbonata* (J. Sowerby)	82
	Schloenbachia subvarians Spath	
2.	*Avellana sp.*	78
	Schloenbachia subvarians	
3.	*Serpula sp.*	75
	Kingena lima (Defrance)	
	Inoceramus pictus J. Sowerby	
	Schloenbachia cf. *subtuberculata* (Sharpe)	
4.	' *Rhynchonella* ' *martini* (Mantell)	73
	Inoceramus crippsi Mantell	
	fish-scale	
5.	*Lima (Plagiostoma) globosa* (J. de C. Sowerby)	70
	Plicatula inflata J. de C. Sowerby	
	Schloenbachia aff. *subplana* (Mantell)	
6.	*Inoceramus crippsi*	67
	Turrilites cf. *mantelli* Sharpe	
	fish-coprolite ?	
7.	*?Serpula sp.*	65
	' *Rhynchonella* ' *martini*	
	Terebratulina striatula (Mantell)	
	Inoceramus cf. *pictus*	
	Pycnodonte vesicularis (Lamarck)	
	Scaphites equalis J. Sowerby	
	Schloenbachia aff. *subtuberculata*	
8.	*Kingena lima*	62
	' *Cardita* ' *sp.*	
	Scaphites equalis	
	Schloenbachia cf. *subtuberculata*	
9.	' *Rhynchonella* ' *martini*	59½
	Inoceramus sp.	
10.	*Inoceramus sp.*	57
	Lima (Plagiostoma) globosa	
11.	' *Rhynchonella* ' *martini*	54
	Terebratulina sp.	
	Inoceramus crippsi	
	Pycnodonte vesicularis	
12.	Phymosomatid radiole	51
	Pycnodonte vesicularis	

Height of base
above top of
Gault
Ft

13.	*'Rhynchonella' martini*	...	...	...	...	...	49
	Pycnodonte vesicularis						
14.	*Inoceramus crippsi*	...	...	...	...	...	46
	Pycnodonte vesicularis						
15.	starfish ossicle (fragment)	...	...	...	...	...	43
	'Rhynchonella' martini						
16.	*Entolium orbiculare* (J. Sowerby)	...	...	...	...	40
	Plicatula inflata						

Osman also presented a series of fossils from " Abbotscliff Trial Pit "; this
is presumed to have been a trial made in a heading 100 yd east of the Warren
end of the Abbot's Cliff Tunnel, in which the Gault surface was at 36 ft below
O.D. The fossils are listed below.

Bed No.

Height above
top of Gault
Ft   in

2.								
(Sponge Bed)	*Schloenbachia subvarians*	...	...	...	...	56	0
	Cymatoceras sp.	...	...	...	...	...	50	0
	'Rhynchonella' martini	...	...	...	...	46	6
	Inoceramus sp. [juv.]							
	Lima (Plagiostoma) globosa							
	Pycnodonte vesicularis							
	Inoceramus crippsi	...	...	...	...	...	38	1
1.	*Plicatula inflata*	...	...	...	...	...	2	7
(Chloritic Marl)								

R.V.M.

East of The Warren the chalk in the foreshore is largely concealed by tumbled
masses of fallen chalk and flint, but at intervals solid chalk is exposed in the
wave-cut platform ; fossils of the *varians* Zone were collected 2060 yd S. 33° W.,
1890 yd S. 30° W. and 1160 yd S. 14° E. of Farthingloe (287388, 289389 and
300394) ; 1220 yd S. 37° E. of Farthingloe (304395) a *subglobosus* Zone fossil
was found ; between the last two localities, some 650 yd E.N.E. of the western
end of the Shakespeare Tunnel, Bed 7 of Jukes-Browne's classification rises in
the cliff from below beach level.   G.B.

Sections showing plenus Marls : Traces of yellowish grey soft marly chalk of
the *plenus* Marls were noted in the banks of the sunken lane 400 yd N. 30° E. of
Westwell church.

A quarry (036481) 200 yd E. 25° S. of Boughton Aluph church exposed soft
but firm grey or greenish grey blocky chalk, seen for 4 ft, overlain by laminated
grey marly chalk in 4 to 6-in seams alternating with bands of harder greyish
blocky chalk, in all 2½ ft thick, beneath Melbourn Rock (p. 132). Jukes-Browne
(1903, p. 48) recorded at this quarry 14 to 16 ft of firm greyish white chalk passing
up to 4¼ to 4½ ft of ' Belemnite Marl '.   B.C.W.

Some 3 ft of yellow marly chalk of the *plenus* Marls are present in a small pit
(063480) 620 yd S. 35° E. of Olantigh Towers. Here the marly chalk has been
bored by burrowing organisms ; the borings show as small circular or spiral
nodules of similar marly chalk. In a pit (073460) 200 yd N. of Amage Farm,
10 ft of shattered partially laminated yellow marly chalk from the same horizon

were noted. In an extensive old pit (083443) 950 yd N. 38° E. of Beddlestone Farm the face is graded and partially obscured by talus (in 1954), but the following section was seen :

	Ft
Rubbly grey chalk, very dark in colour in top few feet	40
Firm light grey to grey chalk with pyrite nodules	24
Grey rubbly firm chalk with streaks and patches of light grey chalk and showing a fault of indeterminable throw	36
Grey shattered firm chalk with 'rust-balls'	6

The uppermost dark-coloured chalk is probably the lowest beds of the *plenus* Marls.

In a working quarry (103423) 700 yd N. of Brabourne church 26 ft of massive greyish white chalk, darkest grey in the lowest 10 ft, pass up into 9 to 10 ft of massive greyish green marly chalk which underlie bedded dark grey and green marl, ¾ to 1 ft, beneath Melbourn Rock. The two uppermost beds are grouped as *plenus* Marls.   J.G.O.S.

The *plenus* Marls, overlain by Melbourn Rock (p. 133), are seen in a chalk-pit (098426) on the north-east side of the road, 1500 yd S. 4° E. of South Hill; firm light greyish white to white marly chalk, seen for 6 ft, passes up imperceptibly into about 7 ft of firm greyish white chalk; above this two seams of greenish grey marl, the lower 2 in thick and the upper 4 in, are separated by 6 in of firm greyish white chalk. The thickness of the *plenus* Marls is indefinite, but may be as great as 8 ft.

The junction between Lower and Middle Chalk is seen in the banks of sunken lanes 250 yd N. 40° W., 380 yd N. 35° E. and 800 yd E. 40° S. of the cross-roads at Ottinge, 650 yd E. 16° S. and 1700 yd S.E. of Lyminge church, 1150 yd N. 39° E. of Danton Farm, 220 yd E. 30° N. of Coombe Farm, 180 yd W. 20° N. and 100 yd E. 35° S. of Lower Standen. The junction can also be examined in a large chalk-pit (225380) beside the Folkestone–Hawkinge road, 250 yd E.N.E. of Sugarloaf Hill, where the exposed beds below the Melbourn Rock are as follows:

	Ft
4. Yollowish grey marl, horizontally streaked, the middle 4 to 6 in being locally firm and whiter	1
3. Firm yellowish white chalk, becoming darker towards the base; yellowish patches of marl grade laterally into white chalk ...	4
2. Firm yellowish grey chalk, very marly near the top, but becoming less so below and grading down into	5
1. Buff-white chalk, firm and massively bedded	35

Beds 4, 3 and 2 constitute the *plenus* Marls; 'borings' containing chalk of contrasting colour are present in patches in beds 3 and 2, and in places in the top 2 ft of bed 1.

A chalk-pit (234377) west of Dover Hill, 850 yd S. 5° W. of Hope Farm, displays firm light yellowish grey marly chalk, seen for 1 ft 6 in, beneath 1 ft 9 in of yellowish grey marly chalk, soft in the top 4 in; this underlies 6 in of dark yellowish grey marl with lenses of firm yellowish grey marly chalk, which is overlain by Melbourn Rock (p. 134). In a pit (241375) 550 yd N. 7° E. of the Roman Villa east of Folkestone, greyish white chalk grades up into firm pale yellowish grey marly chalk, locally harder and whiter (patches show mottled colouration), 7 ft 6 in; below fawn friable marl, 3 in; below soft yellowish grey marl which encloses lenses of firmer light grey marly chalk, 11 in, grading up into nodular chalk of the Melbourn Rock (p. 134).

Eastwards from the last locality the Lower Chalk–Middle Chalk boundary is exposed almost without break in the cliffs which skirt Folkestone Warren and the coast-line, but the contact is rarely accessible until it descends to the shore at Shakespeare Cliff; however, the cliff (258385) 170 yd E. 25° S. of Capel Court shows firm white chalk overlain by the *plenus* Marls, which consist of 6 ft 10 in of yellowish grey massively-bedded marly chalk, friable towards bottom and top, beneath 6 in of hard yellowish grey marly chalk, overlain by 1 ft 2 in of yellowish grey marl, largely friable but firm in part.

On the Abbot's Cliff path (268385), at the east end of Folkestone Warren, some 140 ft of Lower Chalk are accessible, while the section can be continued down a further 45 ft in the cliff below and to the east. The combined section is as follows[1]:

	Ft	in
10. Yellowish grey marl	–	4
9. Firm light yellowish grey marly chalk	–	8
8. Yellowish grey marl	–	3
7. Firm light yellowish grey marly chalk	3	2
6. Yellowish grey marl ... ... 0 to	–	4
5. Firm yellowish grey massively-bedded marly chalk, with occasional small pyrite nodules and rods up to ½ inch in diameter	60	0
4. Firm yellowish grey massively-bedded marly chalk with impersistent beds of harder, coarser chalk	9	0
3. Grey marly chalk	37	0
2. Grey marly chalk with hard bands which stand out prominently in the cliff at intervals of 2 to 4 ft	29	0
1. Grey marly chalk, becoming darker grey downwards: harder pale grey beds occur as follows: 8 in at 16 ft below the top of the bed, 10 in at 21 ft, 1 ft at 33 ft	46	0

In the absence of marker horizons it is difficult to relate the lower part of this section to that of Jukes-Browne, but bed 4 above corresponds to his Bed 7, bed 5 is equivalent to Bed 8 and beds 6 to 10 make up the *plenus* Marls (Bed 9).

In the cliff to the west of the Lydden Spout cliff-path (283388) the following section can be seen:

	Ft
4. *A. plenus* Marls: yellowish grey marl (not readily accessible) 6 ft to	8
3. Greyish white marly chalk, massive and apparently uniform, with only occasional irregular marly wisps	57
2. Greyish white marly chalk, becoming more marly downwards and yellowish grey at the base, with impersistent harder beds ...	8
1. Pale yellowish grey marly chalk, with darker grey streaks and wisps; in bottom 30 ft darker grey bands, 6 to 10 in thick, alternate with the paler chalk, and some beds protrude in places from the cliff face	46

At Aker's Steps (297394), above the western end of the Shakespeare Tunnel, roughly 36 ft of chalk below the Melbourn Rock are obscured by debris and vegetation: however, this part of the succession can be seen in the cliff 100 yd west of the tunnel entrance. The composite section is as follows:

	Ft
5. Yellowish grey marly chalk (uppermost 2 ft not seen)	9
4. Firm pale yellowish grey to greyish white marly chalk, with occasional small pyrite nodules; patchy grey and pale grey colouration in basal foot	57½

[1] This section and those at Lydden Spout and Aker's Steps were measured by Messrs F. G. Dimes and G. Bisson (Fig. 3, p. 119).

Ft

3. Rather rough yellowish grey marly chalk with thin impersistent
beds of harder coarser chalk, which protrude from the cliff face
7 ft to   7½

2. Rather rough pale yellowish grey chalk ; patchy grey and pale
grey colouration in basal 4 in ; resting on an irregular surface ...   1½

1. Rather rough yellowish grey marly chalk, with occasional impersistent darker grey streaks. (The base of this bed is 10 ft above
rail level)   10

Beds 3, 4 and 5 in this section are respectively equivalent to Beds 7, 8 and 9
of Jukes-Browne.

At the end of the sea-wall (308398) 1290 yd E. 29° S. of Farthingloe, firm
greyish white chalk, seen for 4 ft, underlies the *plenus* Marls which comprise firm
gritty greyish white chalk with a 4-inch bed of yellowish grey marl 3 ft 1 in
below the top, 7 ft 2 in, overlain by yellowish grey marl alternating with firm
greyish white chalk streaked with marl, 1 ft 10 in ; this is overlain by the
Melbourn Rock represented by nodular coarse gritty chalk, marly in the basal
6 in.   G.B.

MIDDLE CHALK

In the water borehole (980572) 1100 yd N. 38° W. of Belmont, resistivity
measurements[1] and the borehole records suggest that the Middle Chalk is 233 ft
thick, with its base at a depth of 298 ft (−161 ft O.D.). A thickness of 225 ft
was indicated by the chisel cuttings from the Throwley water borehole (995559),
500 yd E. 30° N. of Throwley church, where the base of the Middle Chalk is
believed to lie at a depth of about 335 ft (−160 ft O.D.). Chisel samples from
Selling No. 1 water borehole (035575), examined by Dr. E. R. Shephard-Thorn,
showed the Middle Chalk to be about 199 ft thick, with its base at about 376 ft
depth (−262 ft O.D.). A Geological Survey borehole (076534) 800 yd E. 20° S.
of Chilham church, drilled in connexion with a hydrological research project
(Ineson 1962, p. 63 ; Davies and others 1964, pp. 59–60), encountered the base of
the Middle Chalk at 189 ft below surface (−116 ft O.D.).   J.G.O.S.

Inoceramus labiatus Zone. At a quarry (036481) 200 yd E. 25° S. of Boughton
Aluph church the Melbourn Rock is unusual both in its facies and its fauna.
Resting with a sharp junction on *plenus* Marls (p. 129), the chalk, seen for about
8 ft, is white, very hard and stone-like, fine-grained and only sparingly nodular.
It is much broken by joints and in its top 4 to 5 ft is rubbly. It yielded
occasional *Inoceramus* and, in some places, numerous moulds of small
ammonoids (Casey 1959). The moulds occur in the body of the chalk, not in
nodules, but otherwise the mode of preservation resembles that of the
Hyphantoceras reussianum fauna. Such an occurrence has not been known
hitherto at so low an horizon and it points to peculiar localized conditions of
deposition.

Beds of similar lithology at an horizon a few feet higher are seen in a roadside quarry (040483) 350 yd N. 11° W. of Boughton Corner, which shows about
12 ft of sparsely nodular, hard and stone-like chalk with occasional Rhynchonelloids. The lowest 1 to 2 ft exposed are more nodular than the chalk higher in
the face and contain abundant *Inoceramus* fragments.   B.C.W.

Resistivity measurements suggest that the base of the Middle Chalk lies at a
depth of 507 ft (−442 ft O.D.) in the water borehole (053594) at Boughton Street,

---

[1] Electrical resistivity measurements in this and some other boreholes mentioned in the text
were made by Messrs. D. A. Gray, T. K. Tate and K. H. Murray of the Water Department
of the Geological Survey. Gamma-ray logs were made by Messrs. D. Ostle and J. Taylor.

860 yd E. 44° S. of Nash Court. The division of Middle from Upper Chalk in this borehole is uncertain.

The Melbourn Rock forms such a well-marked feature on the eastern side of the Stour valley that the minor faulting near Wye is conspicuous. South of Trimworth Manor the rock gives rise to steep river-cliffs and valley sides, while on the west side of the Stour valley the feature is subdued and obscured by drift deposits. In a scar (078453) in the deep coombe (Devil's Kneadingtrough) 950 yd S. 42° E. of Amage Farm 20 ft of coarse greyish nodular chalk are exposed. In a small pit (084448) 920 yd S. 28° W. of Staple Lees are 9 ft of rubbly rust-stained firm chalk with a few harder lumps.

In the working quarry (103423) 700 yd N. of Brabourne church Lower Chalk (p. 130) is overlain by bedded firm white chalk with nodular seams up to 1 ft in thickness, 10 ft, beneath hard nodular coarse white chalk, 9½ ft. The nodules in the uppermost exposed bed are 3 to 4 inches in diameter but are somewhat smaller below.

J.G.O.S.

Melbourn Rock rests on Lower Chalk (p. 130) in the roadside pit (098426) 1500 yd S. 4° E. of South Hill. At the base, 1 ft 9 in of hard, massive, slightly greyish white chalk is overlain by 1 ft 3 in of rough, nodular, greyish white chalk with greenish grey staining along interfaces; this is succeeded by some 14 ft of hard rough gritty greyish white chalk, including a 6-in bed of rough nodular chalk with sponges 9 ft from the top. In another pit, 100 yd to the north-east, the above section is continued after a vertical gap of about 58 ft; some 15 ft of firm to hard, white and greyish white chalk are exposed, with a nodular bed up to 1 ft 2 in thick about 10 ft below the top. The chalk is yellow-stained near the ground surface.

In a small pit (125428) 500 yd E. 11° S. of Park Farm about 6 ft of firm to hard, white Melbourn Rock can be seen, including a 15-in bed of rough, gritty, nodular chalk with sponges. An excavation (149423) at the side of a field 2050 yd N. 38° W. of Lyminge church revealed 4 ft of hard creamy-white unbedded chalk near the base of the Middle Chalk; another exposure (155423) 300 yd N. 32° W. of Longage Farm shows 7 ft of firm blocky white chalk, with poorly-defined bedding: this must be high in the *I. labiatus* Zone; 560 yd W. 39° N. of the cross-roads at Ottinge, on the north side of the lane (164427), a pipe of red sand lined with dark mottled clay is visible in 4 ft of firm to hard, blocky white chalk, and beside the road (173398) 1730 yd E. 42° S. of Lyminge church 4 ft of hard nodular gritty cream chalk are exposed.

G.B.

North of Brockman's Bushes small exposures (162389) in the banks of the lane to Etchinghill show nodular chalk. The outlier capping Summerhouse Hill is indicated by exposures in old trenches of nodular chalk with *Inoceramus* and Rhynchonelloids, down to some 20 ft below the summit. A small old quarry (180392) 1050 yd E. 10° N. of Coombe Farm showed 4 ft of hard, fine-grained rough-fracturing white chalk of the *labiatus* Zone.

B.C.W.

Lucas (1908, p. 461) described "a buff-coloured rock, closely resembling the smooth parts of the Melbourn Rock" which "runs down both sides of the Alkham Valley as far as the vicarage". The base of this rock was 77 ft above the top of a rock which was proved between 70 and 90 ft below surface in the Lower Standen Well of the Folkestone Waterworks Co. (241404), and which Lucas took to be the Melbourn Rock. The shaft is 130½ ft deep and is now brick lined to within 14½ ft of the bottom, where yellow and grey marly chalk can be seen. Chalk of typical Melbourn Rock aspect is dug up in cottage gardens 100 yd N. of the Lower Standen Pumping Station, and nodular chalk was also found at the bottom of the bank on the south-east side of the valley, 240 yd N.E. of the station, and on the surface 160 yd W.S.W. of Drellingore Pumping Station. On the other hand a hole dug 70 yd N. 35° E. of the Lower

Standen station showed grey marly chalk, reminiscent of Lower Chalk. In the north-eastern adit of No. 3 Well (241408), Mr. S. C. A. Holmes found the base of the Melbourn Rock at 132 ft above O.D., while in the Drellingore well-shafts (243412) it was at + 133 ft (Reynolds 1948, pp. 107–8). It is evident from the disparities between the observations at the surface and in the wells that steep dips, or faults with downthrows to east or north-east, are present in the area (p. 13). Very hard pale grey nodular chalk is exposed in the bank (252420) beside the road 560 yd S.W. of Alkham church.

In a large quarry (225380) 250 yd E.N.E. of Sugarloaf Hill the Middle Chalk is represented by some 60 ft of hard, nodular, white to greyish white chalk streaked with marl, the basal 12 ft of which are in massive beds. Sponges occur at two horizons. It is not clear whether the whole of this chalk should be allotted to the Melbourn Rock, owing to the rubbly nature of the chalk near the surface.

In a chalk-pit (234377) west of Dover Hill, 850 yd S. 5° W. of Hope Farm, the *plenus* Marls (p. 130) are overlain in upward sequence by hard off-white chalk without nodules, 6 in ; hard yellowish white chalk with a tendency to be nodular towards the top, 2 ft 8 in ; grey marl, 1 in ; and hard off-white chalk, nodular except in the basal 2 ft, about 25 ft.

In a pit (241375) 550 yd N. 7° E. of the Roman Villa east of Folkestone, 20 ft of poorly-bedded, nodular, gritty chalk are exposed above the *plenus* Marls (p. 130). Occasional wavy yellowish grey marl partings are present throughout and scattered pyrite nodules occur in the basal 3 ft.

On the cliff-path at Abbot's Cliff (268385), the uppermost 40 or 50 ft of the section are grassed over ; the following section can be examined :

	Ft	in
32. Firm blocky white chalk	12	0
31. Irregular line of rather small flint nodules, black with white cortex.		
30. Firm blocky white chalk ; occasional flint nodules 8 ft down	10	0
29. Firm blocky flintless white chalk ; 1-in marl seams 5, 24 and 26 ft down	42	0
28. Impersistent bed of nodular chalk ... ... 0 to	—	6
27. Massive firm white chalk	9	0
26. Grey marl ; wisps of marl persist in basal 6 in of bed above 2 in to	—	3
25. Massive firm white chalk, nodular in top 1 ft 6 in and spongi-ferous in top 4 in	10	0
24. Grey marl ; wisps of marl persist in basal 5 to 6 in of bed above ... ... 2 in to	—	3
23. Massive firm white chalk with scattered nodules, particularly in uppermost 2 ft	17	6
22. Massive nodular white chalk, with marked beds of nodules in uppermost 1 ft and lowest 1 ft 6 in	9	9
21. Grey marl	—	0½
20. Firm white chalk, nodular in top 1 ft 6 in	2	8
19. Impersistent seam of grey marl ... ... 0 to	—	0½
18. Firm white chalk, unevenly nodular	5	0
17. Grey marl ; wisps of marl persist in basal 6 in of bed above 1 in to	—	2
16. Massive firm white chalk	4	9
15. Hard yellow-stained cream chalk nodules in a softer matrix, with sponges	—	4
14. Massive firm white chalk with an irregular line of nodules 1 ft 6 in above the base	4	2

	Ft	in
13. Marl parting.		
12. Massive firm nodular white chalk, markedly nodular in uppermost 6 in	2	0
11. Rough gritty nodular chalk with marly wisps; yellow-stained in top 4 to 5 in	43	0
1–10. Lower Chalk: (p. 131).		

The above section is interrupted between beds 22 and 23 by a fault (p. 13).

On the Lydden Spout path (283388) the following section is exposed:

	Ft	in
34. Chalk (mostly obscured)	13	6
33. Occasional large black thinly-patinated flint nodules.		
32. Massive firm white chalk, poorly exposed	33	0
31. Grey marl	—	6
30. Massive firm white chalk, marly in uppermost 3 in	4	6
29. Grey marl	—	6
28. Massive white chalk with scattered flint nodules	10	0
27. Nodular white chalk	3	0
26. Massive firm white chalk; occasional flint nodules 10 ft below top	25	0
25. Band of flint nodules.		
24. Firm white chalk: beds of grey marl 1 to 2 in thick occur 14 ft 8 in and 16 ft 6 in below the top	27	6
23. Grey marl; wisps of marl continue in basal 3 in of bed above	—	0½
22. Massive firm white chalk, with a 3-in bed of grey marl 1 ft 9 in below top	19	9
21. Massive white chalk, nodular in uppermost 6 in	10	6
20. Grey marl; wisps of marl continue in basal 3 in of bed above	—	5
19. Massive firm white chalk, nodular in uppermost 2 ft ...	11	0
18. Grey marl; wisps of marl continue in basal 5 in of bed above	—	3
17. Massive firm white chalk, nodular throughout but particularly so in the uppermost 2 ft, between 4 and 6 ft below the top and between 9 and 12 ft below the top; sponges occur between 9 and 12 ft down	21	0
16. Yellow-stained nodular chalk with sponges	—	6
15. White chalk with scattered nodules and marly wisps ...	12	0
14. Grey marl seam, irregular and wispy ... ... 2 in to	—	3
13. Massive white chalk with irregular bands of nodules and with a marl parting 5 ft from the top	8	0
12. Grey marl parting.		
11. White chalk, nodular in places and particularly in top 8 to 12 in	4	0
10. Chalk with scattered nodules; very nodular and spongiferous in uppermost 6 to 9 in	2	0
9. Marl parting.		
8. Massive chalk, irregularly nodular	9	0
7. Marl parting.		
6. Off-white, nodular chalk	1	6
5. Nodular chalk, with wisps, streaks and irregular partings of grey marl. Sponges in uppermost 12 to 18 in	41	0
1–4. Lower Chalk: (p. 131).		

Bed 33 of this section may correspond to bed 3 of the section at the cliff top 75 yd to the west (p. 146–7).

K

At Aker's Steps (297394) the section is as follows:

								Ft	in

50–53. Upper Chalk: (p. 147).

	Ft	in
49. Nodular chalk	13	0
48. Grey marl ... ... 4 in to	—	6
47. Chalk which becomes decreasingly nodular downwards, with flint nodules 7 ft below the top and impersistent marl partings 14, 16 and 19 ft below the top	22	0
46. Grey marl ... ... 2 in to	—	4
45. Chalk, somewhat nodular towards the top	2	4
44. Marl parting.		
43. White chalk	4	0
42. Grey marl ... ... 4 in to	—	6
41. Massive off-white chalk with scattered small flint nodules	5	4
40. Impersistent band of large nodules of black flint.		
39. Massive chalk, nodular between 8 and 14 in above the base ...	6	0
38. Chalk with numerous irregular marl streaks	—	7
37. Nodular chalk	—	10
36. Chalk with numerous irregular marl streaks	1	9
35. Firm greyish white chalk: occasional flint nodules occur 3 ft 5 in below the top and scattered below	25	0
34. Band of flint nodules.		
33. Blocky chalk with occasional small flint nodules	4	7
32. Grey marl	—	1
31. Blocky chalk: occasional small flint nodules in uppermost 2 ft	9	5
30. Chalk with many irregular marl partings	—	6
29. Firm off-white chalk: 1 to 2 in grey marl 1 ft 2 in down ...	2	7
28. Irregular marl partings	—	1
27. Soft off-white chalk: 4 in chalk with four irregular marl partings 6 ft down; 4 in grey marl 8 ft 6 in down: cavities filled with loam and gravel 11 ft 6 in and 16 ft down	26	6
26. Marl parting.		
25. Soft massively-bedded off-white chalk	10	0
24. Grey marl, with white lenses: wisps of marl continue in basal 4 in of bed above	—	2
23c. Nodule bed with sponges	—	3
23b. Soft, off-white chalk without nodules, massively bedded. *Terebratulina lata* common	11	0
23a. Marl band consisting of several marl partings close together. Marl partings persist 4 in into bed above	—	2
23. Chalk, nodular in top 2 ft 9 in and basal 2 ft 5 in, and with scattered nodules between	12	2
22. Chalk with occasional nodules and an irregular marl parting 4 in below the top ... ... 6 in to	1	0
21. Nodular chalk with some sponges ... ... 4 in to	—	5
20. Chalk, nodular except in uppermost and lowest 1 ft 6 in ...	7	0
19. Nodular chalk, with sponges in part ... ... 7 in to	—	8
18. Marl parting, consisting of irregular marl streaks close together.		
17. Nodular chalk, spongiferous in top 5 in ... 11 in to	1	0
16. Marl parting marked by cavities.		
15. Chalk with scattered nodules and irregular marl streaks ...	3	8
14. Nodular chalk with irregular marl partings and streaks ...	2	0
13. Chalk with only occasional nodules; many irregular marl streaks	1	9

	Ft	in
12. Chalk, nodular, particularly in basal 5 in	2	5
11. Chalk with rare nodules	1	10
10. Chalk, nodular, particularly in basal 11 in	2	9
9. Nodular chalk with marly streaks. A discontinuous sponge bed is present 1 ft 6 in below the top, the chalk above being only slightly nodular	5	2
8. Marl parting.		
7. Chalk with irregular marl streaks in uppermost and lowest 4 in, and nodular between	1	4
6. Hard gritty nodular chalk with horizontal marly streaks; spongiferous in uppermost 6 in. (Base not seen)	38	0
1–5. Lower Chalk: (pp. 131–2).		

The cliff north of the railway east of Shakespeare Tunnel is in Melbourn Rock (p. 232).   G.B.

Terebratulina lata Zone : Three large overgrown quarries adjoin each other on the north slope of the spur of the escarpment ⅜ mile W. of Dunn Street. At the top of the middle quarry were noted 12½ ft of white chalk, yielding no fossils, with iron-stained chalk nodules in the top 1½ ft, a 4-in marl seam 4 ft below the top, and a line of large, white-rinded black flint nodules 3 ft below the marl seam. The chalk was burnt for lime at kilns in the quarries.

A large disused quarry (029486) in Eastwell Park, 800 yd W. 40° N. of Boughton Aluph church, which is presumably that referred to by Jukes-Browne (1904, p. 158) as the Warren Quarry, provides on its north face a clear section through the junction of Middle and Upper Chalk. A fault on its west face, seen as a 4-in wide rubble-filled fissure dipping 65° N., with a downthrow to the north probably of 20 ft at least, brings these strata against lower beds. The section is as follows:

	Ft	in
10–20. Upper Chalk: (pp. 142–3) ... ... about	28	0
9. Massive chalk with scattered small flints, and with a ¼-in tabular flint seam 2 ft from top on part of west face, seen for	5	0
Cut out by fault ... ...at least	20	0
8. Soft platy chalk ... ... seen for about	3	0
7. Line of spaced-out flint nodules.		
6. Medium to hard, flintless rough-fracturing chalk	6	0
5. Very massive, hard lumpy chalk, the lumps iron-stained and projecting on the weathered face ; no flints	4	6
4. Marl ... ... 1 in to	–	2
3. Massive, lumpy chalk, as above ; no flints	2	3
2. Well-developed seam of soft grey marl	–	6
1. Rather hard, massive white chalk ... ... seen for	2	0

The lithological succession of the upper beds of the Middle Chalk agrees remarkably well with that in the coast sections at Dover. Jukes-Browne also took the base of the Upper Chalk in this quarry at the base of bed 10, conforming with his interpretation of the coast section. His account of the quarry mentioned no fault. He recorded massively-bedded, firm white chalk with a few scattered flints, corresponding to bed 9 above, as totalling 9 ft and resting on a 6-in marl band, which in turn overlay massive, firm white chalk with a few small scattered flints, seen for 16 ft. The lowest 20½ ft thus recorded are presumably now covered by talus and represent strata cut out by the fault.

Beds 1 to 8 inclusive can be again recognized in a quarry (037488) on the west side of the road up White Hill, 400 yd W. 15° S. of Soakham Farm. Bed 1

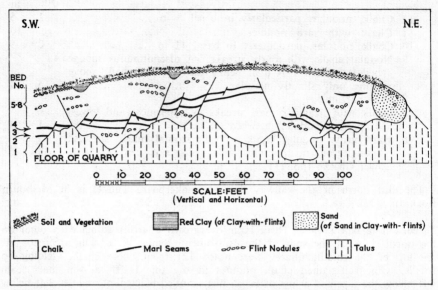

FIG. 4. *Sketch of the face of a quarry in Middle Chalk on White Hill, Boughton Aluph*
The beds are numbered as in the section on Page 137.

is there seen for 8 ft and has flints scattered between 2 and $3\frac{1}{2}$ ft below its top.
The main face of the quarry (Fig. 4) shows a pipe of orange sand and a remarkable series of faults. Sand-filled pipes are numerous in the vicinity of this
quarry, and it is reasonable to assume that the faults were caused by subsidence
connected with the development of the pipes. The faults bounding the relatively
elevated mass of chalk in the centre of the quarry face appear to trend north–
south, obliquely to the face, such that this mass occurs between the pipe at the
east end of the section and a much larger pipe (p. 221) which lies westward of
it, and 20 yd or less north of the quarry.   B.C.W.

In an old quarry (075532) in the river-cliff 1000 yd E. 19° S. of Chilham
Castle, some $3\frac{1}{2}$ ft of hard yellow and white nodular chalk are exposed below
13 ft of Upper Chalk (p. 144). A band of large brown-patinated flints at their
top is considered to be the base of the Upper Chalk. An exposure (079520)
350 yd E. 10° S. of East Stour Farm shows 9 ft of nodular coarse creamy
chalk high in the *T. lata* Zone, beneath 8 ft of loam and Coombe Deposits. The
Godmersham water borehole (073516), sited on the alluvium 540 yd S. 27° W.
of this farm proved $118\frac{1}{2}$ ft of Middle Chalk (base at −55 ft O.D.) below
$19\frac{1}{2}$ ft of drift. Some 130 yd E. of Pope Street Farm (077514), $8\frac{1}{2}$ ft of fine white
chalk with a nodular and lumpy band and two marked marl seams are exposed
below $3\frac{1}{2}$ ft of flinty loam.

In a large old quarry (059505) 370 yd W. of Godmersham church the exposed
section is:

	Ft
8. Upper Chalk: (p. 143)	$4\frac{1}{2}$
7. Firm grey blocky chalk, with one scattered band of large flints ...	$6\frac{1}{2}$
6. Band of very large flints.	
5. Coarse grey marly lumpy chalk with scattered small flints and a widely scattered band of flints near the base ... $7\frac{1}{2}$ ft to	$8\frac{1}{2}$
4. Coarse hard creamy iron-stained nodular chalk (some nodules are fibrous) ... ... $4\frac{1}{2}$ ft to	6

Ft

3. Firm massive coarse grey marly chalk 1½
2. Grey marl seam, 2 inches.
1. Hard coarse grey chalk nodules (some fibrous) in a matrix of soft
    grey marly chalk with dark grey streaks 3¾

Jukes-Browne (1903, pp. 157–8), in describing this section, stated that the junction of the *T. lata* and *H. planus* zones lay between beds 2 and 3. Recent collecting does not support this view and beds 3 to 7 are now also allotted to the *T. lata* Zone. There are two small faults in the face.

At Crundale Limeworks (073497), 700 yd E. 17° S. of Ripple Farm, 19 ft of massive hard grey marly chalk with much ferruginous staining in places are separated by a thin marl seam from 3 ft of massive hard grey marly chalk above; these are overlain in upward order by grey marl, 2 in; soft massive grey marly chalk, 4 ft; thin streaks of grey marl; soft massive grey marly rust-speckled chalk, 8½ ft; and chalky loam, 10 in. A small fault downthrowing 2½ ft to the east is present on the north face of this pit. The marl seams show signs of disturbance where they approach the surface. Some 3 ft of rubbly chalk passing down into soft marly greyish chalk with harder lumps seen for 9 ft, the whole being extensively iron-stained, were noted in a pit (088494) at the road-bend 300 yd E. of Winchcombe Farm. South of the road 210 yd N. 25° E. of Crundale House a small pit (083488) showed 6 ft of massive brittle coarse white chalk. A similar thickness of fine brittle white chalk was noted in another pit (084486) 170 yd E. 19° S. of the house.

A pit (070480) on the spur 620 yd W. 20° N. of Marriage Farm shows 1½ ft of firm coarse grey chalk with a scattered band of large flints at the top, overlain by 4 ft of massive coarse firm greyish chalk with occasional hard nodules and scattered flints, below 1 ft of chalky soil. In 1954 the large quarry (071466) below and 60 yd W. of the crown cut on Wye Downs showed the following section in upward order: massive well-bedded soft grey coarse chalk (drying whiter and harder) with rare hard nodules and impersistent marl seams, one of which was underlain by 4 in of iron-stained nodular chalk, 43 ft; grey marl, 2 to 3 in; firm grey occasionally nodular and iron-stained chalk (drying whiter and harder) with bands (mostly impersistent) of grey marly chalk 2 to 3 ft apart, 26½ ft; a scattered band of small thinly patinated flat flints; and firm grey chalk, 2½ ft. Three small faults with downthrows of under 5 ft were seen in the pit. Jukes-Browne's record (1903, p. 47) of a section "one third of a mile east of Coldharbour Farm" presumably referred to this pit, but in this event he was in error in regarding it as being opened in the *H. subglobosus* Zone.

J.G.O.S.

In a chalk-pit (106439) 150 yd E. 15° S. of Kingsmill Down 4 ft of soft greyish white chalk, probably of the *T. lata* Zone, were exposed, and at the sharp road-bend 400 yd W. 30° S. of Elmsted church similar chalk was seen. Another chalk-pit (116457) on the west side of the road 370 yd N. of Hill Street exposed 4 ft of rubbly firm white chalk with a tendency to be nodular.

G.B.

In a pit (126469) by the road-bend 500 yd N. 19° E. of Dean Farm nodular hard greyish white chalk, 4 ft, is overlain successively by shattered massive greyish white lumpy chalk with dark marly streaks, 4 ft, a band of medium-sized flints with varied thickness of patina, and by rubbly firm greyish white coarse chalk, 5½ ft. At the roadside (125466) 220 yd N. 15° E. of this farm 7½ ft of coarse firm greyish white chalk with occasional hard iron-stained nodules are seen, and in another roadside exposure (119466) 130 yd S. of Ittinge Farm 5 ft of rubbly firm greyish white chalk are exposed. On the valley side (149460)

630 yd S. 25° E. of the windmill at Stelling Minnis 2 ft of soft greyish chalk are overlain by 6 ft of rubbly weathered soft white chalk.

At the road-fork (161478) 930 yd E. 19° N. of Highchimney Farm the following upward succession of beds is exposed: massive, well-bedded marly greyish white occasionally lumpy chalk, 4 ft; band of scattered flints; nodular and lumpy greyish white chalk with streaks of grey marly chalk, 4 ft; very well-marked band of large flints with a variable thickness of patina; well-bedded coarse greyish white lumpy and nodular chalk with some flints, 5½ ft; thin marl seam; soft white blocky chalk, lumpy in places, with a band of flint 2½ ft above base, 5½ ft. The prominent flint band dips 2° and 1° to north and south respectively from the central part of the face; the uppermost flint band probably lies at the base of the *H. planus* Zone.

In a small pit (158470) by the roadside 220 yd E. 25° N. of Fryarne Park 6¼ ft of rubbly firm marly chalk were noted and in another roadside pit (157466) 350 yd S. 15° E. of this farm the following section was seen: rubbly greyish white coarse chalk with a band of large flints, 4 ft, beneath grey marl, 9 in, dull white rubbly lumpy chalk with a few fibrous nodules, 4 ft, a thin seam of grey marl, and soft dull white rubbly lumpy chalk, 2 ft.

There are pits high in the *T. lata* Zone by the roadside 1050 yd S. 33° W. of Dane Farm. In the more northerly pit (171471) the section in ascending order is: coarse creamy lumpy chalk with large thickly-patinated flints, 7 ft; coarse creamy nodular chalk, 2 ft; a band of medium to large flints with thick patinas; coarse creamy lumpy chalk, 2 ft; a band of medium-sized flints; coarse creamy and white nodular chalk, 2 ft; dirty grey marl, 2 to 3 in; rubbly coarse greyish white lumpy chalk, 2 ft. In the southerly pit (171470) the section shows massive greyish white lumpy chalk, 5½ ft, beneath nodular creamy and white coarse chalk with flints, 5 ft, below lumpy white coarse chalk, 1½ ft. In a small pit (168469) 270 yd E. 18° N. of Wildage Farm 5 ft of rubbly nodular white chalk were noted.

J.G.O.S.

Some 8 ft of soft white chalk with a few scattered small flint nodules up to 2 inches long, overlain by dark brown to black Clay-with-flints in pipes, and probably representing the *T. lata* Zone, are seen in a pit (162449) beside Clavertye Wood, 2000 yd W. 36° N. of Elham church. In the same valley, 600 yd N. 37° W. of Upper Parkgate Farm, a pit beside the road (168457) shows about 30 ft of soft white to greyish white rubbly but well-bedded chalk, belonging to the same zone; flint nodules are present at two horizons about 10 ft apart, as well as occasionally at other levels.

In a pit (161433) beside the lane 220 yd S. 16° E. of Mount hard creamy-white chalk, seen for 1 ft, is overlain by 3 in of pale grey and white marly chalk, below 11 ft of firm white chalk which is well-bedded and blocky, and becomes softer towards the top.

G.B.

In pits near the sharp road-bend 420 yd W. 25° N. of Wingmore are sections high in the *T. lata* Zone. On the east side of the road (184467) the section is: coarse massive lumpy greyish white chalk with marly streaks, 3½ ft; overlain successively by hard creamy iron-stained nodular chalk, 1 ft; band of scattered flints; massive coarse soft greyish white chalk, 2 ft; grey marl, 2 in; coarse soft greyish white blocky chalk with harder lumps, 2¼ ft; band of scattered flints; coarse white rubbly chalk with hard nodules, particularly in masses at the base, 3½ ft. On the west side of the road (184467) 5 ft of blocky coarse white chalk with some harder nodules and a marl seam at the base were seen below 3 ft of flinty loam.

J.G.O.S.

In a pit (182437) beside the lane 580 yd E. 18° S. of Elham church the following section in chalk of the *T. lata* Zone is exposed:

	Ft
Firm blocky white chalk, becoming softer, greyish white and marly in the lowest foot; scattered sponges	6
Hard blocky white chalk with scattered fragments of *Inoceramus* ...	3
Light grey marl	0¼
Firm creamy-white blocky chalk with fragments of *Inoceramus* and thin impersistent marl seams	2½
Light grey marl ... ... 0 to	0¼
Firm creamy-white blocky chalk with sponges, especially 1 ft below the top	4
Soft greyish white blocky chalk with *Inoceramus*	7

On the south side of an old pit (192426) 425 yd N. 27° W. of Acrise church extremely hard nodular cream chalk is overlain by about 9 ft of poorly-exposed soft greyish white chalk with a course of large irregular thinly-patinated grey flint nodules some 4 ft above the base. On the north side of the pit, at a somewhat lower level, about 2 ft of similar hard nodular chalk underlie an impersistent bed of flints, the chalk below being soft and white. These beds are probably low in the *T. lata* Zone.

In Limekiln Plantation (198420), beside the road 500 yd E. 28° S. of Acrise church, 4 ft of firm greyish white blocky flintless chalk with hard patches towards the top are overlain by 6 in of chalk with hard cream nodules and sponges, beneath about 7 ft of intermittently-exposed firm white blocky flintless chalk. Topley (in MS.) noted 15 or 20 ft of white chalk without flints in this pit, with a well-marked greyish band 1 ft thick.

At the side of the road (203442) 180 yd E. 8° N. of Rakesole, the upper part of the *T. lata* Zone is represented by about 9 ft of soft to firm blocky white chalk with some marly streaks and rare small flint nodules, gradually becoming more nodular upwards, and overlain by about 7 ft of blocky, nodular white to greyish white chalk, with grey marly patches, in which scattered large flint nodules, with ⅛-in white patinas, tend to be arranged in courses. On the opposite side of the valley, in Craft Shaw (197441), 550 yd S. 12° E. of Dreal's Farm, are seen 6 ft of greyish white chalk, hard with soft patches and nodular towards the top; and at the southern end of Canter Wood (199442), 490 yd S.E. of Dreal's Farm, are exposed 4 ft of firm to hard white chalk with occasional sponges and, at one horizon, large flint nodules. A pit (207459) 50 yd N. of Gatteridge Farm exhibits 5 ft of soft white chalk overlain by 5 ft of hard nodular cream chalk with scattered large irregular flint nodules, probably near the top of the Middle Chalk.

A pit (214391) 150 yd E. of Terlingham displays 9 ft of firm blocky white chalk containing a pipe of red sand with brown and mottled clay with flints. An old pit (229424) 660 yd W. 29° N. of Great Everden Farm exhibits several feet of white chalk with some flints, near the top of the *T. lata* Zone. Chalk lower in this zone was dug in a cesspit (233430) 20 yd S. of Beard's Hall Farm, where it was very white and massive, with a 4 to 5-in marl seam and small elongated flints. Some 10 ft of soft white chalk are intermittently exposed in a large disused pit (231415) 750 yd S. 34° W. of Great Everden Farm, and an impersistent band of black flint nodules with thin white patinas is seen at the south end of the pit.

In a water borehole (224481) 1730 yd N. 29° E. of Denton Court, resistivity measurements indicated that the base of the Upper Chalk was at 104 ft below surface (131 ft above O.D.), while the base of the Middle Chalk was at 314 ft, but by gamma-ray logging (see footnote, p. 131) the latter was put at 319 ft. The Lower Chalk–Middle Chalk contact in a borehole (251442) in Lyoak Wood,

1060 yd E. 36° S. of Swanton Court Farm, was found by resistivity logging to be at about 410 ft below surface (21 ft above O.D.), although the driller's, log recorded hard chalk to 417 ft. In a borehole at Lyoak Pumping Station, 440 yd to the N. 23° E., samples indicated that the contact was at about 268 ft below surface (12 ft below O.D.).

In the yard of Norton Farm (256424), 150 yd E. 9° N. of Alkham church, about 12 ft of compact white chalk without either flint or marl seam are exposed, while in a partially degraded roadside pit (257417) 650 yd S. 9° E. of Alkham church the following section is seen in upward order : firm blocky creamy-white chalk, becoming greyish white and marly upwards, with sponges in the lowest foot exposed and in the top foot, 12 ft ; soft grey to yellowish grey marl, 2 to 5 in ; soft blocky greyish white marly chalk, 1½ ft ; firm blocky greyish white chalk, occasionally nodular with interstitial grey marl and sponges and other fossils, 2 ft ; soft blocky greyish white marly chalk, 7 ft.

In a pit (243400) in Cuckoo Wood, 200 yd W. of Hockley Sole, about 20 ft of firm white blocky chalk, with a 6-in bed of harder chalk containing occasional sponges near the base, are overlain successively by 6 in of hard rocky chalk with yellow sponges and by 3½ ft of firm white blocky chalk with irregular near-horizontal partings of grey marl in the basal 6 in. No flint is to be seen.

At the western end of a pit (237377) on Dover Hill, 950 yd S. 20° E. of Hope Farm, some 32 ft of firm, blocky white to off-white chalk, probably in the *T. lata* Zone are seen : two seams of grey marl 2 to 3 in thick are 11 ft apart, and sponges occur in an impersistent 0 to 4-in bed 6 in below the lower seam and in the 6 in of chalk immediately beneath the upper seam.

In a pit (273404) on Lowslip Hill, 550 yd E. 5° N. of the windmill at West Hougham, soft white to greyish white blocky chalk with scattered white-patinated grey flint nodules is seen for 5 ft and is overlain by 5 in of light grey marl and marly chalk and by 2 ft of chalk rubble and soil.   G.B.

UPPER CHALK

Holaster planus Zone. A quarry (985496) near the valley bottom, 690 yd N. 34° E. of Frittenfield, showed 4 ft of massive, soft, medium to coarse-grained chalk with a band of large flint nodules, near the base of the *planus* Zone. Coarse-grained soft white chalk, seen for 10 ft, with a band of very large flints about 6 ft below the top of the face and a few scattered smaller flints, occurs in a disused quarry (994494) 850 yd N. 42° E. of Dean Court. Fossils from here, and from a similar but overgrown quarry (995491) ¼ mile to the south indicate a horizon high in the *planus* Zone or low in the *cortestudinarium* Zone.

Loose fragments of nodular chalk with fossils of the *reussianum* fauna occurred near the valley bottom (006493) 600 yd W. 22° N. of Challock church. Similar fragmental chalk was used as a guide in mapping the Middle Chalk–Upper Chalk boundary between there and the Warren Quarry (029486), where the section of Upper Chalk is:

		Ft	in
20.	Rather soft, coarse-grained, white platy chalk, with hard lumps or nodules with moulds of fossils, ammonites being common, and with scattered small flints ... ... seen for	2	0
19.	Fairly hard massive chalk with harder lumps, many iron-stained and some with fossils, including ammonites. Numerous flints near the top almost form a definite band, and small flints are scattered throughout	7	0

Ft in

18. Massive, lumpy to nodular chalk, the nodules standing out in
relief on the weathered face. Scattered small flints occur.
Micraster is common, and some of the nodules contain
ammonites   6   6
17. Continuous band of flints ...   .   –   3
16. Hard, coarse-grained massive chalk   1   6
15. Marl parting   ...   . up to   –   0½
14. Hard, lumpy massive chalk   1   6
13. Continuous flint band (similar to bed 17)   ...   .   –   6
12. Hard, massive chalk   3   6
11. Continuous seam of flaky grey marl   ...   ...   ½ in to   –   2
10. Massive chalk with iron-stained nodules. Large flints scattered
in lowest 1 ft   5   0
1–9. Middle Chalk: (p. 137)   ...   .about   23   6

The description of this quarry by Jukes-Browne (1904, p. 158) differs slightly
from that given above; the latter, however, matches the coast section he
recorded at Dover (Jukes-Browne 1904, p. 138). Bed 11 corresponds to the marl
band there taken as the base of the Upper Chalk by Rowe.   B.C.W.

A pit (974541) 550 yd W.7°N. of the road fork at Tong Green exposed 4 ft
of soft coarse yellowish white blocky chalk with hard iron-stained nodules,
and in another exposure (975525) 310 yd W.32°N. of Rushmere 5 ft of blocky
firm yellowish white lumpy chalk were noted. Fossils collected from the above
two localities were not zonally diagnostic: they are assigned to the *H. planus*
Zone.

Resistivity measurements made in the water borehole (980572) 1100 yd N.
38° W. of Belmont suggest that the base of the Upper Chalk lies at a depth of
65 ft (+ 72 ft O.D.). In samples from Throwley water borehole (995559) (p. 132)
Micraster praecursor was found from a depth of 85 ft and *Solariella* from 115 ft.
This latter level is probably below the base of the Upper Chalk, which is thought
to be at about + 65 ft O.D. Similar chisel samples from the Selling No. 1 water
borehole (035575) yielded *reussianum* fossils from a depth of 162 ft and indicated
that the base of the Upper Chalk was at a depth of 177 ft (– 63 ft O.D.).

Near Chilham hard nodular chalk beds at the base of the zone give rise to steep
slopes on the valley sides. The highest beds of the zone, forming part of the Top
Rock, are seen at the southern tip of Park Wood (042523), 1400 yd E. 13° N. of
Coppins Farm, where 3½ ft of hard creamy nodules set in coarse soft yellowish
chalk underlie 7 ft of soft white lumpy chalk with occasional hard yellow nodules
and a scattered band of flints. In the same pit, west of a minor fault not shown
on the map, 7 ft of soft coarse white blocky chalk with two scattered bands of
flints are exposed. A fine section in a pit (051528) 300 yd W. 28° N. of Young
Manor Farm shows 25 ft of well-bedded hard greyish white nodular chalk with
a well-marked band of flints. In a down-faulted part of the face in a pit (059505)
370 yd W. of Godmersham church, 4½ ft of firm greyish white blocky chalk with
hard coarse greyish white nodules, overlying a scattered band of flints, are considered to be the lowest beds of the Upper Chalk. The *reussianum* fauna was
found in the bank of a trackway above the face of this pit. (See also pp. 138–9.)

A large pit (075539) 600 yd W. 34° N. of the station at Bagham (Chilham
Station) is in Top Rock, comprising 6 ft of hard yellow fibrous-nodular chalk with
flints in massive 2-ft beds, separated by a well-marked band of large flints from
the overlying 5 ft of hard yellow sparsely nodular chalk with many flints; these
are overlain by 15 to 20 ft of soft white chalk with occasional nodules which
represent the *M. cortestudinarium* Zone. In the old quarry (075532) in the river

cliff 1000 yd E. 19° S. of Chilham Castle the section at the base of the Upper Chalk is similar to that exposed in the Dover cliff-section (White 1928a, p. 30) : the Middle Chalk (p. 138) is overlain in upward sequence by hard yellow or white nodular chalk with a band of flints at the base, 2½ ft ; marl seam, 2 in ; firm greyish white chalk, 1½ ft ; nodular chalk with a well-marked band of large flints, 1½ ft ; firm greyish white chalk, 1½ ft ; thin marl seam ; rubbly firm greyish white chalk with a band of large flints near the base, 6 ft. As a result of frost action the beds become increasingly rubbly as the ground surface is approached, the marl seams are disturbed, and the chalk between them becomes very rubbly and attenuated, the whole having the appearance of being drawn out down the slope. Immediately north of Whitehill a pit (082542) exposes about 25 ft of firm coarse yellowish grey to white well-bedded chalk with bands of flints. Beds high in the zone, and perhaps including some chalk of the overlying *M. cortestudinarium* Zone, are exposed in a roadside quarry (091493) 650 yd E. 10° S. of Winchcombe Farm. In the section given below none of the beds is sufficiently nodular to warrant the term Top Rock : coarse grey lumpy chalk, 5 ft, is overlain succes- sively by a scattered band of flint nodules ; hardy greyish lumpy chalk, 2 ft ; a course of large flint nodules ; massive hard white lumpy chalk in beds 1 to 2 ft thick, 7 ft ; coarse white lumpy chalk with a band of flints, 2½ ft ; rubbly lumpy white chalk with two bands of flints, 4½ ft ; and soil, ½ ft.

In the northern tip of the wood 300 yd E. 5° N. of Huntstreet a pit (093486) exposes 1½ ft of coarse hard iron-stained nodular chalk overlain by 10 ft of soft white rubbly coarse chalk with firmer bands. Some 1300 yd W. 17° S. of Waltham church a roadside pit (102481) exposes soft greyish white chalk seen for 1 ft, beneath 7½ ft of bedded greyish white coarse chalk containing a variable quantity of cream and white iron-stained chalk nodules.

Soft massive white chalk 5 ft in thickness, with a band of flints and associated tabular flint, is seen in a roadside quarry (114484) 160 yd E. 32° S. of Waltham church. On the eastern margins of woodland 450 yd E. 34° N. of Sarness Farm a pit (114492) shows 4 ft of hard white nodular chalk which is separated by a layer of percolated brickearth and some small flints from the overlying 8½ft of blocky soft white chalk with a band of large flint nodules 4½ ft up ; the section is capped by 4 ft of rubbly chalk, the fragmentation having been caused by frost, which also produced attenuation of the beds towards the valley.

In a pit (125476) in woodland 965 yd S. 23° W. of Wadden Hall 7 ft of nodular coarse greyish white chalk were noted and 6 ft of massive similar chalk were seen in a pit (128473), also in woodland, 1325 yd S. of this Hall. About 25 ft of soft white well-bedded chalk with some harder brown-stained lumps, becoming firmer and more rubbly at the top, were seen in a section (128489) 480 yd N. 6° E. of Wadden Hall. There were four scattered flint bands in the face and tabular flints along the bedding in places.

In woodland 380 yd W. 20° N. of Broadwaygreen Farm a pit (120511) shows 7 ft of firm greyish white coarse slightly lumpy chalk separated by a band of flint nodules from an overlying 7 ft of softer coarse white slightly lumpy chalk containing scattered flints and one band of flints. Nearby a smaller pit (121512) 300 yd N. 42° W. of this farm exposes 4 ft of massive yellow lumpy chalk at an horizon high in the zone. On his map F. H. Edmunds recorded 15 ft of chalk with flints in a pit (127530) 350 yd W. 27° N. of Swarling Farm, representing an horizon very high in the zone or perhaps in the overlying *M. cortestudinarium* Zone. On the west side of the road (121519) 200 yd N. 32° W. of Wootton Farm the following beds high in the zone were exposed: sparsely nodular and lumpy greyish white chalk with scattered flint nodules, 3½ ft ; below greyish white soft chalk, 2 ft ; hard coarse white nodular chalk, 1 ft ; soft coarse white chalk with scattered small flint nodules, 3½ ft ; nodular chalk associated with flints, 0 to 6 in ;

overlain by soft coarse white chalk with a band of flint nodules, 5 ft. On the east side of the road (121520) 300 yd N. 32° W. of this farm 15 ft of coarse yellow and white chalk are exposed. Chalk of the *H. planus* Zone is again seen near the head of the valley north-west of Bossingham. A pit (147491) south of the road 350 yd W. of the inn in the village, shows the following upward sequence : soft dull white, well-bedded marly chalk with a band of flint nodules and scattered flints, 8 ft ; passing up into massive hard yellow nodular chalk, 2½ ft ; impersistent band of percolated brickearth and clay ; coarse soft white massive chalk with a few hard nodules, 2¼ ft ; flinty loam (dump), 1½ to 2 ft. A further section (149494) 480 yd N. 38° W. of this inn shows 4 ft of hard white chalk with discontinuous bands of flint nodules underlying 6 ft of soft white chalk with a scattered band of flints.

A pit (157490) 690 yd E. 13° S. of the inn at Bossingham exposes 6 ft of blocky soft coarse greyish white marly chalk below 2 ft of Clay-with-flints. On the roadside (158490) 580 yd W. 15° N. of Lynsore Court, 6 ft of hard white and cream chalk nodules set in soft greyish white coarse chalk with occasional grey marly streaks and scattered flints were noted. A pit (161489) on the spur 230 yd W. 7° N. of the Court shows 12 ft of lumpy coarse blocky chalk and on the roadside (162509) 250 yd W. 32° S. of Bursted Manor 4 ft of soft coarse greyish white chalk with large flint nodules are exposed. Another pit (162511) 220 yd W. 40° N. of this Manor shows 8 ft of hard coarse white chalk with three bands of flint nodules, below 2 to 3 ft of flinty loam.

In a pit (157477) at Stelling Minnis, 490 yd E. 20° N. of Highchimney Farm, 4½ ft of well-bedded soft coarse grey marly chalk underlie 2 ft of hard creamy nodular chalk with a scattered band of flints at the base. On the valley side 200 yd E. 10° N. of Wildage Farm, a pit (167469) shows coarse cream to white chalk nodules in beds of 1 ft thickness set in 15 ft of coarse creamy white soft chalk with many scattered flints. Eastwards of a fault in the face the chalk is nodular at the base, but the nodules become rarer upwards and are set in massive fine soft greyish white chalk. Another pit (170463), 320 yd W. of Jacques Court, shows 7 ft of grey to white nodular and lumpy chalk below 3 to 4 ft of Clay-with-flints.   J.G.O.S.

In a pit (186499) on the north side of the valley 450 yd S. 32° W. of Marley, 3 ft of soft white chalk are overlain by 2 ft of hard nodular cream chalk, fibrous in places, below 4 ft of soft white chalk. The hard chalk in this section may be at or near the horizon of the Top Rock.

The whole of the *H. planus* Zone, together with parts of the zones below and above, is present in the sides of the sunken road (202485) east-south-east of South Barham, between the disused Elham Valley railway and the road from Breach to Barham, but the section is too poorly seen to permit of detailed description.

About 760 yd E. 8° N. of Wingmore (195467) some 8 ft of chalk yielded fossils indicative of a position high in the *H. planus* Zone or low in that of *M. cortestudinarium*. This chalk was yellow-stained white, and soft, becoming harder towards the base, with flint nodules dispersed and in bands.

At the edge of a small wood (191460), 310 yd W. 34° N. of Wingate Farm, 4 ft of white to creamy white nodular chalk were found, with scattered flint nodules and a *reussianum* fauna. Near the head of the valley south of Breach (199459), 550 yd E. 11° N. of Wingate Farm, a pit shows 5 ft of well-bedded soft rough chalk, with hard nodular patches in the uppermost 2 ft and flint nodules throughout. This chalk is near the base of the *planus* Zone.

Fossils of the *H. planus* Zone occur in 6 ft of firm rather gritty greyish white chalk, in a roadside pit (209465) 250 yd N. 26° W. of Tappington Hall :

in the pit there are two bands of widely-spaced flint nodules 2 ft apart, and 6 in to 1 ft below the lower of these there are occasional orange-coloured sponges associated with hard cream patches.

In a chalk-pit (214460) in Denton Wood, on the east side of the Denton–Selstead road, 430 yd E. 38° S. of Tappington Hall, the following strata are exposed in upward order : soft white chalk, 1 ft ; band of flint nodules ; soft greyish white chalk with harder yellow-stained patches, 2½ ft ; tabular flint band, 1 to 3 in ; soft blocky gritty white chalk with occasional flint nodules and bands of widely-spaced large irregular thickly-patinated black flints 6 ft and 10 ft above the base, the chalk near the surface being broken up, 13 ft. Palaeontological evidence suggests that the tabular flint band, which is at about 365 ft above O.D., is at or near the junction of the zones of *H. planus* and *M. cortestudinarium*.

Chalk high in the *planus* Zone is seen in a pit (211449) in Biggin Wood, 1050 yd N. 20° E. of Brandred Farm : greyish white gritty chalk with hard cream patches and rare nodules, in which flints are few but tend to occur at one or two horizons, 10 ft, is overlain successively by a band of widely-spaced flint nodules ; greyish white gritty chalk with hard cream patches in basal 1½ ft, 3½ ft ; a band of widely-spaced flint nodules ; greyish white gritty chalk with occasional scattered flint nodules, 3 ft ; a band of large irregular thickly-patinated black flint nodules ; soft greyish white gritty chalk with hard cream patches, 4 ft ; and reddish brown Clay-with-flints in pipes up to 5 ft deep. Rough nodular chalk is present in an old pit (216447) 275 yd W. 13° S. of the public house at Selstead and the *reussianum* fauna was found in similar chalk in the wood (214442) 550 yd to the S. 31° W. It follows that the base of the Upper Chalk is close beneath the bottom of the valley in which these pits are situated.

In a roadside exposure (238436) 320 yd S. 6° W. of North Court some 5 ft of rough nodular chalk of the *planus* Zone are visible. A pit (239431) 800 yd S. 5° E. of North Court is in soft creamy-white chalk with hard patches, representative of the top of the *planus* Zone or the base of the *cortestudinarium* Zone. From the pits around the well at Lyoak Pumping Station (253446), 1060 yd E. 11° S. of Swanton Court Farm, and in the farmyard (260454) 70 yd S. 21° W. of the milestone at the bottom of Lydden Hill,. came faunas indicative of a position high in the *planus* Zone : these localities are 1200 yd apart and the inclination of the strata along this line must coincide closely with the slope of the valley floor.

Chalk of the *planus* Zone, crisp, white, yellow-stained and with one band of large, white-patinated flints, is exposed in a 5-ft face in a pit (288404) in the wood 1000 yd W. of Farthingloe.

Near the cliff-top (282388) 75 yd west of the Lydden Spout path the following beds can be seen :

	Ft	in
12. White chalk	1	0
11. Band of flint nodules.		
10. Greyish white chalk, with yellowish patches, not notably nodular	2	0
9. Grey marl	–	2
8. Chalk (obscured)	2	6
7. Band of large white-patinated black flint nodules.		
6. Rough white chalk with hard cream patches and nodules and occasional scattered small flint nodules	3	8
5. Grey marl ... ... 2 in to	–	3
4. Rough white chalk with hard cream patches and nodules ...	4	0

Ft in

3. Band of large white-patinated black flint nodules.
2. Rough white chalk with bands of flint nodules 2½ ft and 5½ ft
below top ... ... about 10 0
1. Grey marl.

This succession closely resembles that described by Jukes-Browne (1903, p. 377) straddling the Middle Chalk-Upper Chalk boundary, the marl seam, bed 5, being the horizon chosen by Rowe (1900) to be the common limit of the zones of *T. lata* and *H. planus*. In places elsewhere along the top of this cliff the presence of flint bands and marl seams suggests that a thin layer of Upper Chalk is present, often penetrated by pipes of the overlying Clay-with-flints. At Aker's Steps (p. 136) the Upper Chalk section is as follows in upward order: Bed 50, chalk with scattered flint nodules in basal 3 ft, 7 ft; Bed 51, grey marl, 4 in; Bed 52, rough white chalk, 3 ft 8 in; Bed 53, band of flint nodules. The same sequence can be made out near the top of Shakespeare Cliff. G.B.

Micraster cortestudinarium Zone. In the valley north of Westwell, 1050 yd N. 20° W. of Dean Court are two small quarries (986497), one above the other, in beds about the junction of the *cortestudinarium* and *coranguinum* zones. The lower quarry exposed the following beds in ascending order: soft, white fine-grained chalk, with a line of spaced-out flints 2 ft from top, seen for 3 ft; overlain in turn by a band of medium to large-sized flints, 6 in; massive, hard, white or greyish chalk, with some iron-staining, 1 ft 6 in; a band of fairly large, spaced-out flints, 6 in; rubbly weathered chalk, mostly extremely hard and stony, 3 ft; capped by a flint band with 2 ft of hard white chalk, weathered platy, on top. The upper quarry, which yielded no fossils, showed 7 ft of soft chalk with two lines of flint nodules, 3 ft apart. Chalk of the *cortestudinarium* Zone probably has a continuous outcrop from here eastward through Eastwell Park, below the edge of the Clay-with-flints plateau. B.C.W.

A pit (977520) 300 yd S. 35° W. of Rushmere shows: hard white lumpy chalk with large scattered flint nodules, 2 ft 6 in; overlain by soft white coarse chalk, 1 ft; lumpy white chalk with scattered flint nodules, 2 ft 4 in; soft greyish white lumpy chalk with marly streaks, 2 ft; hard nodular chalk with many large flint nodules, 2 ft; hard white and yellow nodular chalk, 3 ft 6 in; firm white lumpy chalk, 3 ft, overlain by a band of large flint nodules, upon which rests white coarse hard nodular chalk, 3 ft, capped by flinty clay, 2 ft.

In another pit (976518) 650 yd S. 28° W. of this place, the following beds are exposed in upward order: coarse dull white firm chalk, ½ ft; band of large flint nodules; firm coarse white chalk, 3 ft; band of large flint nodules; firm white chalk, 1 ft; band of large flint nodules; soft lumpy chalk, 2 ft; band of flint nodules; firm blocky dull white chalk, 1½ ft; scattered flint nodules; bedded firm white chalk with two bands of flint nodules, 3 ft; blocky chalk with a band of flint nodules (inaccessible), 6 ft.

In the valley (987532) 700 yd W. 30° N. of Cadmans Farm 3 ft of blocky coarse yellow and white lumpy chalk underlie 3 ft of blocky firm white coarse chalk. In woodland (986540) 370 yd W. 43° S. of the inn at Throwley Forstal 6 ft of white and yellow-stained lumpy blocky chalk were noted, and on the valley side (985542) 350 yd W. 4° S. of this inn were seen 5 ft of rubbly coarse white chalk with a band of large flint nodules below 2½ ft of rubbly chalk with pockets of loam. In the north tip of woodland (979554) 210 yd W. 39° S. of Hockley 9 ft of soft white yellow-stained lumpy chalk with 'cannon-ball' flints underlie 4 ft of soft white massive chalk with an impersistent flint band. The zonal horizon of this section is not certain. Near the roadway (995528) 530 yd N. 44° E. of Snoadstreet 20 ft of soft white blocky chalk with four bands of scattered flint nodules are exposed, and in woodland (998520) 880 yd E. 40° S. of Snoadstreet 20 ft of coarse white blocky chalk with a well-marked band of flint nodules were noted. A roadside pit (001530) 300 yd N.W. of Broomfield shows soft coarse yellow to white chalk with a few firmer nodules and three bands of flints, seen for 6 ft, overlain by a thin marl seam on which rest 5 ft of soft coarse yellow to white blocky chalk with two bands of flints. Another pit (998524) low in the valley 760 yd W. 40° S. of Broomfield shows 8 ft of soft coarse dull white blocky chalk with streaks of marl and hard creamy nodules.

Hard creamy nodular chalk is seen for 10 ft in the cutting (064537) of the road to Shottenden, 380 yd W. 40° N. of Chilham Castle, and a pit (068542) near the roadside 800 yd N. 11° E. of this Castle shows 12 ft of white chalk with hard yellow nodules which are referred to this zone. F. H. Edmunds noted on his map 15 ft of chalk with scattered flints at the southern tip of the wood (115535) 700 yd S. 10° W. of the Mental Hospital at Chartham, and 10 ft of soft highly-weathered white chalk in a pit (114530) 180 yd N. 10° W. of Sappington Court.

In the valley 500 yd S. 28° E. of Buckholt Farm a pit (106496) shows 6 ft of hard creamy chalk nodules in a matrix of soft coarse chalk with scattered flints and a persistent tabular sheet of flint at the summit, overlain, by 10 ft of soft coarse white lumpy chalk with a band of flint nodules. In a pit (111490) 180 yd E. of Sarness Farm, soft white massive chalk, seen for 4 ft, is separated by a band of large flint nodules from the overlying 2 ft of nodular and lumpy chalk with 'rust balls', which are succeeded in turn by 10 ft of soft white massive chalk with a band of flint nodules, and by Clay-with-flints.

A roadside pit (130487) 350 yd N. 30° E. of Wadden Hall reveals 8 ft of coarse white nodular chalk and lumpy chalk with 'cannon-ball' flints and scattered nodules of flint, which are referred to this zone, as are the 10 ft of soft massive coarse white chalk seen in another pit (134493) 750 yd S. 18° W. of Dane Chantry. Immediately east of Dane Chantry (136500) 14 ft of massive coarse firm white lumpy chalk with several courses of flint nodules are seen. Fossils from this pit indicate that the horizon is high in the *M. cortestudinarium* Zone or perhaps low in the *M. coranguinum* Zone. In another pit (137501) 150 yd N. 30° E. of Dane Chantry white lumpy chalk with impersistent flint bands passing upwards into soft coarse white chalk is seen for about 20 ft.

A pit (149507) south of the roadway 380 yd W. 32° N. of Upper Hardres Court shows 12 ft of soft white chalk with three bands of flint nodules which are believed to be at a horizon high in this zone. Some 850 yd E. 40° S. of the Court a degraded large pit (158502) straddles the junction of this zone with that of *M. coranguinum*. An exposure showing 7 ft of coarse soft white rubbly chalk with large flint nodules in the north-east corner of the pit (159501) yielded a *Micraster* of the *M. cortestudinarium* Zone, while in chalk (p. 157) in the south-west corner of the pit a *Micraster* of the *M. coranguinum* Zone was found.

On the south side of the road (158513) 50 yd S. of Broxall Farm hard nodular creamy chalk with large flint nodules, seen for 4 ft, is overlain by 10 ft of soft coarse white chalk with flint nodules in scattered bands, below 2 to 5 ft of Clay-with-flints. In the grounds of cottages (157520) 460 yd W. 10° S. of the T road-junction at Pett Bottom, 9 ft of soft creamy white rubbly chalk with one course of flint nodules are exposed, and in woodland north of the roadway 1050 yd N. 30° W. of Dane Farm, a pit (170487) shows 10 ft of massive soft white coarse chalk with flint nodules.   J.G.O.S.

In the disused railway cutting (189516) that skirts the south-western margin of Charlton Park the following succession is exposed:

Ft

11. Rough white chalk, with scattered flint nodules which are sometimes arranged in loose straggly bands. The chalk is massive, but becomes blocky and rubbly towards the surface   12

10. Irregular band of flint nodules.

9. Coarse gritty white or greyish white chalk with scattered flint nodules ; a band of sparse flint nodules is present 3 ft below the top   9

8. Band of flint nodules with violet-stained patinas.

7. Coarse gritty white or greyish white chalk with hard patches in a softer matrix, occasional small streaks of grey marl and occasional scattered flints ; sponges occur between 1½ and 2 ft below the top 4½

6. Band of widely-spaced flint nodules.

5. Coarse gritty massive white or greyish white chalk with hard spongiferous patches, especially between 1½ and 3 ft below the top 4½

4. Band of flint nodules.

3. Coarse gritty massively-bedded white or greyish white chalk, with occasional scattered flint nodules and sponges ; a band of flint nodules is present 4 ft below the top   7

2. Irregular band of flint nodules.

1. Coarse gritty massively-bedded white or greyish white chalk ; occasional scattered flint nodules ; many sponges between 2 and 3 ft below the top   5

Obscured   5

The flint nodules in the bands are of dark grey to black flint, large and irregular with white patinas from ⅛ to ⅜ inch in thickness. The section comprises the upper part of the *M. cortestudinarium* Zone and the lower part of the *M. coranguinum* Zone, but the position of the junction was not located. No trace of either the *M. cortestudinarium* or the *M. coranguinum* tabular flint bands was found. In this cutting (190515) 900 yd W. 18° N. of Kingston church, Whitaker (in MS.) described lenticular patches of brown sandy earth, one of which was finely bedded ; the patches occurred along a marly layer, in places continued for a long way, and some were connected by oblique planes to the surface. These were evidently ' sheet pipes '.

The basal part of the zone is poorly exposed in roadside banks (203485) east-south-east of South Barham farm, while chalk higher in the zone is seen in a pit (207490) beside the road 350 yd S. of the inn at Derringstone ; here about 20 ft of firm white well-bedded chalk are visible, with four strong bands of flint nodules ; other flints occur in less well-defined bands. The apparent dip of the beds is about 2° towards the north. A pit (203472) in the wood on Lodge-lees Down, 750 yd S. 43° E. of Breach, shows 6 ft of crisp white chalk with many flints which may be in the *M. cortestudinarium* Zone.

In a large pit (238437), now mostly overgrown, 160 yd S. 12° W. of North Court, chalk of the *M. cortestudinarium* Zone is exposed at intervals ; at the base 2 ft of hard, rocky coarse gritty greyish white chalk with sponges are overlain by some 6 ft of softer, white gritty chalk with occasional irregular nodules of black flint with ⅛-in white patinas ; about 16 ft above this, 3 ft of soft white gritty chalk, with hard patches, occasional sponges and a course of flints, are seen ; the intervening chalk is white, hard, gritty and nodular, but poorly exposed.

A pit (253427) beside the road 500 yd N. 33° W. of Alkham church shows
the following beds in upward order: coarse gritty creamy white chalk, 4 ft;
tabular flint band, 0 to 1 inch; coarse creamy white chalk with hard patches
and occasional flint nodules, 2 ft; band of large irregular flint nodules with
white patinas up to ¼ in thick; rough greyish white chalk, 2 ft; massive, well-
bedded, hard gritty pale cream chalk, nodular in patches, with impersistent
bands of flint nodules 6½ ft and 2 ft below the top, 9½ ft. Higher up the valley
side (255427), near the Clay-with-flints line and 420 yd N. 11° W. of Alkham
church, 8 ft of white chalk, with poorly-defined bedding and only occasional
flint nodules, represent the upper part of the *M. cortestudinarium* Zone or the
base of that of *M. coranguinum.*   G.B.

Micraster coranguinum Zone. A small outlier of chalk of the *M. coranguinum* Zone near Tong Green is mostly concealed by Clay-with-flints. At the
road-fork in Tong Green (980540) 7 ft of massive firm white chalk with two
courses of flint nodules are exposed below hard yellowish white nodular rubbly
chalk, 7 ft, below soft coarse yellowish white rubbly chalk with scattered flint
nodules and thin tabular sheets of flint, 4 ft. Near the roadside in a small
wood (984537) 500 yd S. 42° E. of the above road-fork 16 ft of coarse white
slightly lumpy chalk with a scattered course of flint nodules are seen.

On the valley side 450 yd E. 38° S. of the T road-junction at Hockley a pit
(983552) shows 14 ft of massive soft white chalk with three courses of flint
nodules and a little tabular flint, underlying 0 to 3 ft of Clay-with-flints, while
in another pit (975559) 750 yd W. 39° N. of the same place 11 ft of massive
white iron-stained soft chalk with two courses of flint nodules are seen, below
2 ft of Clay-with-flints. The chalk at the last locality may lie within the
M. cortestudinarium Zone, but its horizon could not be determined from the
single fossil collected.

In a pit (983562) 350 yd W. 29° S. of Belmont, 12 ft of blocky soft white
chalk streaked with yellow were noted, while 400 yd W. 40° N. of this house,
7 ft of soft white massive chalk are exposed (983566). A further 8 ft of
similar chalk are seen (983569) 700 yd N. 21° W. of the house. At Scooks
(982578) 7 ft of white soft chalk are exposed, with many creamy bands and
streaks and courses of flint nodules; 520 yd N. 28° E. of this place, 8 ft of
similar chalk were noted below 1½ ft of Clay-with-flints (984580).

In a small wood (003533) 560 yd N. 2° W. of Broomfield 11 ft of soft clean
white chalk with one course of flint nodules were noted, and 14 ft of similar
chalk were seen below 2 to 6 ft of Clay-with-flints in another pit (998532)
700 yd W. 33° N. of this farm.   J.G.O.S.

A quarry (997513) 1000 yd W. 40° S. of Pested Farm shows 24 ft of roughly
fracturing chalk with some flint bands, unfossiliferous except for occasional
thin-shelled *Inoceramus* in the upper part. Similar chalk, yielding few fossils,
is visible in old pits 350 yd due west of here, on the east edge of Longbeech
Wood (993513), 660 yd W. 40° S. of Pested Farm (999517); 760 yd S. 35° W.
of Pested Farm (000515); and 950 yd S. 10° W. of Pested Farm (002511).

A quarry (019515) just north of the Canterbury road, 100 yd N. of Tower Farm,
shows 10 ft of fine-grained white chalk with a few small flints.   B.C.W.

On the valley side 340 yd N. 20° E. of Hegdale Farm a pit (011526) shows
18 ft of soft white chalk with flint courses, while 480 yd E. 42° N. of this farm
20 ft of soft bedded white chalk with scattered flint nodules and one well-defined
course of large flint nodules are exposed (014526). A pit (017519) 280 yd S.
10° E. of Howlett's Farm exposes 18 ft of coarse white soft to brittle chalk
with what is probably the same well-marked band of flint nodules, and the

band is possibly seen again in woodland (014535) 400 yd N.W. of Dryland
Farm, where 8 ft of soft slightly blocky chalk are exposed; fossils collected
from a similar thickness of chalk exposed in the road cutting at the north end
of the wood suggest a horizon low in this zone. On the valley side (009535)
560 yd S. of the inn at Leaveland, $9\frac{1}{2}$ ft of clean white chalk with scattered
courses of flint and common thin vertical tabular sheets of flint are seen below
9 ft of loam and Clay-with-flints. A pit (004545) 390 yd S. of Leaveland Court,
low in the valley, exposes 16 ft of soft white chalk with two courses of flint
nodules. A roadside pit (003547) 200 yd S. 38° W. of the same Court shows
25 ft of rather blocky, bedded, clean white chalk with occasional impersistent
beds of firm chalk 1 to 2 ft thick; a planed-off echinoid observed on one of
the bedding planes indicates a slight break in the deposition of the chalk at
this horizon. Immediately south of Throwley Valley (999550) 6 ft of yellow
iron-stained soft chalk were noted, as were also 6 ft of clean white massive
soft chalk with a prominent course of flint nodules in a pit (996561) 300 yd N.W.
of Parsonage Farm, and 5 ft of firm white chalk just west of the road-fork
at North Wilderton (994570). A pit (995576) 600 yd N. 23° E. of North
Wilderton displays 8 ft of firm white chalk with scattered flints below 2 ft
of flinty loam, and a roadside pit (004577) 700 yd W. 28° S. of Littles Farm,
shows 8 ft of massive white chalk with one course of flints; 5 ft of massive
yellow and brown-stained chalk were noted in woodland (997581) 700 yd E.
29° S. of the road junction at Kennaways.

In the road cutting (994589) 100 yd E. of Painter's Forstal, a total of 11 ft
of rubbly yellow-stained white chalk with scattered flints and yellow chalk
nodules is exposed, and in parkland 370 yd E. 20° N. of this place a pit
(995590) shows 11 ft of hard yellow silicified chalk. Some 9 ft of greyish white
rubbly chalk with scattered flints were seen in a section (002589) 100 yd N.
of Plumford, and 6 ft of blocky white soft chalk with two courses of flints in a
wood-side pit (000596) 580 yd W. 8° N. of Brogdale.

At Littles Farm (010580), 14 ft of massive soft white chalk with rare harder
chalk nodules and scattered flints are exposed; 6 ft of massive white chalk
with scattered bands of flint nodules are present in a section (009572) 450 yd N.
32° W. of Sheldwich church.

At the corner of the road (021554) 840 yd S. 7° E. of Lees Court, 30 ft of
soft white massive chalk with six rather poorly-marked courses of flint are
exposed; 420 yd E. 15° S. of Badlesmere Court another pit (019550) shows
14 ft of clean white soft blocky chalk with three bands of flint nodules. A pit
(024552) 630 yd N. 18° W. of Woods Court shows 7 ft of soft white chalk with
a well-marked course of flints, underlying 2 ft of Clay-with-flints; 180 yd S.
15° W. of Stringmans Farm is a pit (025542) with 11 ft of similar chalk in
which the same flint band is exposed. This prominent flint band is at about
the stratigraphical level of Whitaker's Three-inch Band. It is believed to be
seen again in a pit (032546) 500 yd E. 5° S. of Woods Court, in 23 ft of soft
white chalk with six other more scattered bands of flint nodules; again in a pit
(033549) 760 yd E. 20° N. of this Court, in 20 ft of massive, white chalk with
rather scattered flint nodules; and also south of the road (037542) 820 yd W.
of the inn at Shottenden, in 20 ft of soft white chalk with other nodular flint
courses and scattered flints.

On the side of a minor valley 630 yd E. 4° N. of Dryland Farm, a pit (023532)
shows 15 to 20 ft of soft blocky chalk with some harder nodules and scattered
flints underlying 1 to 2 ft of Clay-with-flints. Another pit (024532) by the
roadside 750 yd E. of this farm, shows 11 ft of similar chalk, which are largely
inaccessible.

L

In the eastern tip of the wood (035553) 1540 yd N. 39° W. of the inn at Shottenden 7½ ft of clean white chalk with two courses of flint nodules are seen, together with a small pipe of Thanet Beds, and 9 ft of similar chalk are exposed in a pit (031567) 220 yd W. 20° S. of Harefield Farm. In the corner of the road 350 yd W. of Hogben's Hill a pit (028568) shows 30 ft of soft white chalk ; courses of flint nodules are present particularly in the lower half of the section, and one prominent band 18 ft below the top is believed to be Whitaker's Three-inch Band. Whitaker (in MS.) noted a 'sheet pipe', consisting of a horizontal layer of buff clayey sand, about 3 in thick and full of pebbles at one part, which extended for nearly 20 yd in chalk in the lane cutting a little over ¼ mile south-west of the church near Hogben's Hill (Selling church). Some 400 yd W. 32° N. of this church 15 to 20 ft of soft clean white massive chalk with sporadic flint nodules are exposed. In parts of the face, sponges preserved in limonite occur immediately beneath chalk in which *Conulus* is common and the section is believed to be opened at the horizon of Barrois's Sponge Bed and the *Conulus* Band, at the base of the *Uintacrinus* Zone, though *Uintacrinus* has not been found.

East of Owens Court, chalk high in this zone is exposed. A pit (030579) 350 yd E. of this Court exposes 8 ft of soft white chalk and 490 yd E. 21° S. of the same 10 ft of similar chalk are seen (031577). Both pits show scattered flints and impersistent courses of flint nodules. On the east side of the road (032590) 300 yd S. 41° W. of Colkins 20 ft of soft white massive chalk with scattered flints are exposed ; a prominent but impersistent flint band 8 ft below the top of the section may represent Whitaker's Three-inch Band.

Some 350 yd W. of Brenley Corner, in a pit (035600) south of the road, 30 ft of clean white soft chalk with very ill-defined bands of flint nodules are exposed below 2 to 4 ft of flinty loam and rubbly chalk. A large flint band near the base of the pit probably represents Whitaker's Three-inch Band, and *Conulus* is common in the face about 20 ft above. The uppermost 10 ft or so of chalk exposed are therefore believed to lie within the *Uintacrinus* Zone, but the index fossil has not been found. North of the road 20 ft of similar chalk .re exposed. In the railway cutting (040585) 580 yd S. 15° E. of Brenley House, ft of massive dull white chalk with large flints were noted below 3 ft 4 in of Thanet Beds. In a pit (044587) 450 yd W. 43° S. of Wellbrook Farm, 18 ft of well-bedded dull white chalk with a few scattered flints at a horizon about that of the *Conulus* Band underlie 6 to 12 ft of piped Thanet Beds. In a pit (047586) 360 yd N.W. of Boughton Church Farm, 6 ft of firm white rubbly chalk with flints at about this same horizon were noted. A large pit (051583) 250 yd E. 30° S. of this farm shows 20 to 25 ft of massive well-bedded dull white firm chalk with occasional flints overlain by 3 ft of Thanet Beds. This pit is believed to be opened in chalk above Whitaker's Three-inch Band and to include a few feet of chalk of the *Uintacrinus* Zone. A pit (050578) in the road-fork 200 yd S. 30° W. of Crouch shows 13 ft of white rubbly chalk with scattered flints and occasional harder yellow nodules. Whitaker (1872, p. 29) referred briefly to a prominent flint band here, which is no longer visible ; it was probably his Three-inch Band. A pit (040581) 100 yd W. 40° N. of Dane Court shows 14 ft of soft white chalk with perhaps the same flint band and other courses of more scattered flints.

In pits 300 yd E. 35° S. of Gushmere 14 ft of massive white chalk with scattered flints were seen north of the road (048572) and 6 ft south of the road. A deep pit (046565) by the road 200 yd E. 35° N. of Norham Farm, exposes 21 ft of white massive chalk, with a prominent flint band near the base of the section. Another pit (049566) 630 yd E. 25° N. of this farm shows 20 ft of similar chalk including the flint band, and 20 ft of massive white chalk at about the same horizon are exposed (053560) some 580 yd W. 19° S. of Rhodecourt Farm.

In a pit (049554) 750 yd E. of the earthwork in Perry Wood 10 ft of firm white well-bedded chalk with a few scattered flints were noted, and immediately east of Stonestile Farm (059553) 12 ft of massive hard white chalk with poorly-defined bands of flint nodules are exposed. Chalk considerably lower in this zone is present north-east of Molash. In a pit (034527) 100 yd S. of Bower Farm 20 ft of massive soft white chalk with occasional hard yellow chalk nodules and impersistent scattered courses of flint nodules are exposed below 1 to 6 ft of Clay-with-flints and in a section (035525) 750 yd E. 32° N. of Coppins Farm 20 ft of soft white chalk with courses of flint nodules and small tabular sheets of flint were noted. At the west tip of Park Wood (041525) 1150 yd E. 26° N. of Coppins Farm, 11 ft of soft bedded chalk with ill-defined flint nodule courses are exposed below 2 ft of chalk rubble. At the north-west tip of this wood (041529) 470 yd S. 29° W. of Denne Manor Farm 15 ft of clean white chalk with courses of flint nodules were noted. Some 10 to 12 ft of soft white rubbly chalk with scattered courses of flint nodules and pipes of Thanet Beds and Drift were observed at a locality (048538) 580 yd S. 38° E. of the inn at Shottenden, as were also 12 ft of similar chalk (060540) 940 yd N. 28° E. of Dane Court.

At the road corner (063556) 530 yd E. 26° N. of Stonestile Farm 7 ft of soft white chalk with one course of flint nodules are present below 5 ft of rubbly chalk.

In the roadside (065553) and roadside pits (067553) about 800 yd W. 33° N. of the cross-roads at Old Wives Lees some 40 ft of chalk of the *M. coranguinum* Zone are discontinuously exposed. The composite section is:

	Ft
Soft clean white chalk with small scattered flints at probably a higher horizon than that described below in the road cutting	8
(Above exposed in pit at entrance to orchard)	
Soft clean white massive chalk with one band of flint nodules ...	7
(Above exposed in pit below entrance to orchard)	
Soft clean white massive chalk with prominent rust staining	3
Course of flint nodules.	
Massive soft yellowish chalk	4
Scattered band of flint nodules.	
Soft clean white chalk	3
Scattered band of flint nodules.	
Soft clean white massive chalk ... ... 4 ft to	5
(Above exposed north of the road)	
Blocky white firmer chalk	3
Band of large flint nodules.	
Blocky white firm chalk	2½
Band of large flint nodules.	
Coarser and softer blocky white chalk	3
Band of flint nodules.	
20 yd of the road bank obscured.	
Band of flint nodules.	
Soft fine blocky chalk with scattered flints ... ... about	6
(Above exposed in the low bank on the south side of the road).	

At Old Wives Lees (072550), on the north side of the road 200 yd N. 22° W. of the cross-roads, 20 ft of soft clean white chalk with occasional harder yellow nodules and impersistent bands of flint nodules were noted, their lithology suggesting a horizon high in the zone. A large pit (074558) 700 yd E. 15° S. of Upper Ensinge, showed 25 ft of massive white chalk with scattered small flints and a few large flint nodules at the top of the section. In a pit (087549) 280 yd W. of the bridge over the river at Shalmsford Street, hard white or yellow nodular chalk with many flints, seen for 7 ft, passes up into soft white chalk with widely spaced bands of flint nodules, 15 ft. A pit (087534) near a road-fork 900 yd W. 20° S. of Mystole House, shows 12 ft of white soft chalk with a few hard nodules and scattered bands of flint nodules.

Some 8 ft of soft massive white chalk with no flints were noted in a pit (088558) 580 yd W. 30° S. of Nickle Farm. Immediately north of the railway a pit (091560) 220 yd W. 30° S. of the farm shows 12 to 15 ft of soft white rubbly chalk developed in massive 3-ft. beds with a few small flint nodules. At Nickle Farm (093561) 15 ft of rubbly white chalk with a 6-in band of hard yellow nodular chalk were noted,

The three railway cuttings to the east show very good sections in the chalk of this zone. In the cutting (095558) 400 yd S. 35° E. of Nickle Farm firm white chalk with scattered impersistent bands of small flint nodules (not con-tinuously exposed but showing a gentle easterly dip), 22 ft, is overlain by massive firm white chalk with occasional hard rust-stained nodules mostly in the lowest 2 ft and uppermost 1 ft, 8ft ; below white firm chalk, 6 ft ; and chalky soil, 6 in. Immediately south of this cutting (095557) 8 ft of massive firm white chalk were noted.

In the cutting east of the level crossing (106558) (Chartham Siding, some ½ mile north of Chartham Station) the section on the north side is: white soft chalk, well-bedded, with some of the bedding planes marked by yellow and brown stains, 10 ft ; overlain by massive yellow-stained white chalk with impersistent and irregular bands of flint nodules spaced up to 3 ft apart, becoming more rubbly in the upper part with more regularity in the flint bands, 20 to 25 ft ; below clayey loam and gritty sands with water-worn and angular flints and siltstones (p. 218), up to 3 ft in pipes. The flint bands on the south side of the cutting have an apparent westerly dip. The fall in level of the beds is counteracted by gentle anticlines which are usually associated with rubbly chalk and are of superficial aspect. ' Sheet pipes ' of laminated fine sand occur in the chalk of the cutting: the largest measured 2 ft by 24 ft and had a thin bed of chocolate-coloured clay at the base. A similar chalk section is seen in a roadside quarry immediately south of the level-crossing. In the cutting west of the level-crossing there is considerable piping of gravel and brickearth. Some 360 yd N. 34°·W. of this level-crossing (104559) 5 ft of rubbly hard white chalk with some harder nodules and scattered flints were noted. In a roadside pit (105559) 200 yd N. 10° W. of the crossing 25 ft of firm white chalk with yellow impersistent hard bands of chalk and some courses of flint nodules are exposed, and a pit (107560) in woodland 430 yd N. 43° E. of the level-crossing shows 15 ft of soft white chalk with three impersistent bands of flint nodules. A pit (116569) 950 yd N. 4° W. of Howfield Farm reveals 5 ft of blocky chalk with scattered flint nodules underlying 6 ft of Thanet Beds and brickearth ; a 1-inch layer of tabular flint below the Bullhead Bed in this section is probably of secondary origin. Immediately south of the railway (130573) 740 yd E. 24° N. of Tonford Manor 6 ft of blocky, well-bedded chalk with pipes of Eocene sand were noted.

Chalk of the *M. coranguinum* Zone has an extensive outcrop on Chartham Downs between the Stour valley and the Petham nailbourne valley ; however, some of the highest chalk present on this ridge is within the *Uintacrinus* Zone. In the road cutting (127559) 450 yd W. 23° S. of Cockering Farm the 8 ft of firm white chalk exposed are rubbly as a result of frost action, and the two prominent flint courses present dip sharply towards the minor valley to the east. Near the Ashford–Canterbury road (127566) 520 yd W. 28° S. of Thanington church, 10 ft of firm white chalk with irregular flint courses and pipes of river gravel and sand were noted ; on the roadside (123563) 1000 yd W. 28° S. of this church, 20 ft of massive, well-bedded chalk, becoming blocky towards the top and with indistinct bands of flint nodules, were present. On the east side of a footpath (101547) 470 yd E. of the road-rail bridge at Shalmsford Street 18 ft of chalk (recorded on his map by F. H. Edmunds) yielded fossils which indicate a position very low in the zone. At the roadside (106541) 600 yd N.E. of Thruxted 8 ft of chalk were noted and a similar thickness of soft white chalk with occasional nodular flints was seen in a pit (102533) at the west end of the wood 400 yd S. of this place.

Some 12 ft of chalk were seen in a pit (131533) north of the road 520 yd N. of Swarling Farm. A section (129542) to the east of the Mental Hospital and 800 yd S. 22° W. of Newhouse Farm shows 2 ft of soft white chalk separated by a layer of large flint nodules from the overlying 12 ft of soft white chalk with scattered flints, below 1 to 2 ft of soil. The layer of large flint nodules is believed to be Whitaker's Three-inch Band. In a pit (140538) 675 yd W. 15° S. of Heppington House, on the east side of Stone Street, the 18 ft of almost flintless, well-bedded, soft, white chalk exposed lie wholly within the *Uintacrinus* Zone. This is the only place where *Uintacrinus* has been found west of the Little Stour River, but much of the chalk capping the ridge north-eastwards is probably at or about the same horizon. On the south side of the road (147554) 350 yd W. 40° N. of Merton Farm 35 ft of soft evenly bedded chalk with bands of nodular flints low on the face are visible. A band of large nodular flints towards the top of the section is believed to be Whitaker's Three-inch Band. The section does not extend high enough for any part of the face to lie within the *Uintacrinus* Zone. A temporary section (152555) 375 yd N. E. of this farm exposed 12 ft of soft white chalk with scattered flints below piped sandy loam and clay. In a pit (154558) in playing fields 750 yd N. 39° E. of the same farm, 12 ft of soft occasionally lumpy grey to yellowish white chalk with scattered flints were noted. In the cutting (142560) of the most northerly part of Stone Street (Hollow Lane) about 450 yd N.W. of Stuppington Farm there is an extensive section where Whitaker's Three-inch Band is easily accessible. The northerly dip of the strata is only slightly less than the slope of the roadway, and although the section is about 600 yd in length only the following strata in the *M. coranguinum* Zone can be seen: soft white chalk with scattered flint nodules, $2\frac{1}{2}$ ft; overlain by soft white chalk with bands of ferruginous staining and a few scattered small flints, 4 to 5 ft; below a well-marked band of semi-tabular flint (Whitaker's Three-inch Band); below very rubbly soft white chalk with some ferruginous staining, 4 ft. At the northern end of the exposure are 20 ft of soft white chalk with two bands of flint nodules. The correlation between the two sections is uncertain but the uppermost flint band here is believed to represent the lowest flints of the road section. There is some thin local silicification of the chalk at the contact with benched drift deposits exposed in this cutting (pp. 237–8). In one pit (138559) west of the above roadway, 850 yd E. 20° S. of Cockering Farm, 9 ft of soft blocky white chalk with one band of flint nodules were noted, and another (138561) exposed 18 ft of similar chalk with two bands of flint nodules and scattered flints. Both these pits are opened in chalk at a lower horizon in the zone than the above road cutting.

In Canterbury, Denne's Limeworks Pit (147567), near the disused Canterbury–Folkestone railway and 700 yd S. 10° E. of the principal railway station in the city, provides the finest inland section in *M. coranguinum* chalk in the district; the details are:

							Ft	in
Surface soil ...	...	...	...	...	...	...	1	0
Thanet Beds: (p. 186)	...	...	...	...	...	...	14	0

	Ft	in
Soft, clean white, massive, well-bedded chalk with scattered flint nodules and abundant *Conulus*	21	0
Chalk as above but more thinly bedded, with three scattered courses of flint nodules	11	0
Flint band, tabular and up to 2 in thick at north end of pit, semi-tabular and up to 6 in thick in south (Whitaker's Three-inch Band).		
Soft white massive chalk with some scattered flint nodules ...	12	6
Band of scattered medium-sized flints.		
Soft white massive chalk with a well-marked band of small to medium-sized flints 4½ ft down	7	6
Band of medium to large-sized flints.		
Soft white chalk with a well-marked band of large flints 5 ft down	6	6
Band of large flints.		
Soft white chalk with occasional scattered flints	3	6
Thin prominent band of flints.		
Soft white chalk with a scattered band of medium to large-sized flints 2 ft down and a sparse impersistent band of flints 7½ ft down	9	10
Scattered band of medium to large flint nodules.		
Massive soft white chalk with scattered small and large flint nodules	5	6
Scattered band of medium-sized flints and a small oblique tabular sheet of flint.		
Massive soft white chalk with a well-marked band of very large flints 3 ft 8 in below the top, a band of medium to small-sized flints 9 ft 4 in below the top and scattered small flints between the bands	10	10
Total	103	4

A very slight northerly dip can be observed at the western end of the pit. Some of the flint courses near the base of the pit are associated with brown loams and clays infilling fissures and cavities and deposited from percolating waters. The largest mass of loam measures 3 by 1½ ft. The uppermost 12 ft or so of the section are referred to the *Uintacrinus* Zone, although the zone fossil has not been found. *Conulus*, however, is very common about 20 ft above the tabular flint band and this horizon is believed to be the inland representative of the *Conulus* Band at the base of the *Uintacrinus* Zone.

North of the railway (154572) 870 yd E. 10° S. of the principal railway station in Canterbury, 20 ft of soft white rather rubbly chalk with random flint nodules were noted below 2 to 3 ft of rubbly chalk, and 3 to 6 ft of brickearth. Some of this chalk probably lies within the *Uintacrinus* Zone.

In the disused cutting (172548) 400 yd S. 30° W. of Renville, 20 ft of horizontal, well-bedded, massive white chalk with scattered thinly-patinated flints are exposed below 0 to 6 ft of brickearth (deeper in pipes). Fossils from this cutting suggest the presence of the *Conulus* Band. A pit (165549) 750 yd W. 10° S. of Renville shows 15 ft of well-bedded massive white chalk with occasional large flints, and in a further pit (175556) 640 yd N. 28° E. of Renville 10 ft of chalk were seen. At Milestone Farm (170558) about 30 ft of massive well-bedded soft white chalk were seen, with tabular sheets and scattered impersistent bands of flint in the lower part of the section and scattered flint nodules in the upper, rather rubbly part, underlying 4 ft of piped drift (p. 219). North of Bridge and 1000 yd E. 10° N. of Renville, 6 ft of chalk noted in an overgrown exposure (182553) below 1 to 2 ft of loam are perhaps in the *Uintacrinus* Zone.

South of the ridge at Chartham Downs chalk of the *M. coranguinum* Zone is restricted in extent. At the edge of the wood (102533) 350 yd S. of Thruxted 8 ft of soft white chalk with flints yielded fossils suggesting a horizon low in the zone. Some 9 ft of soft clean massive yellowish chalk with courses of scattered flints seen in a pit (089518) 720 yd W. 42° N. of Upper Thruxted Farm, and 11 ft of firm coarse white rubbly chalk with ill-defined scattered courses of flint nodules exposed below 3 ft of rubbly chalk and flinty clay in a pit (092511) 300 yd S.W. of this farm, are referred to the *M. coranguinum* Zone. In a pit (117499) 230 yd E. of Haults Farm, soft white lumpy chalk with a very thin marl seam exposed for 9½ ft below 1 ft of flinty loam must lie very near the base of this zone.

West of the road (147519) 100 yd W. 30° S. of Young's Farm at Upper Hardres, 8 ft of soft white chalk with one course of flint nodules and scattered flints are exposed below 2 ft of rubbly chalk. Part of an old large pit (158502) 850 yd E. 40° S. of Upper Hardres Court shows 10 ft of coarse white soft rubbly chalk with a band of large flints at a horizon at the base of the zone. Chalk of the underlying *M. cortestudinarium* Zone is also exposed nearby (p. 148). Near the road-fork (157506) 460 yd E. 15° S. of this Court 4 ft of coarse yellowish nodular chalk with a band of flint nodules are overlain by 11 ft of soft white chalk with two scattered bands of flints. This section is probably near the base of the zone, but diagnostic fossils were not found.

High on the valley side (169502) 250 yd N. 40° W. of Westwood Farm, fine massive white chalk with bands of flint nodules and one course of tabular flint is exposed for 12½ ft below 2 ft of rubbly chalk; another section (178498) some 850 yd E. 20° S. of this farm shows 9 ft of coarse soft white chalk with scattered bands of flint nodules. At Park Farm (177512), 27 ft of soft white well-bedded chalk with persistent and impersistent courses of flint nodules are exposed below 2 ft of rubbly chalk; fossils indicate a low position in the zone. At a pit (167528) 1050 yd E. 41° N. of Pett Bottom, 10 ft of soft chalk with scattered flints were seen, and a similar thickness of chalk with nodular flints was exposed in a pit (166536) 1900 yd W. 9° S. of Bridge church. In a pit (175529) 690 yd W. 21° S. of the house in Bourne Park 10 ft of massively bedded chalk with few flints were noted.

Some 25 ft of massive, well-bedded, white, blocky chalk with many courses of large flints were recorded, with overlying Head (p. 228), in the railway cutting (175537) 850 yd N. 42° W. of the house in Bourne Park. About 8 ft of chalk were noted in the cutting (183526) 620 yd S. 20° E. of this house.   J.G.O.S.

The lowest beds of the *M. coranguinum* Zone are exposed in the disused railway cutting south-west of Charlton Park (p. 149). The cutting (187522) ¼ mile south of Bishopsbourne church is about 40 ft deep, steep sided and much obscured by vegetation and talus. The section comprises the following strata in upward order: massive coarse gritty white chalk, 4 ft; band of closely-spaced flint nodules; massive white chalk, less gritty than that below, 6 ft; sparse band of flint nodules; coarse white chalk with scattered flint nodules, 12 ft; chalk with bands of flint nodules at intervals of about 3 ft, 9 ft. The flint nodules here are large and black, with thick white patinas.

In a pit (188493) close to the base of the Clay-with-flints, 1100 yd W. 33° S. of Heart's Delight, 8 ft of firm white massively-bedded chalk with occasional large flint nodules are exposed, and are referred to the *M. coranguinum* Zone.

In an old pit (195498) and in excavations for the foundations of a barn (196498), respectively 175 yd W. 14° S. and 50 yd S. of Heart's Delight, firm white chalk with flints yielded fossils younger than the *H. planus* Zone. At the former locality 5 ft of chalk are exposed, with one course of rather widely-spaced small black flint nodules with thick white patinas, and occasional flints scattered elsewhere. This chalk is probably within the *M. coranguinum* Zone, as are the 6 ft of similar chalk with occasional flint nodules visible in the small pit (197497) 300 yd S. 35° E. of Heart's Delight. Chalk of this zone in a pit (206515) on the north-east side of the Canterbury–Dover road, 900 yd E. 19° N. of Kingston church, is traversed by a 1-in thick tabular flint band transverse to the bedding; the uppermost 3 ft of chalk is rubbly due to frost action, and within this zone the tabular flint is dragged downhill, demonstrating the occurrence of hill-creep.

A pit in Bankyfield Shaw (210474), 900 yd N. 27° E. of Snodehill Farm, exposes 12 ft of firm white massively-bedded chalk, with large thickly-patinated black flint nodules both scattered and in impersistent bands. This chalk, which is referred to the *M. coranguinum* Zone, is not particularly coarse, but it contains some hard patches and occasional sponges.

Near the 400-ft contour 350 yd E. 15° S. of Denton Court, a pit (220466) shows about 15 ft of chalk, which may belong to the *M. coranguinum* Zone. In the pit face 9 ft of soft blocky white chalk with poorly-defined bedding are seen; flint nodules are not very numerous, but large ones tend to be concentrated into sporadic courses, with flints in groups or widely spaced, some $3\frac{1}{2}$ ft and $5\frac{1}{2}$ ft up; an additional 6 ft of similar chalk is exposed below the face in steeply inclined tunnels. Another pit (219469), at a lower level than the last, is situated 300 yd N. 37° E. of Denton Court: here some 10 ft of massively-bedded rough creamy-white chalk, with a tendency to be nodular towards the base, are poorly exposed; flint nodules are common, large and black, with $\frac{1}{8}$-in white patinas, and appear to be arranged in courses. This chalk is probably high in the *M. cortestudinarium* Zone or low in the zone above.

The pit (226459) 150 yd N.W. of Hill House Farm is obscured except for a section in a tunnel and above the tunnel entrance. The tunnel is some 40 yd long, 8 ft wide, 9 ft high at the entrance, and slopes downward into the hillside. In all about 25 ft of well-bedded soft white chalk are exposed; a strong band of large irregular black flint nodules, with $\frac{1}{16}$-in to $\frac{1}{8}$-in white patinas, is seen near the bottom of the section in the tunnel; elsewhere flints are rare and scattered. . No fossil was found.

A pit (228465) 330 yd E. 15° S. of Wootton church is in the *M. coranguinum* Zone. The lowest 3 ft of chalk are massively bedded, hard and gritty, with a $\frac{1}{2}$-inch parting of grey marl at the middle; these are overlain by 5 ft of massively bedded firm creamy-white chalk, with scattered small flint nodules in the lowest $1\frac{1}{2}$ ft, an impersistent 6-inch sponge-bed $1\frac{1}{2}$ ft up and some sponges in the uppermost foot; below $5\frac{1}{2}$ ft of rather rough white chalk with harder creamy-white patches, flint nodules scattered in the lowest foot and flint courses $1\frac{1}{2}$ ft and $4\frac{1}{2}$ ft up; below 5 ft of firm blocky white chalk with a flint course at the base and scattered large flints above; below 3 ft of rubbly soft white chalk with a course of widely-spaced flints at the base. The flint courses mentioned are more or less impersistent, while the section is traversed by a steeply inclined tabular flint vein $\frac{1}{2}$ to 2 inches in thickness.

At the edge of the wood 830 yd E. 17° S. of Wootton church, a pit (232464) shows 4 ft of chalk with fragments of large *Inoceramus* and a course of flint nodules. Soft white chalk is exposed for 6 ft in a pit (244466) beside the lane 630 yd E. 21° N. of Geddinge.

A pit (224484) 560 yd E. 18° N. of Broome shows 14 ft of well-bedded massive firm white chalk of the *M. coranguinum* Zone: flint nodules here are large, irregular in shape, and are widely spaced in even courses, 1, $4\frac{1}{2}$, 6 and 10 ft above the bottom of the section, the lowest two courses being impersistent: well-developed near-vertical joints strike north-west.   G.B.

A pit (222537) west of Adisham, 1000 yd E. 24° N. of Woodlands, exposing 10 ft of soft white well-bedded chalk with many flints, was recorded on his map by F. H. Edmunds and must lie close to the top of the zone.   J.G.O.S.

In a pit (221518) 1380 yd E. 23° S. of Ileden fossils of the *coranguinum* Zone were found in 7 ft of crisp white chalk, the uppermost 2 ft of which are rubbly due to frost action. Well-marked bands of large irregular black flints with white patinas up to ¼ in thick are present 5 ft and 2 ft below the top of the pit, with scattered nodules elsewhere. A pit (220504) 770 yd W. 17° S. of Womenswold church displays 5 ft of soft white chalk with occasional flint nodules, which belong to the *coranguinum* Zone or possibly to the *Uintacrinus* Zone, while 400 yd N. 38° E. of Dennehill Farm 16 ft of crisp white well-bedded blocky chalk, probably of the *coranguinum* Zone, can be seen in a pit (227500) and in tunnels into the quarry face. One strong band of flint nodules is present 1 ft above the base of the section; above this flints are scattered and become increasingly rare. Joints in the chalk here strike approximately east–west.

A roadside pit (250483) 500 yd N. 30° E. of West Court Farm shows 12 ft of soft white and creamy-white blocky chalk, with three fairly well-developed courses of flint nodules and scattered smaller flints. In a pit at Coxhill (252472) 10 ft of crisp white blocky chalk, rubbly in the top 4 ft, are exposed; major joints are about vertical and strike north-west; flint nodules are widely spaced in two poorly-developed courses and scattered elsewhere. 'Sheet pipes', consisting of beds of brown loam up to 4 inches thick, are present at three levels about 2 ft apart, connected by vertical fissures filled with similar loam; these beds are discontinuous, the lowest being traceable for ten yards; lumps of chalk are enclosed in the loam in places. Some 9 ft of firm white chalk with occasional sponges are exposed in a pit (254472) 150 yd E. 19° S. of Coxhill and in a tunnel into the pit-face, now partly collapsed. Isolated large flint nodules occur at certain horizons, forming bands about 3 ft apart in parts of the pit, and scattered smaller flints are also present. This chalk is thought to include the *Conulus* Band.

At the side of the levelled site (262482) 400 yd E. 10° S. of Shepherdswell Station 5 ft of firm white chalk of the *M. coranguinum* Zone were exposed. In Limekiln Plantation, 775 yd S.E. of Fredville, the crisp white chalk exposed in a large pit (262507) is believed to be at the *Conulus* Band horizon: 5 ft of chalk with rare flints near the base and a sparse band of flints 3 ft up, are overlain by 6 in of chalk (3 in above and 3 in below a well-defined bedding plane) with sponges and flint nodules over a 10-ft horizontal span, below 6 ft of chalk with occasional scattered flints and clusters of *Conulus,* below a fairly well-marked band of flints at the top of the pit.

The railway from Shepherdswell to Canterbury passes through a succession of cuttings in which chalk of the *M. coranguinum* and succeeding zones is exposed. Here Whitaker (1872, p. 29) noted that clay had in places filtered into the chalk along lines of jointing and of bedding. Beneath the road-bridge just south of Shepherdswell Station (258481) are exposed some 20 ft of massive coarse white chalk in almost horizontal beds with four prominent, more or less evenly spaced, nodular flint bands. Nothing resembling Bedwell's columnar flint band is to be seen. In similar chalk, 250 yd N. 18° W. of Shepherdswell Station (257484) two tabular flint bands are present some 6 ft apart: the lower is impersistent and up to 1 in thick; the upper is up to 2 in thick, occasionally swelling out into nodules 6 in or more in diameter, and passing laterally into a nodular flint band. About 4 ft above the upper tabular flint band, a nodular flint band with nodules scattered through about 1 ft of chalk is present ; 120 yd farther north a ½-in tabular flint seam occurs 2 ft above this, while other tabular flint lenses, scattered flint nodules and steeply-inclined tabular flint bands are present in the chalk above. The chalk here is closely jointed in a north-west direction.

A strong 2 to 3-in tabular flint band, locally swollen into big nodular masses, is believed to be Whitaker's Three-inch Band. It is first seen beneath drift about 20 ft above the cutting floor (255488), 720 yd N. 18° W. of Shepherdswell Station ; it reaches the floor 300 yd farther north, where the cutting is about 35 ft deep, in powdery white massively-bedded blocky chalk, with flints in sparse impersistent bands, of which one, better developed than the rest, is about 20 ft above the tabular band.

At the road-bridge (254492) 1160 yd N. 18° W. of Shepherdswell Station, two ill-defined flint bands are present, 2½ ft apart ; the upper one includes nodules up to 4 ft long and 1 ft thick, and above it sponges occur associated with *Conulus* in a 6-in bed about 13 ft above the floor of the cutting. The sponge-bed was not easily traced in either direction, but might represent Barrois's Sponge Bed. The cutting (251500) between 1850 and 2400 yd N. 18° W. of Shepherdswell Station is about 25 ft deep, the chalk being white, uniform, blocky and massively bedded, with flints rare and usually small ; traces of sponges are occasionally present, and some are associated with *Conulus,* which occurs scattered through the chalk and possibly represents the *Conulus* Band in a dispersed form.

North of Snowdown Halt the cutting (245512) is about 20 ft deep, in blocky soft white chalk, which is more massively bedded downwards. Flint is represented by isolated nodules. This chalk may be high in the *M. coranguinum* Zone or low in the succeeding *Uintacrinus* Zone.

In a pit in Callow Wood (246447) 480 yd S. 35° E. of Swanton Court Farm 10 ft of massively-bedded crisp white chalk with courses of flints are visible. Chalk of the *coranguinum* Zone is probably present in pits (252451 and 252454) respectively 800 yd S. 43° E. and 550 yd E. 20° S. of Wickham Bushes, where the chalk is poorly fossiliferous: at the latter locality 10 ft of soft white and cream chalk with flint courses are seen.   G.B.

Uintacrinus socialis Zone and Marsupites testudinarius Zone. On Barham Downs (197529) 1000 yd S. 13° E. of Highland Court 8 ft of soft white well-bedded chalk with few flints yielded *Uintacrinus socialis* Grinnell and some 200 yd E. of Highland Court (195538) F. H. Edmunds noted on his map a further 8 ft of similar chalk. Near the road at Chalkpit Farm (199553) 12 ft of soft white chalk with many flints are believed to lie wholly within the *Uintacrinus* Zone. On the valley side at the north-east buildings of Garrington Farms (206563), 4 ft of soft white rubbly chalk are exposed below 4 ft of Thanet Beds and made ground, while 15 ft of similar chalk were seen in an adjacent temporary exposure.

In 1958 the chalk-pit (211552) near the road-bridge over the railway 1560 yd E. 38° N. of Shepherd's Close showed 40 ft of soft white blocky chalk with many closely-spaced joints ; one discontinuous band of large irregular black flint nodules occurred 9 ft above the base, and a few very large flint nodules were scattered in the lower half of the face. The uppermost 20 ft of chalk exposed in this quarry are seen again together with succeeding beds in the adjacent railway cuttings. The highest chalk in these cuttings has yielded *Marsupites testudinarius* and what is probably an outlier of this zone caps the ridge thereabouts. The base of this zone is not marked by any lithological feature and the extent of the outlier is indeterminable.

At the road-fork (225561) 800 yd south of Bramling 25 ft of bedded soft fine white chalk of the *Uintacrinus* Zone with rare flint nodules were noted below 2 to 4 ft of weathered chalk and soil ; 12 ft of massive fine white chalk were seen in a pit near the T road-junction (225558) on Bramling Downs, 250 yd farther south. At Bramling, 200 yd S. of the milestone (226567), 3 to 5 ft of rubbly white yellow-stained chalk with *Uintacrinus* were seen.

Near Adisham (210541), 950 yd N. 21° W. of Woodlands, 10 ft of soft white well-bedded chalk with very few flints yielded *Uintacrinus,* and 6 ft of similar chalk in the road cutting (229545) 100 yd N. of Adisham church are also referred to this zone.

The railway cuttings (226547) west of the point 420 yd N. 11° W. of Adisham church reveal 35 ft of soft white blocky chalk, with a few small or very. large flints and oblique tabular flints. *Marsupites* plates were found as low as 13 ft above the floor of the cutting, but again the base of the *Marsupites testudinarius* Zone could not be defined in respect of a lithological feature. The chalk of this zone passes northwards under Drift and Thanet Beds, so that a connexion with the similar chalk near Wingham cannot be proved, though such a link probably exists.

Near the cross-roads (231547) 600 yd E. 25° N. of Adisham church 25 ft of soft white chalk with large nodular flints include the *Conulus* Band at the base of the *Uintacrinus* Zone. A pit (235555) 500 yd N. 20° E. of Bossington House exposes 6 ft of soft white chalk with a few flints and yielding *Marsupites,* below 1½ ft of loam with ' Bullhead ' flints. In the cutting (235538) 250 yd S. 37° E. of Adisham Station 18 ft of soft white thinly-bedded and closely-jointed chalk of the *Uintacrinus* Zone are exposed. No zonally diagnostic fossil was found either in 12 ft of well-bedded chalk with occasional very large flints exposed immediately south of Ratling Court (240535) or in 8 ft of similar thinly-bedded chalk with very large flints seen at the north-east corner of buildings at Old Court Farm (246533) ; the two pits are, however, considered to lie within the *Uintacrinus* Zone.   J.G.O.S.

The railway cutting (241529) immediately north of Aylesham Halt shows about 25 ft of soft white blocky chalk of the *Uintacrinus* Zone, with rare small flint nodules ; midway between the station and the bridge 250 yd to the north, two steeply-inclined tabular flint bands are to be seen ; between this bridge and a place 550 yd farther north, the cutting is lower and much obscured by vegetation and grime. South of Aylesham Halt the cutting is again badly obscured, but it shows about 35 ft of chalk similar to that to the north of the halt, with occasional isolated flint nodules ; the bedding is not discernible.

The chalk rubble in a shallow degraded pit (208518) 620 yd S. 11° W. of Ileden yielded a fauna including *Uintacrinus socialis,* which was also found in chalk dug from graves in Aylesham Cemetery (220508) 750 yd W. 17° N. of Womenswold church. The *Uintacrinus* Zone was also represented by 7 ft of firm white chalk, with very few flints, seen in a small pit (227512) 750 yd N. 5° W. of Womenswold church.   G.B.

The chalk at the foot of the Eocene escarpment north of Wingham probably lies wholly within the *Marsupites testudinarius* Zone, but it is nowhere exposed. South of Wingham, 520 yd N. of Dene Farm (242563), 4 ft of rubbly firm yellowish chalk with few flints and yielding *Uintacrinus* are exposed in the roadside. A plate of *Marsupites* found in rubble below the Thanet Beds thereabouts indicates the presence of the higher zone in the bank above this exposure. F. H. Edmunds noted on his map 8 ft of soft white chalk yielding *Marsupites,* below 2 ft of pockety loam, with flints, exposed in a farmyard (250559) 970 yd E. 6° N. of Dene Farm. At the Waterworks (243553) 500 yd S. 7° E. of Dene Farm, 5 ft of thinly-bedded soft white chalk at about the same horizon are exposed, and in a large old pit (249569) 1100 yd E. 44° S.

of Wingham church, 11 ft of soft white chalk were seen. In a pit (262551)
950 yd E. 36° N. of Goodnestone church, 20 ft of thinly-bedded very soft
white chalk of the *Marsupites* Zone, with a few sporadic flint nodules, were
described on his map by F. H. Edmunds : a well-marked sponge-bed was present
4 ft above the base of the pit, while the uppermost 6 ft of chalk were highly
weathered and disintegrated, but had been recemented in part to form a breccia.
On the south of the road (253538) 100 yd E. 20° S. of Bonnington Farm,
F. H. Edmunds recorded 6 ft of soft white chalk and he noted a similar
thickness of chalk in a section (248542) 850 yd W. 24° S. of Goodnestone
church ; both sections are believed to lie within the *Uintacrinus* Zone. Some
10 ft of well-bedded soft white chalk, almost devoid of flints and yielding
Marsupites, were noted on F. H. Edmunds's map at a pit (256530) 1050 yd
S. 28° E. of Bonnington Farm ; this chalk is presumably near the base of the
Marsupites Zone because *Uintacrinus* was found in 8 ft of similar chalk in
another pit (261530) at a slightly lower level, 1400 yd E. 31° S. of the same
farm.   J.G.O.S.

Chapter VII

PALAEONTOLOGY OF THE CHALK

THE FOSSILS occurring in the Chalk in the district described in the present memoir have not hitherto received any special study, apart from the work of Rowe (1900) on the zoning of the cliff-sections in east Kent, and the present account seems to be the first in which Rowe's zonal methods have been applied to the inland parts of the district.

Two features of the palaeontology of the Chalk call for mention in the present context. The first concerns the limits between the zones recognized in the Lower and Middle Chalk respectively. Jukes-Browne (1903, pp. 32–7) reviewed previous work and divided the Lower Chalk into the zone of *Schloenbachia varians* below (?67 ft) and the zone of *Holaster subglobosus* above (126 ft including the Belemnite Marls). He took the boundary between these zones 19 ft above the base of a bed of grey marly chalk, 79 ft thick, approximately at the highest level of occurrence of species of *Schloenbachia* (Fig. 3). This boundary has not been closely tested by collecting during the present work, but the occurrence of *Holaster subglobosus* below the base of Jukes-Browne's Bed 7 at Abbot's Cliff (p. 131) sets a limit to any possible raising of the horizon of the zonal junction. In the Middle Chalk, Rowe (1900, pp. 317, 319–20) drew particular attention to the difficulty of finding a satisfactory palaeontological criterion for distinguishing the zones of *Inoceramus labiatus* [*Rhynchonella cuvieri*] and *Terebratulina lata*, and concluded that the most satisfactory junction lay at a "flint-line" 70 ft above the base of the Middle Chalk. Jukes-Browne (1903, pp. 374–5) also attributed a thickness of 70 ft to the lower zone, though without referring to a flint band, and indeed no such band was seen during the re-survey (p. 122). At about the horizon mentioned, however, a marl seam, 2 to 3 in thick (bed 24 at Abbot's Cliff), separates underlying hard, nodular chalk with irregular marly wisps and abundant fossils from firm, smooth, white, sparsely fossiliferous chalk above. *Inoceramus labiatus* (Schlotheim) was not found above this marl seam in the cliff sections, and recent work on the Rhynchonelloid brachiopods (Pettitt 1954) and on the sea-urchins of the genus *Conulus* (Popiel-Barczyk 1958) suggests that the horizon may mark a more definite palaeontological break than was apparent to Rowe. *Orbirhynchia cuvieri* (d'Orbigny) and the other species said by Pettitt to be characteristic of the *labiatus* Zone were found only below the marl seam, while *O. heberti* Pettitt, a species of the *lata* Zone, was found within 12 ft above it. It is too early as yet to apply Popiel-Barczyk's work on Turonian *Conulus* from Poland unreservedly to the British fossils, but forms with straight sides and with little or no contraction of the adoral surface, some of them approaching *C. subrotundus* Mantell *subglobosus* Popiel-Barczyk, were found only above this marl seam (see also Rowe 1908, pp. 294, 296). Intensive collecting of the poorly

163

fossiliferous lower beds of the *lata* Zone will be necessary to test more precisely the validity of this junction.

Secondly, the boundary between the Upper Chalk zones of *Micraster coranguinum* and *Uintacrinus socialis* is the same as that fixed by Rowe (1900) in the coast sections of the Isle of Thanet. Although Barrois's Sponge Bed could not be identified in inland sections of weathered chalk, the occurrence of small nests of crushed specimens of *Conulus* at a definite level 20 ft above a continuous three-inch bed of tabular, or continuous nodular flint, and the increasing abundance of *Uintacrinus* plates at about 20 ft above the former level make possible a confident correlation with Whitaker's Three-inch Band and the *Conulus* Band of the coast.

LIST OF FOSSILS

In the subjoined lists there is given first a systematic table of the Chalk fossils collected by Survey officers in the district dealt with in the present memoir, but no attempt has been made to relate individual species to the localities where they were found, because all the species here recorded from each zone, and others besides, might well be found by further collecting at any locality in that zone. The subzone of *Actinocamax plenus* is regarded as a part of the Turonian *Inoceramus labiatus* Zone. Species of foraminifera, polyzoa and ostracoda are not listed. There follows a separate list of species illustrated in the literature by specimens from the area.

The Hexactinellid sponges have been determined by Mr. R. E. H. Reid, and some of the Lower Chalk ammonites by the late L. F. Spath. The remaining fossils have been determined in the Palaeontological Department of the Geological Survey.

Species	Zone								
	Schloenbachia varians	*Holaster subglobosus*	*Inoceramus labiatus*	*Terebratulina lata*	*Holaster planus*	*Micraster cortestudinarium*	*Micraster coranguinum*	*Uintacrinus socialis*	*Marsupites testudinarius*
	1	2	3	4	5	6	7	8	9
Porifera (Hexactinellida)									
cf. *Calyptrella bertae* Schrammen	.	.	.	.	.	6			
Cameroptychium campanulatum (T. Smith)	.	.	.	.	5				
Cephalites capitatus T. Smith	.	.	3						
cf. *C. capitatus*	.	.	.	4					

	1	2	3	4	5	6	7	8	9
Porifera (Hexactinellida)–*contd.*									
Cephalites longitudinalis T. Smith				4					
C. spp.				4	5				
cf. *Coscinopora infundibuliformis* Goldfuss							7		
cf. *C. spp.*						6	7		
Eurete spp.			3	4					
Exanthesis labrosus (T. Smith)	1								
Guettardia stellata Michelin					5				
G. or *Coscinopora sp.*							7		
Hexactinosan indet., *? Hexactinella* or *Strephinia sp.*			3						
Hexactinosan indet., *? Eurete sp.* or internal lamellae of *Cincliderma quadratum* Hinde			3						
Laocetis fittoni (Mantell)		2							
Lychniscosan indet., perhaps '*Laocetis*' *fittoni* (Mantell)			3						
Paraplocia labyrinthica (Mantell)			3						
Plocoscyphia elegans T. Smith					5				
Porochonia simplex (T. Smith)					5				
Sporadoscinia alcyonoides (Mantell)				4	5				
S. sp., two conjoined examples, probably *S. alcyonoides*						6			
S. or *Ventriculites sp.*					5				
Strephinia convoluta Hinde			3						
Synolynthia subrotunda (Mantell)						6			
Toulminia benettiae (Mantell)				4					
Tremabolites cf. *perforatus* (T. Smith)				4	5				
Ventriculites chonoides (Mantell)				4					
? V. impressus T. Smith					5				
V. sp.					5				
'*V.*' cf. *muricatus* (Roemer) (*non* T. Smith)					5				
Ventriculitid indet., possibly *V. chonoides* or *Cephalites longitudinalis*					5				
Ventriculitid Lychniscosan, cf. *Leiostracosia angustata* (F. A. Roemer)					5				
Porifera (Calcarea)									
Pharetrospongia strahani Sollas				4					
Porosphaera globularis (Phillips)	1	2	3	4			7	8	9
P. nuciformis (von Hagenow)							7	8	
P. patelliformis Hinde			3				7	8	
Anthozoa (Zoantharia Scleractinia)									
Caryophyllia cylindracea (Reuss)					5				
Diblasus gravensis Lonsdale							7		
Micrabacia coronula (Goldfuss)	1								
Onchotrochus serpentinus Duncan		2	3	4					
Parasmilia centralis (Mantell)			3	4	5		7	8	
Crinoidea									
Bourgueticrinus bacillus Griffith & Brydone							7		
B. ellipticus (Miller) *sensu lato*			3	4	5		7	8	

	1	2	3	4	5	6	7	8	9
Crinoidea–*contd.*									
Bourgueticrinus fritillus Griffith & Brydone ...							7	8	
B. granulosus Peron							7	8	
B. cf. *granulosus*							7		
B. utriculatus (Valette)							7	8	
Isocrinus agassizi (von Hagenow)			3	4	5				
Marsupites testudinarius (Schlotheim)									9
Uintacrinus socialis Grinnell								8	
Asteroidea									
Calliderma smithiae (Forbes)	1	2	3	4			7		
Crateraster quinqueloba (Goldfuss)			3	4			7	8	
Metopaster exsculptus Spencer							7	8	
M. parkinsoni (Forbes)			3	4			7	8	
M. uncatus (Forbes)							7		9
Pycinaster angustatus Spencer							7	8	9
P. humilis Spencer			3	4					
P. magnificus Spencer								8	
Stauranderaster bulbiferus (Forbes)			3				7		
S. coronatus (Forbes)				4			7		
Echinoidea									
Camerogalerus cylindricus (Lamarck)		2							
Cardiaster pygmaeus (Forbes)			3	4					
' *Cidaris* ' *bowerbanki* Forbes, radioles... ...		2							
' *C.*' *dissimilis* S. P. Woodward, radioles ...		2	3						
' *C.*' *hirudo* Sorignet, test and radioles ...				4					
' *C.*' *perornata* Forbes, radioles	1		3					8	
' *C.*' *serrifera* Forbes, radioles			3		5				
Conulus albogalerus Leske							7	8	
C. castanea (Brongniart) (Desor 1842, pl. iv, figs. 14–16)			3						
C. castanea cf. *planus* Popiel-Barczyk			3						
C. rothomagensis (Sismonda)			3						
C. subrotundus Mantell (typical form, Popiel-Barczyk)			3						
C. subrotundus (Wright 1875, pl. liii, fig. 2) ...			3						
C. aff. *subrotundus* (supernormal form, Popiel-Barczyk)			3						
C. cf. *subrotundus*				4					
C. subrotundus conoideus Popiel-Barczyk ...			3						
C. subrotundus cf. *conoideus*			3						
C. subrotundus subglobosus Popiel-Barczyk ...				4					
C. subrotundus cf. *subglobosus*				4					
C. subrotundus (d'Orbigny) *non* Mantell ...							7	8	
Discoides dixoni (Forbes)			3	4	5				
Echinocorys scutata Leske					5	6	7		
E. scutata elevata Griffith & Brydone									9
E. scutata aff. *elevata*								8	
Gauthieria radiata (Sorignet)			3	4	5				
G. wetherelli (Forbes)							7		
Glyphocyphus radiatus (Hoeninghaus)			3						

	1	2	3	4	5	6	7	8	9
Echinoidea–*contd.*									
Hemiaster (Peroniaster) nasutulus Sorignet ...	.	.	3	4					
Holaster gregoryi? Lambert	1								
H. subglobosus (Leske)	.	2							
H. subglobosus cf. *depressus* Cotteau	.	2							
H. cf. *trecensis* Leymerie (Wright 1881, pl. lxxii, fig. 2)	.	2							
H. (Sternotaxis) placenta J. L. R. Agassiz ...	.	.	.	.	5	6	7		
H. (S.) planus (Mantell)	.	.	3	4	5				
Infulaster excentricus (S. Woodward)	.	.	.	.	5				
Micraster coranguinum (Leske)	.	.	.	.	.	.	7	8	
M. coranguinum latior Rowe	.	.	.	.	.	.	7		
M. corbovis Forbes	.	.	.	4					
M. aff. *corbovis*	.	.	.	.	5	6			
M. cortestudinarium (Goldfuss)	.	.	.	.	5	6			
M. sp., between *M. corbovis* and *M. praecursor* Rowe	.	.	.	4					
M. leskei (Desmoulins)	.	.	.	4	5				
M. praecursor Rowe	.	.	.	.	5	6			
M. (Isomicraster) senonensis Lambert ...	.	.	.	.	.	.	7		
Salenia granulosa Forbes	.	.	3						
Stereocidaris sceptrifera (Mantell)	.	.	.	.	.	.	7	8	
Tylocidaris clavigera (Koenig)	.	.	.	.	5	.	7		
Annelida									
Flucticularia fluctuata (J. de C. Sowerby) ...	.	.	.	.	.	6			
Neomicrorbis (Neomicrorbis) crenatostriatus (Münster)	.	.	.	4					
Proliserpula (Proliserpula) ampullacea (J. de C. Sowerby)	.	2	.	.	5	6			
Serpula carinata S. Woodward, *non* Bosc ...	.	.	.	.	.	.	7		
S. macropus J. de C. Sowerby	.	.	.	.	.	.	7		
S. plana S. Woodward	.	.	.	.	.	.	7		
S. umbonata (J. Sowerby)	.	2							
Brachiopoda									
Concinnithyris albensis (Leymerie)	.	2	3	.	5				
C. albensis minor (Leymerie)	.	.	3	4	5				
C. burhamensis Sahni	.	2	3						
C. obesa (J. de C. Sowerby)	.	.	3	4	5				
C. cf. *obesa*	.	.	.	4					
C. protobesa Sahni	.	.	3						
C. cf. *protobesa*	.	.	.	.	5				
C. subundata (J. Sowerby)	.	2							
Crania parisiensis Defrance	.	.	.	4					
Cretirhynchia cuneiformis Pettitt	.	.	.	.	5				
C. aff. *cuneiformis*	.	.	.	.	.	.	7		
C. minor Pettitt	.	.	.	.	5				
C. octoplicata (J. Sowerby)	.	.	.	.	.	.	7		
C. aff. *octoplicata*	.	.	.	.	5	6			
C. subplicata (Mantell)	.	.	.	.	.	6			
C. cf. *subplicata*	.	.	.	.	5	6			

M

	1	2	3	4	5	6	7	8	9
Brachiopoda–*contd.*									
Gibbithyris ellipsoidalis Sahni					5		7		
G. cf. *ellipsoidalis*					5				
G. grandis Sahni					5				
G. media Sahni				4	5				
G. merensis Sahni					5				
G. pyramidalis Sahni					5				
G. semiglobosa (J. Sowerby)						6			
G. subrotunda (J. Sowerby)			3	4	5				
Kingena lima (Defrance)	1	2			5		7	8	
Lacazella (*Bifolium*) *wetherelli* (Morris)							7		
Orbirhynchia compta Pettitt			3						
O. aff. *compta*			3						
O. cuvieri (d'Orbigny)			3						
O. dispansa Pettitt				4	5				
O. cf. *dispansa*				4					
O. heberti Pettitt				4					
O. cf. *mantelliana* (J. de C. Sowerby)	1								
O. obscura Pettitt	1								
O. orbignyi Pettitt			3						
O. cf. *parva* Pettitt	1								
O. cf. *pisiformis* Pettitt							7		
O. cf. *praedispansa* Pettitt			3						
O. reedensis (Etheridge)					5				
O. aff. *reedensis*				4	5				
O. cf. *wiesti* (Quenstedt)			3						
O. sp. nov.?				4					
' *Rhynchonella* ' *martini* (Mantell)	1								
Terebratulina lata Etheridge			3	4	5				
T. nodulosa Etheridge		2							
T. rowei Kitchin							7		
T. striatula (Mantell) *sensu lato*			3	4	5		7	8	
T. triangularis Etheridge	1								
Gastropoda									
Avellana cf. *cassis* d'Orbigny	1								
Bathrotomaria perspectiva (Mantell)							7		
Eutrochus schlueteri (Woods)					5				
' *Solariella* ' *gemmata* (J. de C. Sowerby)				4	5				
' *Turbo* ' *sp.*					5				
Scaphopoda									
Dentalium turoniense Woods					5				
Bivalvia									
?Astarte angulata Guéranger					5				
Atreta nilssoni (von Hagenow)					5	6	7		
Bakevellia sp.					5				
Barbatia sp., cf. *B. galliennei* (d'Orbigny)	1								
' *Cardita* ' *cancellata* Woods					5	6			
' *Cardium* ' cf. *cenomanense* d'Orbigny					5				
' *C.*' *turoniense* Woods					5				

	1	2	3	4	5	6	7	8	9
Bivalvia—*contd.*									
Chlamys cretosa (Defrance)				4			7		
C. (Aequipecten) aff. *campaniensis* (d'Orbigny)					5				
C. (A). pexata (Woods)				4					
Corbis morisoni Woods					5				
Cuspidaria caudata (Nilsson)					5				
C. pulchra (J. de C. Sowerby)					5				
Entolium orbiculare (J. Sowerby)	1								
Freyia similis (Münster pars, Holzapfel)					5				
Gryphaeostrea canaliculata (J. Sowerby)		2		4			7		
Inoceramus cordiformis J. de C. Sowerby							7		
I. costellatus Woods					5				
I. cf. *costellatus*			3						
I. crippsi Mantell	1	2							
I. crippsi reachensis Etheridge		2							
I. cuvieri J. Sowerby				4	5		7		
I. cf. *etheridgei* Woods	1	2	3						
I. inconstans Woods			3	4	5		7		
I. involutus J. de C. Sowerby							7		
I. labiatus (Schlotheim)			3						
I. aff. *labiatus*					5				
I. labiatus latus J. de C. Sowerby					5				
I. lamarcki Parkinson			3	4	5				
I. cf. *lamarcki*					5				
I. lamarcki apicalis Woods				4	5				
I. lamarcki, approaching *websteri* Mantell					5				
I. lamarcki websteri						6			
I. pictus J. Sowerby	1								
I. cf. *pictus*			3						
Lima (Plagiostoma) globosa (J. de C. Sowerby)	1	2							
L. (P.) hoperi (Mantell)				4	5		7		
L. (Ctenoides) divaricata Dujardin							7		
Limatula wintonensis (Woods)				4	5	6			
Limopsis sp. (Woods 1897, pl. xxvii, figs. 7, 8)					5				
Lopha semiplana (J. de C. Sowerby)					5		7	8	9
Neithea quadricostata (J. Sowerby)		2							
N. quinquecostata (J. Sowerby)		2							
N. sexcostata (S. Woodward)							7	8	
Nucula sp.					5				
Nuculana [*Palaeoneilo?*] cf. *semilunaris* (von Buch)					5				
Ostrea boucheroni Coquand							7	8	
O. incurva Nilsson			3	4			7	8	
Plicatula barroisi Peron		2		4	5	6			
P. inflata J. de C. Sowerby	1								
Pseudoptera coerulescens (Nilsson)					5				
Pycnodonte vesicularis (Lamarck)	1	2	3	4					
Spondylus dutempleanus d'Orbigny				4					
S. latus (J. Sowerby)				4	5		7		
S. spinosus (J. Sowerby)				4	5	6			
' *Trapezium* ' *trapezoidale* (Roemer)					5				

	1	2	3	4	5	6	7	8	9
Cephalopoda (Ammonoidea)									
Acanthoceras aff. *evolutum* Spath	1								
A. cf. *rhotomagense* (Defrance MS.), Brongniart sp.	1								
Allocrioceras sp.					5				
? *Allocrioceras sp.*			3						
Calycoceras aff. *newboldi* (Kossmat)	1								
Hyphantoceras reussianum (d'Orbigny)					5				
Hyphoplites falcatus (Mantell)	1								
Hypoturrilites mantelli (Sharpe)	1								
Lechites cf. *gaudini* (Pictet & Campiche)	1								
Lewesiceras mantelli Wright & Wright					5				
L. sharpei (Spath)					5				
Mantelliceras cantianum Spath	1								
M. aff. *hyatti* Spath	1								
M. mantelli (J. Sowerby)	1								
Metoicoceras sp.			3						
Parapuzosia sp., group *P. leptophylla* (Sharpe)							7		
Scaphites equalis J. Sowerby	1								
S. geinitzi d'Orbigny				4	5				
S. obliquus J. Sowerby	1								
Sciponoceras baculoides (Mantell)	1								
S. bohemicum (Fritsch)				4					
S. cf. *gracile* (Shumard)			3						
Schloenbachia intermedia (Mantell)	1								
S. subplana (Mantell)	1								
S. subtuberculata (Sharpe)	1								
S. subvarians Spath	1								
S. subvarians aperta Spath	1								
S. ventriosa Stieler	1								
Subprionocyclus branneri (Anderson)					5				
S. hitchinensis (Billinghurst)					5				
S. neptuni (Geinitz)					5				
S. cf. *neptuni* (Schlueter 1872, pl. xvi, fig. 2)					5				
ammonoid aptychus							7		
Cephalopoda (Belemnoidea)									
Actinocamax plenus (Blainville)			3						
A. cf. *primus* Archangelsky			3						
A. verus Miller							7	8	
Crustacea (Cirripedia)									
Cretiscalpellum glabrum (Roemer)		2		4	5		7		
Pisces									
Corax falcatus J. L. R. Agassiz		2							
Isurus mantelli (J. L. R. Agassiz)				4					
Ptychodus rugosus Dixon					5				
Scapanorhynchus rhaphiodon (J. L. R. Agassiz)									9
S. subulatus (J. L. R. Agassiz)			3						

FOSSILS FIGURED FROM THE CHALK OF THE DISTRICT

The following list includes not only species recorded from localities certainly within the one-inch sheets dealt with in the present memoir, but also those recorded as from the Lower Chalk ('Grey Chalk', 'Chalk Marl') of Dover or near Dover. The specimens on which these records are based must in fact have come from the area of the Folkestone (305) and Dover A (306) sheets since no Lower Chalk is exposed at the surface within the Dover (290) Sheet. Species recorded from the Middle or Upper Chalk of Dover are not listed, because they might have come from Sheet 290, in which Dover itself is situated.

Porifera (Demospongia)

Jerea cordiformis Hinde, Hinde 1883, pl. xv, figs. 3, 3a. Grey Chalk, near Dover.

Nelumbia tuberosa Hinde, Hinde 1883, pl. xvi, figs. 1, 1a, b. Grey Chalk, near Dover.

Polyjerea arbuscula Hinde, Hinde 1883, pl. xvi, figs. 2, 2a. Grey Chalk, near Dover.

P. lobata Hinde, Hinde 1883, pl. xvi, fig. 3. Grey Chalk, near Dover.

Siphonia ficus Goldfuss, Hinde 1883, pl. xiii, figs. 3, 3a. Grey Chalk, Dover.

Stachyspongia spica (Roemer), Hinde 1883, pl. vi, figs. 2, 2a. Grey Chalk, near Dover.

Thamnospongia? reticulata Hinde, Hinde 1883, pl. xviii, figs. 3, 3a, b. Grey Chalk, near Folkestone.

Porifera (Hexactinellida)

Exanthesis labrosus (T. Smith), Hinde 1883, pl. xxix, fig. 2. "Upper Greensand" [i.e. Chloritic Marl], near Folkestone.

'Laocetis' fittoni (Mantell), Hinde 1883, pl. xxiii, fig. 2. Grey Chalk, near Dover.

Sestrocladia furcata Hinde, Hinde 1883, pl. xxvii, figs. 1, 1a, b. Grey Chalk, near Dover.

Stauronema planum Hinde, Hinde 1883, pl. xxiv, figs. 2, 2a–c. Grey Chalk, near Folkestone.

Strephinia convoluta Hinde, Hinde 1883, pl. xxiii, figs. 3, 3a, b. Grey Chalk, near Dover.

S.? reteformis Hinde, Hinde 1883, pl. xxiii, figs. 4, 4a. Grey Chalk, near Dover.

Asteroidea

Calliderma latum (Forbes), Sladen *in* Sladen & Spencer 1891, pl. iii, fig. 2. Chalk Marl, Dover.

C. smithiae (Forbes), Sladen *in* Sladen & Spencer 1891, pl. vii, fig. 2. Grey Chalk, Folkestone.

Trachyaster rugosus (Spencer), Spencer *in* Sladen & Spencer 1907, pl. xxix, figs. 7, 7a. Lower Chalk, Dover and Folkestone.

Ophiuroidea

Amphiura cretacea Spencer, Spencer *in* Sladen & Spencer 1907, pl. xxviii, figs. 6, 6a. Lower Chalk, Folkestone.

Ophiotitanos laevis Spencer, Spencer *in* Sladen & Spencer 1907, pl. xxviii, figs. 3, 3a, 4, 4a. Lower Chalk, Dover.

O. magnus Spencer, Spencer *in* Sladen & Spencer 1907, pl. xxviii, figs. 5, 5a, pl. xxix, fig. 13. Lower Chalk, Folkestone.

O. tenuis Spencer, Spencer *in* Sladen & Spencer 1907, pl. xxviii, figs. 1, 1a, 2, 2a. Lower Chalk, Folkestone and Dover.

Echinoidea

Allomma normanniae (Cotteau), Wright 1868, pl. xxi, figs. 3a–d. Grey Chalk, near Folkestone.

'*Cidaris*' *bowerbankii* Forbes *in* Dixon, Dixon 1850, pl. xxix, fig. 4. Grey Chalk, Dover.

'*C.*' *dissimilis* S. P. Woodward, Wright 1864, pl. iii, fig. 2, pl. iiiA, figs. 1–5. Grey Chalk, Dover.

'*C.*' *dixoni* Cotteau, Wright 1868, pl. xi, fig. 4. Grey Chalk, Dover.

Discoides subuculus (Leske), Wright 1874, pl. xlv, figs. 5, 6. Grey Chalk, near Folkestone.

Glyptocyphus difficilis (J. L. R. Agassiz), Wright 1870, pl. xxii, figs. 1, 2, 4. Grey Chalk, near Folkestone.

Holaster trecensis Leymerie, Wright 1881, pl. lxxiv, fig. 2. Grey Chalk, near Folkestone.

Salenia austeni S. P. Woodward, Wright 1871, pl. xxxvii, figs. 1, 2. Grey Chalk, near Folkestone.

S. clarkii S. P. Woodward, Wright 1871, pl. xxxviii, fig. 1, pl. xxxix, fig. 1, 1872, pl. xlii, figs. 1, 2, 5. Grey Chalk, near Folkestone.

Tetragramma brongniarti (J. L. R. Agassiz), Wright 1868, pl. xx, figs. 2a–c, pl. xxiA, figs. 2–4, pl. xxiB, figs. 3a–e. Grey Chalk, near Folkestone.

Tiaromma schlueteri (de Loriol), Wright 1868, pl. xvi, figs. 4a–e, pl. xix, figs. 1a–e, pl. xxiA, figs. 1a, b. 2a, b. Grey Chalk, Folkestone.

Brachiopoda

Concinnithyris latifrons (Leymerie), Sahni 1929, pl. viii, fig. 3. Turonian?, Folkestone.

C. protobesa Sahni, Sahni 1929, pl. ii, figs. 1–3. Turonian, Folkestone.

Kingena lima (Defrance), Davidson 1852, pl. v, fig. 2. Grey Chalk, between Folkestone and Dover.

Orbirhynchia mantelliana (J. de C. Sowerby), Davidson 1855, pl. xii, figs. 20, 21. Lower Chalk, Folkestone.

O. obscura Pettitt, Pettitt 1954, pl. iii, fig. 2. Lower Chalk, Folkestone.

'*Rhynchonella*' *martini* (Mantell), Davidson 1855, pl. xii, figs. 15, 15a, b. Grey Chalk in the vicinity of Folkestone.

'*Terebratula*' *squamosa* Mantell, Davidson 1852, pl. v, figs. 6, 6a, b. Grey Chalk, between Folkestone and Dover.

'*Terebratula*' *carteri* Davidson, Davidson 1855, pl. vii, fig. 3. Grey Chalk in the vicinity of Dover.

Terebratulina striata auctt. angl. *non* Wahlenberg, Davidson 1852, pl. ii, fig. 25. Grey Chalk, Dover.

Gastropoda

'*Rostellaria*' *pricei* H. Woodward, H. Woodward 1872, pl. iii. Grey Chalk, Folkestone.

Scaphopoda

Dentalium major Gardner, Gardner 1877a. pl. xvi. 'Cast bed', Grey Chalk, Dover.

Bivalvia

Camptonectes dubrisiensis (Woods), Woods 1902, pl. xxix, fig. 8. Chalk Marl, Dover.

Chlamys elongata (Lamarck), Woods 1902, pl. xxxii, fig. 1. Grey Chalk, Dover.

C. (Aequipecten) arlesiensis (Woods), Woods 1902, pl. xxxvii, figs. 9–11. Chalk Marl, Folkestone.

Entolium orbiculare (J. Sowerby), Woods 1902, p. 151, text-fig. 1. Chalk Marl, Folkestone.

Exogyra conica (J. Sowerby), Woods 1913, p. 411, text-figs. 227, 230, 231. Chalk Marl, Folkestone.

Inoceramus lamarcki Parkinson, Woods 1912, pl. lii, fig. 6. Upper Chalk, *planus* Zone, Shakespeare Cliff, Dover.

I. undulatoplicatus Roemer var. *digitatus* Schlueter, Woods 1912, p. 307, text-fig. 62. Upper Chalk, *coranguinum* Zone, Snowdown Colliery Shaft.

Lima subovalis J. de C. Sowerby, Woods 1904, pl. ii, fig. 6. " Greensand bed at base of Chalk ", Folkestone.

Lopha diluviana (Linnaeus), Woods 1913, p. 351, text-fig. 122, 126–130. Chalk Marl, Folkestone.

Neithea sexcostata (S. Woodward) form α, Woods 1903, pl. xli, figs. 6–8. Chalk Marl, Dover.

Pseudolimea elongata (J. de C. Sowerby), Woods 1904, pl. vi, figs. 5–7. Chalk Marl, Folkestone.

Cephalopoda (Ammonoidea)

Acanthoceras sherborni Spath, Sharpe 1857, pl. xvii, figs. 1a, b. Grey Chalk (? *varians* Zone, Wright & Wright 1951, p. 37), Dover.

Lewesiceras peramplum? (Mantell), Sharpe 1857, pl. xxi, figs. 1a–c. Grey Chalk (? *labiatus* Zone, Wright & Wright 1951, p. 38), Dover.

Crustacea (Malacostraca)

Glyphea willetti (H. Woodward), Woods 1928, pl. xvii, fig. 5. Lower Chalk (probably *varians* Zone, Woods, *loc. cit*.), Dover.

Crustacea (Cirripedia)

Cretiscalpellum glabrum (Roemer), Withers 1935, pl. xvi, figs. 4–5 (*varians* Zone), 7 (*subglobosus* Zone), 8 (? *subglobosus* Zone), Dover

Scalpellum (Arcoscalpellum) hastatum (Darwin), Withers 1935, pl. xxix, fig. 1. (Darwin 1851, pl. ii, fig. 13), Grey Chalk, Dover.

S. (A.) lineatum (Darwin), Withers 1935, pl. xxviii, fig. 15. ? *subglobosus* Zone, Dover.

S. (A.) trilineatum (Darwin), Withers 1935, pl. xxiv, fig. 8 (Darwin 1851, pl. ii, fig. 5), Grey Chalk, Dover ; fig. 10, *varians* Zone, West Cliffs, Dover.

Pisces

Apateodus lanceolatus A. S. Woodward, A. S. Woodward 1907, p. 105, text-fig. 32. Lower Chalk, probably *varians* Zone, Dover.

Belonostomus cinctus J. L. R. Agassiz, A. S. Woodward 1908, pl. xxx, fig. 1. Lower Chalk, *subglobosus* Zone, Dover.

? Cimolichthys lewesiensis Leidy, A. S. Woodward 1902, pl. xii, fig. 5 (Gunther 1864, pl. vi). Lower Chalk, *subglobosus* Zone, Folkestone.

Corax falcatus J. L. R. Agassiz, A. S. Woodward 1911, pl. xlii, figs. 25, 26. Lower Chalk, *subglobosus* Zone, Dover.

Isurus crassidens (Dixon), A. S. Woodward 1911, pl. xliv, fig. 6. Lower Chalk, *subglobosus* Zone, Dover.

I. mantelli (J. L. R. Agassiz), A. S. Woodward 1911, pl. xliii, fig. 12. Lower Chalk, *varians* Zone, Dover.

Lamna appendiculata J. L. R. Agassiz, A. S. Woodward 1911, p. 207, text-fig. 63. Lower Chalk, *subglobosus* Zone, Dover.

Lepidotus? pustulatus A. S. Woodward, A. S. Woodward 1909, pl. xxxv, fig. 1 (A. S. Woodward 1895, pl. viii, fig. 1). Lower Chalk, *varians* Zone, Folkestone.

Macropoma praecursor A. S. Woodward, A. S. Woodward 1909, pl. xxxviii, figs. 8–10. Lower Chalk, *varians* Zone, Folkestone.

Portheus sp., A. S. Woodward 1912, p. 249, text-fig. 78. Lower Chalk, *subglobosus* Zone, Dover.

Ptychodus decurrens J. L. R. Agassiz, A. S. Woodward 1912, pl. li, fig. 14. Lower Chalk, *subglobosus* Zone, Dover.

Saurodon intermedius (Newton), A. S. Woodward 1907, p. 105, text-fig. 32 (Newton 1878, pl. xix). Lower Chalk, probably *varians* Zone, Dover.

Scyllium dubium A. S. Woodward, A. S. Woodward 1911, pl. xlii, figs. 5–6. (A. S. Woodward 1889, pl. xvi, figs. 7, 8). Lower Chalk, *subglobosus* Zone, Dover.

Synechodus dubrisiensis (Mackie), A. S. Woodward 1911, pl. xlv, fig. 6. (Mackie 1863, pl. xiii). Lower Chalk, *subglobosus* Zone, Dover.

Reptilia

Acanthopholis horridus Huxley, Huxley 1867, pl. v. Lower Chalk (8 ft above base of Chalk Marl, see Etheridge 1867), immediately east of Copt Point, Folkestone.

R.V.M.

Chapter VIII

TERTIARY: EOCENE

GENERAL ACCOUNT

THE INTERVAL between the deposition of the highest Chalk and that of the lowest Thanet Beds spans the major geological discontinuity between the Mesozoic and Tertiary eras, during which the mammals became dominant over the reptiles, and several groups of animals became extinct, including the dinosaurs on land and the ammonites in the sea. Yet the angular discordance between the Chalk and the lowest Eocene strata can be demonstrated here only by palaeontological zoning of the former, and it has been found that only about 110 ft of the Chalk represented within this district are cut out by the Thanet Beds in an 18-mile east–west traverse across their outcrop.

It appears that after its deposition the Chalk was uplifted and gently warped, tilted and eroded. The erosion probably reduced the Chalk uplands to nearly a plane surface before the advancing Eocene sea trimmed away the remaining irregularities, and upon the resultant platform the Eocene sediments were deposited. Relics of this trimmed surface extend beyond the present outcrops of Eocene strata and are known as the Eocene bench. Southwards this plane is itself cut off by a later sub-Pliocene surface (Wooldridge and Linton 1955, pp. 48–9). West of the Great Stour valley the Eocene bench extends almost to the crest of the Chalk escarpment before it is truncated by the Pliocene bench. East of this river, truncation occurs on Chartham Downs and thence somewhat north of the A.2 Dover Road on Barham Downs. The Eocene strata must formerly have extended at least over all the ground north of this line of truncation. As the youngest chalk is found in the north-east, the post-depositional tilting of the Chalk must have been in that direction, and if the Eocene bench is used as an indication the thickness of chalk removed before Eocene deposition (relative to the youngest chalk preserved in the district) would have amounted to about 300 ft in the south-west. Thus the magnitude of the tilt between roughly Wye and Wingham was of the order of $\frac{1}{4}°$.

Earth movements similar to the above continued during deposition of the early Eocene strata, with the result that the latter are of diverse nature. The more pronounced movements caused partial erosion of the sediments.

The Tertiary strata of the district comprise:

	Whitaker's classification	European Stages	
	London Clay	} Ypresian	
Lower	⎧ Oldhaven (and Blackheath) Beds		
London	⎨ Woolwich (and Reading) Beds	} Landenian	⎧ Sparnacian
Tertiaries	⎩ Thanet Beds		⎩ Thanetian

The Eocene beds below the London Clay were called the "Plastic Clay" formation in the early 19th century. They were systematically described, together with the London Clay, by Prestwich (1850; 1852; 1854a; 1854c), who also referred to the work of earlier observers; on these foundations Whitaker (1866; 1872) erected the classification of the Eocene used in this account. Stamp (1921) applied the stage-names Landenian and Ypresian to the deposits and discussed the classification, as also did Wrigley (1949). A plea for the simplification of the European terminology was made by Davis and Elliott (1958, p. 273), who wished to call the Thanet Beds and Woolwich "bottom bed" the Palaeocene, and the remaining strata, up to and including the Lower Bracklesham Beds, the Lower Eocene.

The Eocene strata, as described by Stamp (1921), were deposited in a shallow sea, offshore from a large estuary centred in Hampshire and including the London area, Kent and Sussex. The centre of the present Weald was undergoing erosion, but it is uncertain how far, if at all, land actually emerged. The flora and fauna of the period indicate that its climate ranged from temperate to sub-tropical.

THANET BEDS

The lowest beds of the Eocene comprise a variable series of marine sediments up to about 120 ft in total thickness. They were named Thanet Sands by Prestwich (1852), who described their occurrence in Hampshire, London and Kent; in particular he noted localities within the present district where he saw the beds and their fossils. Whitaker (1866, pp. 406–9) renamed the strata Thanet Beds.

The five general subdivisions of the Thanet Beds in Kent (Whitaker 1872, p. 56) occur within the present district, where they comprise:

    e. Fine grey to golden-brown glauconitic sand, about 40 ft thick.
    d. Grey to greenish grey sandy clay and stiff clay, very glauconitic and loamy at base, about 25 to 30 ft thick.
    c. Fine greyish brown sand, locally clayey and with some ironstone; large borings (p. 186) and an occasional pebble at the top; up to about 10 ft thick.
    b. Alternating beds of greenish brown sand and sandy or stiff clay, about 10 ft thick.
    a. Intensely glauconitic green loamy sand up to 3 ft or more thick with, at its base, the 'Bullhead Bed', consisting of unworn but occasionally split green-patinated flints up to 1 ft long but usually smaller.

Bed *a* is ubiquitous. Bed *b* has been seen only in the railway cuttings south of Canterbury, although Whitaker recorded it at Upper Ensinge Tunnel, near Selling church and at Bekesbourne. Bed *c* is exposed in the railway cuttings south and east of Canterbury, and it is prominent in the long outlier south-east of Wingham and in the escarpment north-east of that village. It appears to be absent in the Blean and Selling districts. However, it readily breaks down to form brickearth, and may be concealed by that deposit in many places. Beds *d* and *e* are present everywhere and the contact between them at outcrop is commonly marked by a series of small springs. Some of the sediments previously recognized as beds *c* or *e* are perhaps, as Whitaker (1872, p. 408) realized, no more than weathered portions of bed *d*. Such weathering at Pegwell Bay, Isle of Thanet, has been described by Pitcher, Shearman and Pugh (1954, pp. 310–1).

M. I. Gardner (1888, p. 755) studied the matrix of the Bullhead Bed as found at Pegwell Bay, Chislet, Upnor and Chislehurst—localities outside the present district—and found that it was a very fine sand formed of about equal quantities of dark and light grains with more or less clayey matter. The sand was composed of 45 per cent subangular quartz grains ; 15 per cent glauconite in rounded grains, made up of aggregates of smaller wedge-shaped grains, the cracks between which were filled with iron oxide ; 20 per cent of flint in light grey grains or sharp angular transparent or opaque chips ; the remaining 20 per cent consisted of heavy minerals and some twinned feldspar crystals. Grain size in the bed was up to about 1 mm.

Beneath the green patinas of the flints in the Bullhead Bed is an inner brown layer ; each of the layers is about $\frac{1}{16}$ inch thick. The rounded black flint pebbles which are a feature of all the succeeding Eocene pebble beds of the district are absent. The unabraded condition of the flints and their green colour have caused much speculation on the origin of the bed. Prestwich (1852, p. 253) referred to a " powerful but transient action necessary to uproot these flints from the chalk ". Hughes (1866) believed that the bed was formed by even solution of the underlying chalk after deposition of the Thanet Beds. He thought that the solution was retarded by lines of tabular flint. Dowker (1866) considered that the Bullhead Bed flints were the remains from subaerial solution of the chalk before deposition of the Eocene sediments. He also thought that an underlying tabular flint was formed after deposition of the Thanet Beds, by the solidification of soluble silica derived from those beds. This is probably a reliable explanation of the origin of the tabular flint and furthermore Wrigley (1949) and Haynes (1958, p. 87) expressed opinions on the formation of the Bullhead Bed similar to those of Dowker. Wrigley regarded it as a " gentle redisposition, by an advancing sea, of a clay-with-flints which had accumulated on a long exposed, chalk land-surface ". J. S. Gardner (1883), Boswell (1917, pp. 538–40) and Wooldridge (*in* Dewey and others 1925, pp. 266–7) regarded the bed as a basal conglomerate. Gardner considered it to have formed among much seaweed in the Laminarian zone of 0 to 15 fathoms depth.

Foraminifera found in the Thanet Beds in East Kent were described by Haynes (1956–8), who only referred to one fossiliferous locality—the Upper Ensinge (Selling) Tunnel—within the present district. He found (1958, p. 87) that the Bullhead Bed contained a derived Cretaceous fauna but no foraminifera of Thanetian age, and later (Haynes and El-Naggar 1964) considered it probable that all the planktonic foraminifera in the Thanet Beds were derived. He believed that the beds were deposited in a cool shelf sea of shallow to moderate depth and (Haynes 1955) were upper Palaeocene in age.

It is from bed *d* that many of the fossils recorded in the literature have been obtained, either as fragile shells or as moulds in the hard brown loams to which it weathers (Whitaker 1866, p. 408). The following list of fossils found in the Thanet Beds of this district has been assembled mainly from

lists given by Prestwich (1852 ; 1854a) and Whitaker (1872): where possible
equivalent fossil names in current use have been supplied by Mr. C. J.
Wood: *Schizaster* cf. *cuneatus* Gregory, *S.?*; "*Actaeon*", *Aporrhais sowerbyi*
(Fleming), "*Fusus tuberosus*, Sby", "*Natica* (small sp.)", "*Scalaria sp.*",
Sigatica abducta (Deshayes), "*Solarium*", "*Turritella*", "*Dentalium sp.*",
Arctica [*Cyprina*] *morrisi* (J. Sowerby), *A. planata* (J. de C. Sowerby),
"*Artemis* or *Cytherea spp.*" (including the "*C. orbicularis* of Edwards")
[? = *Dosiniopsis bellovacina* (Deshayes)], *Astarte tenera* Morris, "*Corbula
(Arnouldi?)*, Nyst." [? = *C. plateaui* Cossmann], "*Corbula longirostris*,
Desh.", *Corbula regulbiensis* Morris, *Cucullaea decussata* Parkinson,
Cyrtodaria rutupiensis (Morris), *Dosiniopsis bellovacina*, *Eutylus cuneatus*
(Morris), *Garum edwardsi* (Morris), *Glycimeris terebratularis* (Linné), "a
small *Leda*", "*Modiola sp.*", *Nemocardium plumstedianum* (J. Sowerby),
"*Nucula sp.*", "*Pholadomya Koninckii*, Nyst?" [? = *P. oblitterata* Potiez
and Michaud], *Pholadomya oblitterata*, "*Pholas, sp.* (? *Pholadomya*)",
"*Teredo, sp.* (Borings)", *Thracia oblata* J. Sowerby; "*Nautilus*";
"*Hoploparia sp.*", "*Palaeocorystes*", "*Lamna* (? *sp.*) teeth".

The Bullhead Bed rests evenly upon the Chalk, except where disturbed
by piping, which is locally extensive. Piped masses of Thanet Beds are
encountered in many places beyond their mapped outcrop.

WOOLWICH BEDS

In the London and Hampshire basins strata which include an estuarine
facies of vivid mottled clays, and beds of sand and clays mainly of marine
origin, are called the Woolwich and Reading Beds.[1] They were originally
called the Woolwich and Reading Series by Prestwich (1854a, p. 75), but
were renamed by Whitaker (*in* Hull and Whitaker 1861). Only the Woolwich
Beds, mainly of marine origin, occur within the present district.

Collins and Gill (1923) coupled the Woolwich Beds with the underlying
Thanet Beds in the belief that the division between the two could not be
recognized with certainty and used the European stage-name Landenian
(Stamp 1921) for the combination. Wooldridge (*in* Dewey and others 1925,
p. 268) also declared that the two beds were frequently indistinguishable
and even Whitaker (1872, p. 171) found the distinction difficult east of the
Great Stour River. However, throughout the present district the beds are
quite distinct, though it is evident that the difficulty of separating them
increases somewhat towards the north-east, perhaps because the earth movements which caused the erosion at the junction had less effect in that
direction. Within this district, and also much of the area to the north,
movement was sufficient to cause a break in deposition, during which
rounded or occasionally sub-rounded black flint pebbles were scattered on
the sea floor. These pebbles occur at, or within inches above, the uneven
junction of fine and coarse sands, respectively Thanet and Woolwich Beds.

The base of the formation is everywhere argillaceous, the clay occurring
either mixed with the sand or as thin impersistent clay seams or both. The
clay seams found in the Winterbourne Sand-pit (p. 191) extend about 20 ft
above the base of the beds, but this is exceptional for usually only a few

---

[1] The stratigraphy and palaeogeography of the Woolwich and [Reading Beds were reviewed
by Hester (1965).

impersistent streaks of clay occur in about the lowest 3 ft. Green colours are normally associated with this clayey 'bottom-bed', as it was called by Whitaker (*in* Hull and Whitaker 1861), though in places they are replaced by brown, either completely or in bands. The main mass of the Woolwich Beds consists of coarse, generally current-bedded, brown or grey sands speckled with dark-coloured glauconite, giving the sands a distinct 'pepper and salt' appearance. The sands are usually ferruginous and many thin ironstone lenses occur. A bed of hard nodular ironstone up to 4 ft thick is developed near the top in many places and is commonly fossiliferous. Prestwich recorded that this bed was once used for iron-ore, but Whitaker found that it was only quarried for roadstone. Whitaker erroneously stated that this ironstone was locally developed in the base of the Oldhaven Beds, but in the present area only derived ironstone occurs incorporated within the basal pebble bed of that formation (p. 181). It was almost certainly from this bed at Boughton Street that Prestwich (1850, p. 264) collected fossils.

The Woolwich Beds tend to be banded in colours of light greyish brown and brown. The bands reach an individual thickness of about 4 inches and are usually, but not invariably, irregular and poorly defined. This colour banding is a secondary effect induced after deposition of the sands, but a prominent crimson staining or spotting commonly found near the summit of the beds was regarded by Wooldridge (*in* Dewey and others 1925, p. 269) as being at least partly penecontemporaneous.

Conspicuous among the sands are tubular, almost segmented, ferruginous concretions, many of which combine to form complicated shapes. The central tube or core of the nodule is composed of leached sand and the skin is formed of limonitic sand, rather fragile to the touch. They reach approximately 1 ft in length, up to about 2 inches in breadth, and are roughly vertical. Constrictions in the tubes coincide with the bedding outside the nodules and the resulting almost segmented shape has an intestinal aspect, except that lateral protrusions and extensions give rise to grotesque shapes. Locally the nodules interlock and ramify through considerable thicknesses of the sand. Wooldridge (*in* Dewey and others 1925, p. 269) stated that they bore resemblance to tubular cavities found at the present time in the Laminarian zone of the sea coast, through the decay of the root-stocks of algae, but he also considered that they might have been formed by the work of Terebelloid worms. In either case, their presence implies that the beds were deposited in shallow water. Essentially similar nodules occur in Recent marine sands on Romney Marsh (pp. 283, 285) but there they appear to be of inorganic origin.

Locally above the ironstone bed, and forming the top of the beds where pre-Oldhaven Beds erosion has not cut too deeply, there is a succession of clay seams, for which Collins and Gill (1923, p. 307) recorded a total thickness of 6 ft; this appears to be the maximum proved in this district. These authors considered them to be of estuarine origin and this is confirmed by the subsequent discovery of *Corbicula cuneiformis* (J. Sowerby) in the uppermost clay seam which at Winterbourne (p. 190) is 1¼ ft thick and contains small sand lenticles. It forms the summit of the Woolwich Beds thereabouts. A poor exposure north of the Upper Ensinge Tunnel

(p. 192) showed two thin pipeclay seams amongst coarse and fine sands seen for 6 ft. It is not known whether these pipeclays are higher in the succession than the seam at Winterbourne or whether one of them is its representative.

The thickness of the Woolwich Beds at any locality depends upon the amount eroded prior to the deposition of the Oldhaven Beds. The 50 ft at Winterbourne is the greatest thickness known within this district. The beds are 27 to 38 ft thick at Canterbury and 15 ft in Trenleypark Wood (p. 193) and north-east of Wingham.

Most of the fossils collected from the Woolwich Beds have been recovered from the nodular ironstone seam near the top, but ghost fossil impressions, which show as light-coloured streaks in the sand, are commonly present below. The following list of selected fossils from the ironstone in the vicinity of Boughton Street and Winterbourne is largely based upon lists published by Prestwich (1850) and Whitaker (1872); for the most part the specimens have not been preserved, and according to their location and horizon (not known in detail) they may include an admixture of Oldhaven forms; where possible equivalent names in current use have been given by Mr. C. J. Wood: *Aporrhais sowerbyi*, *Brotia melanioides* (J. Sowerby), "*Calyptraea trochiformis*, Lam.", *Calyptraea sp.* (G.S.M. 44702–3), *Cerithium funatum* Mantell, *Euspira glaucinoides* (J. Sowerby), "*Fusus latus*, Sby.", "*Natica sp.?*", "*Pyrula, sp.*", *Arctica morrisi*, "*Astarte sp.*", *Corbicula cordata* (Morris), *C. cuneiformis*, "*Corbula longirostris*, Desh.", *Corbula regulbiensis*, *Glycymeris brevirostris* (J. de C. Sowerby), *G. plumstediensis* (J. Sowerby), *Nemocardium nitens* (J. Sowerby), *N. plumstedianum*, "*Ostrea sp.*", *Panope intermedia* (J. Sowerby).

OLDHAVEN BEDS

Prestwich (1850; 1854a) included part of the Oldhaven Beds in his basement bed of the London Clay. Whitaker (1866, p. 413), however, realized that the strata were separable from the London Clay, and called them after Oldhaven Gap on the Herne Bay coast. The name Blackheath Beds was given by Whitaker to the pebbly facies of the formation, but he did not extend the term to the two patches of pebble beds at Selling, in this district, as they were so small and so remote from other occurrences. This practice is continued here.

The Oldhaven Beds of this district have not been found to exceed $22\frac{1}{2}$ ft in thickness, except near Selling, where Prestwich put the thickness at 30 to 40 ft. Their outcrop is narrow and tends to be partly obscured by slips and clayey Head deposits derived from the overlying London Clay. An unconformity, which is more prominent than any other within the Eocene, occurs at their base. Its associated erosion appears to have been greatest in the southern and north-eastern parts of the Eocene outcrop; although at Canterbury and to the north-east some of the thinning of the underlying Woolwich Beds may be due to reduced deposition, the higher Woolwich Beds strata of the western Blean are apparently missing. The erosion may have been due to early movements of the Isle of Thanet anticline or an associated fold bringing the deposits within reach of wave activity.

At the base of the Oldhaven Beds there is a pebble bed composed largely of smooth, round or discoid, black flint pebbles, together with some fragments of ironstone up to about 3 inches long and usually angular, and abraded fragments of mauve or lilac-coloured friable sandstone similar to mauve sands in the Woolwich Beds northwards of the present district. Large blocks of ironstone from the Woolwich Beds are also incorporated in the pebble bed and these have been found together, so that they retain the appearance of a continuous but jointed bed about 3 ft in thickness. The Woolwich Beds sand formerly surrounding these blocks has been washed away by wave action which was insufficiently powerful to remove the blocks, which consequently have in some cases been lowered vertically rather than transported laterally. The flint pebbles have been washed under and over the blocks, as well as into fissures and cracks which traverse the ' bed ', and into cavities bored in the blocks. These cavities, which are commonly found near Canterbury, have about the same diameter (2 to 3 in), but not the length, as those in the Thanet Beds (p. 186), and as they have been made in hard ironstone, they cannot have been formed by burrowing worms, but were probably made by large boring molluscs.

The derived ironstone formed a much-jointed ' bed ', which could be traced for nearly 200 yd at the entrance to the tunnel on the Whitstable Branch railway. If this was originally developed about 40 ft above the base of the Woolwich Beds, as it is at Winterbourne (p. 190), then it must have been lowered about 13 ft by the winnowing process. Large blocks of ironstone in the pebble bed have, in every instance, been derived ; any ironstone that is indigenous occurs as small flattened biscuits above the pebble bed level. Whitaker (1866, p. 414 ; 1872, p. 266), and probably also Collins and Gill (1923, p. 301), thought the large ironstone blocks were indigenous, although the last-named authors recorded the pebble bed set in the base of the ironstone ; even when Whitaker found the ironstone *in situ* (1872, fig. 72, p. 268) he incorrectly placed the base of the Oldhaven Beds at its level. Incorrect interpretations of the horizon of the ironstone have therefore been made, particularly near Winterbourne and in sections (where the ironstone is often described as sandstone) recorded by Trimmer, Prestwich and Whitaker. On occasions the pebble bed has been noted in two bands, but in the absence of ironstone blocks it only exceeds a thickness of 2 ft in the outliers at Selling.

Except at Selling, beds of marine current-bedded firm greyish white to buff fine sharp sand occur above the basal pebble bed ; the bedding is well defined in the lower part, in contrast to the underlying Woolwich Beds. Impersistent seams of greyish brown clay and sandy clay, displaying changes in thickness, are common.

The preservation of the Eocene outlier at Selling appears to be due to the occurrence there of the Blackheath type of pebble beds, but the reason for the presence of the latter is uncertain, although it is connected with the history of the pebbles in the bed generally. Stamp (1921, p. 73) considered that the unusual roundness of the pebbles was the result firstly of abrasion on the shore of an island in the centre of the present-day Weald, from which they were derived, and secondly of continual agitation during deposition as off-shore bars, such as those described by Bromehead (*in* Whitaker and

Davies 1920, p. 30). It appears unlikely, but is admittedly not impossible, that an off-shore bar would have formed at Selling only (Wooldridge 1926, p. 176), but the original speculation of Bromehead (1922, p. 325) that the pebble beds formed opposite gaps cut in the Chalk during Eocene time, and which remain today as the river gaps, may perhaps have some substance.

The fauna of the beds is sparse.

LONDON CLAY

The term London Clay dates back to the days of William Smith. The uniformity and thickness of the formation over wide areas of southern and eastern England reveal that the intermittent earth movements which had affected deposition in the Eocene below now changed to a steady sinking. The Eocene depositional basin was extended and within it the London Clay accumulated as marine muds (Davis and Elliott 1958).

Within the present district the formation mainly caps the Blean upland, where its outcrop is forested. The Oldhaven Beds were eroded so that, at least near Winterbourne, their surface was gently channelled to a depth of about 18 inches before the basement bed of the London Clay was laid down. This basement bed, which is distinct from that recognized by Prestwich, consists of up to 2 ft of grey sandy and slightly glauconitic clay, with small rounded pebbles scattered very sporadically throughout, and some ferruginous nodules. The mass of the clay above is bluish grey in colour, shaly and tenacious; it contains pyritized fossil wood and selenite. Septarian nodules (cementstones) are rarely seen. At the surface the clay weathers to a reddish brown colour. Some 250 ft of London Clay occur within this district out of a total for East Kent of about 480 ft.   J.G.O.S.

DETAILS

THANET BEDS

Hanslett's Forstal—Lees Court—Brenley Corner.

Two boreholes for water at Elverland (978583), 1150 yd S. 15° W. of Hanslett's Forstal, reached Chalk at recorded depths of 42 ft and 33 ft. The latter figure appears to be the more reliable. Both boreholes are sited upon a patch of Head Gravel, the thickness of which (about 4 ft) is included in the above figures. The beds in the outliers at and north-north-west of Painter's Forstal are in a much weathered condition.

A tongue of Thanet Beds extending southwards beside the Faversham–Ashford road to near Copton is mostly concealed by brickearth. A well at the Faversham Waterworks (013596), 850 yd W. 13° N. of Westwood Court, after passing through $4\frac{1}{2}$ ft of brickearth, proved $25\frac{1}{2}$ ft of Thanet Beds with the Bullhead Bed at the base, before entering the Chalk. The small outliers south of Copton (017588) and near the roadway north of North Street (013584) appear to be piped masses. At Lees Court (023562) the long outlier, composed of grey clays and brown and green fine loamy sand, passes southwards and northwards into Clay-with-flints and north-westwards into Brickearth. The outlier appears to have contributed material to all these drifts. The margin of the outlier at Owens Court is extensively piped. Some $3\frac{1}{2}$ ft of grey and buff mottled sand and greenish speckled sands with a northerly dip, seen (028578) 240 yd E. 40° S. of Owens Court, do not seem to be entirely in place.

The Bullhead Bed and underlying chalk in the railway cutting at the road-bridge (029602) 1100 yd W. 17° N. of Brenley Corner were described by Whitaker (1872, p. 85). The Bullhead Bed, containing many large flints, lies just below the level of the railway lines at the bridge and it rises irregularly in a south-easterly direction. The clayey sands and loams overlying it are poorly exposed for a thickness of 15 ft. The beds exposed in the cutting tongue southwards nearly to Sole Street House, but are almost wholly concealed by Brickearth.

In the outlier at Brenley Corner only the lowest glauconitic loams and the underlying Bullhead Bed remain, except in pipes.

The Thanet Beds were noted by Whitaker (1872, p. 85) in the railway cutting south-east of the level-crossing near Colkins. He recorded that initially only a little light-coloured sand and the Bullhead Bed intervene between chalk and drift (p. 237), but near the road-bridge (038587) 700 yd S. 30° E. of the level-crossing the junction with the Chalk is extensively piped and the overlying beds warped by collapsing into the pipes. South-eastwards the base suddenly becomes more regular and the flints of the Bullhead Bed, which are mostly of a small size, rest upon a band of tabular flint 3 to 6 inches in thickness. Hughes (*in* Whitaker 1872, p. 85) noted that where this tabular flint was unbroken it appeared to have protected the underlying chalk from solution, for it capped bosses of chalk protruding into the Thanet Beds. Where the tabular flint was broken, it sank into the decomposing surface of the Chalk. In 1951 only a tiny section (040585) could be seen 200 yd south-east of the road-bridge, 580 yd S. 15° E. of Brenley House, where a marked flint band in the Chalk (p. 152) was overlain by the Bullhead Bed, 4 inches in thickness, at the base of 3 ft of green glauconitic loam.

Pipes of Thanet Beds, 6 to 12 ft deep, are seen in the chalk in a pit (044587) 450 yd W. 43° S. of Wellbrook Farm (p. 152). These pipes are filled with greyish green and buff mottled sandy loam and the Bullhead Bed, and between the Bullhead Bed and the Chalk there are 2 in of very dark brown manganiferous clay which are associated with pieces of tabular flint and were presumably formed of the residue from the chalk dissolved during the formation of the pipe. The mass of Thanet Beds here is assumed to be connected, below the Brickearth cover, with the main outcrop at Nash Court. At this locality the dip steepens slightly northwards, so that the sands near the top of the beds crop out in the fields surrounding the house. Sands at much the same horizon occur again in Boughton Street, where they lie mostly to the north of the roadway and are occasionally exposed for a few feet only in the banks of the trackways, underlying up to 4 ft of washed sand.

The water borehole (053594) 860 yd E. 44° S. of Nash Court proved, below Brickearth (p. 237), 10 ft of "green silty clay" of the Thanet Beds, before entering the Chalk. In the well at the disused waterworks (054592) 1170 yd S. 44° E. of Nash Court, Brickearth and Thanet Beds totalled 45 ft, their individual thicknesses being unknown.

Only the lowest 3 ft of the beds, resting evenly on the Chalk (p. 152), are exposed in the pit (051583) 250 yd E. 30° S. of Boughton Church Farm, but 7 ft of well-bedded greyish green and buff mottled sandy loam, with a single line of green-coated flints at the base, were noted in the pit (052581) 480 yd E. 43° S. of this farm. There were 1½ in of clay between the Bullhead Bed and the Chalk.

Selling—Shottenden—Upper Ensinge—Chartham Hatch.

The Thanet Beds are well represented in the outliers near Selling but they are indifferently exposed. Whitaker (1872, p. 86) saw the base of the beds in the roadway south-east of Gushmere. The base was also seen at the north end of the tiny outlier south of Hogben's Hill and in a chalk-pit (048554) south of the

N

road 1050 yd W. 10° N. of Stonestile Farm, where it occurs in a pipe and also at the extreme top of the chalk face. Greenish and buff mottled loam near the base was noted for 4 ft in a temporary digging (048561) 540 yd S. 43° E. of Norham Farm. In the lane cutting (047545) 340 yd E. 38° N. of the inn at Shottenden 1½ ft of dark green loam with the Bullhead Bed at the base are separated by a thin clay seam from 2 ft of pale greyish green loam above.

The uppermost sands of the Thanet Beds, overlain by later deposits, occur towards the two highest points of the outlier at Selling. An old sand-pit (045555) 150 yd E. 43° S. of the cross-roads in Perry Wood shows 7½ ft of golden and light brown, grey-mottled, medium to fine-grained, well-bedded, glauconitic sand below 1 to 2½ ft of pebbly wash.

The beds were better exposed in the middle of the last century. Whitaker (1872, p. 86) recorded a section which he stated was "less than a quarter of a mile southward from the church", and is presumed to be 100 yd S. of the church (Selling church) at Hogben's Hill (038567): the base-bed (clayey greensand) was seen for about 5 ft and was overlain by about 5 ft of alternations of sandy and clayey beds, below 4 or 5 ft of light grey, fine, soft sand.

The uppermost 3 ft of the beds, consisting of medium to fine golden brown and grey mottled sand, underlying Woolwich Beds (p. 189), are exposed behind a cottage (044550) 365 yd N. 10° W. of The Mount. It is doubtful, however, if these sands are in place.

Very little remains today of the sections in the railway cuttings at either end of the Upper Ensinge Tunnel. Whitaker (1872, p. 88) described the western cutting as follows, the letters in brackets referring to his subdivisions of the beds (p. 176):

"At the lower (western) end there is grey clayey sand, somewhat greenish, with crumbling fissile sandstone near the top; the lowest part probably belonging to the base-bed (a), and the rest to the alternations of sand and clay that succeed it in the eastern part of Kent (b). Where the level of the ground rises, by the bridge, sand comes on above; it is fine, buff, with thin layers of green earth near the top, and most likely is the same as the bed that forms the whole of the formation in the western part of the county (c). East of the bridge the level of the ground still rises and the upper part of the sand gets more clayey, passing up into the marly division (d), which at the eastern end of the cutting is capped by fine grey and brown sand (e) from which much water is thrown out in parts."

This section is notable as one of the few records of the presence of bed (c) in the western part of the Thanet Beds outcrop. Whitaker recorded that a crushed *Nautilus* and a fish skeleton were obtained from this cutting, and *Cucullaea,* in silicified form, from the garden of Upper Ensinge itself; he also stated that the cutting at the eastern end of the tunnel showed a similar section. In 1952 only 6½ ft of light brown to greyish buff mottled loamy sands, with about 8 in of finely laminated dark green and light grey sands at the base, were exposed behind a plate-layer's hut in the eastern cutting and other beds were indifferently exposed in drainage channels which descended the cutting sides. Haynes (1958, p. 86) named foraminifera collected from one channel, 30 yd E. of the tunnel.

The beds tongue southwards on the crest of the ridge to Old Wives Lees, the upper sands occurring on the higher ground, with a series of springs at their base. The sands present at the highest point of the ridge are thought to lie close to the top of the beds.

The railway cutting (086561) 700 yd W. of Nickle Farm is also now obscured, but Whitaker (1872, p. 88) noted "12 feet of green-grey marly sand, rather hard, with a layer of stony concretions in the middle, and a layer of pinkish-grey sand (c?) at top. Above this are clayey and loamy beds (d)." At the eastern end of this cutting the Bullhead Bed rested irregularly upon chalk. The sands in the higher part of the Thanet Beds occur hereabouts and extend eastwards, with Woolwich Beds in places, below the gravel-capped ridge at Chartham Hatch. The beds also crop out northwards to Denstead Farm. An old sand-pit (094561) beside a woodland trackway 150 yd E. of Nickle Farm, exposes 7 ft of thinly-bedded fine khaki sand with an irregular buff mottling and occasional small ($1\frac{1}{2}$-in) hard dark brown ironstone lenses, underlying 2 ft of sandy wash with flints. A thin impersistent band and pocket of decalcified sandy limestone at the eastern end of the face yielded abundant *Arctica morrisi,* and *Arctica planata, Cucullaea decussata* and *Dosiniopsis bellovacina.*

The sands are exposed, again with overlying Woolwich Beds (p. 192), in a section (105570) 300 yd N. 20° E. of the cross-roads at Chartham Hatch, where 12 ft of fine gold-brown speckled sand were noted. The sands are stained here and there by manganese and in places there is a dark reddish brown mottling. Solution holes 320 yd E. 40° N. of Petty France (108575) indicate that the Chalk is at no great depth. McDakin (*in* Whitaker 1908, p. 49) found these holes to be 25 to 29 ft deep and the Chalk to be 25 ft down.

Harbledown—Canterbury—Fordwich—Bekesbourne.

The extensive outcrop of Thanet Beds in the valley leading eastwards to Harbledown is covered by a variety of soils ranging from heavy clays to sands. Some of the sands in the upper parts of the beds are exposed in the road cuttings east of Bigbury Camp and at Upper Harbledown. They were also noted up to 4 ft in thickness, buff and mottled brown in colour, in a section (113582) 250 yd W. 40° N. of Poldhurst Farm. In an old chalk-pit (116569) 950 yd N. 4° W. of Howfield Farm the Bullhead Bed is overlain by 3 ft of greenish grey dark speckled sandy loam beneath 3 ft of brickearth ; below the Bullhead Bed there is a 1-inch layer of tabular flint, which, although much broken, is of secondary origin, because impressions of small fissures in the chalk are preserved on one of its surfaces.

The railway cuttings (131573) 900 yd S. of the church with tower at Harbledown are now obscured but some details were recorded by Whitaker (1872, p. 89).

On the north bank of the Stour at Canterbury the entire outcrop is obscured by drift. A borehole in nurseries (144587) 400 yd N. 20° W. of the northern station in Canterbury (Canterbury West Station) encountered $20\frac{1}{4}$ ft of Thanet Beds between the drift (p. 238) and the Chalk. According to the borehole record the beds consist of clay and sand, 8 ft, resting upon sandy clay, $12\frac{1}{4}$ ft. In a digging for foundations (144594) 120 yd E. 20° S. of the entrance to the tunnel on the disused Canterbury–Whitstable railway some 5 to 6 ft of fine buff and grey sand with a few small ironstone nodules were noted. Whitaker (1872, fig. 44) recorded about $4\frac{1}{2}$ ft of fine pale grey sand, iron-shot at top, almost white in parts and bedded, below Woolwich Beds (p. 192), at the mouth of this tunnel (143595). A borehole at a derelict monastery (152599) 900 yd N. 29° E. of the church south of Hales Place encountered Thanet Beds at a depth of 98 ft, after passing through London Clay, Oldhaven Beds (p. 197) and Woolwich Beds (p. 193). The Thanet Beds were recorded as follows, in descending sequence: " Live sand ", 26 ft ; " Green sand with clay ", 25 ft ; " Blackish bed ", $3\frac{1}{2}$ ft ; " Green sand ", $55\frac{1}{2}$ ft ; and " ? Flints ", $\frac{1}{4}$ ft. These total $110\frac{1}{2}$ ft and their assignment to the Thanet Beds is based upon a reinterpretation of the published details (Whitaker 1908, p. 96). Another borehole (157595), south of the railway, 1020 yd E. 19° N. of the last-mentioned church, proved 31 ft of " blue clay " of the Thanet Beds, below drift and resting on chalk.

The beds in the outlier south of Canterbury are well exposed in the top of Denne's Limeworks Pit (149567) near the abandoned railway line 700 yd S. 10° E. of the southern railway station. Here the Bullhead Bed at the base rests irregularly upon the Chalk ; it ranges up to above 1 ft in thickness and it is overlain in upward order by 4½ ft of fine greenish brown and buff mottled glauconitic sand, 2½ ft of brown and buff mottled fine loamy sands with clay patches and ferruginous streaks, 6 ft of fine greenish brown sand and sandy clay in alternating beds about 1 ft thick, and 1 ft of surface soil.

In the disused railway cutting to the east of the above pit, 900 yd S. 31° E. of the same station, 6 ft of golden and khaki mottled glauconitic sandy clay with clay seams are overlain successively by 7 ft of greyish brown slightly clayey fine sand with a few ferruginous streaks and ironstone lenses, becoming less sandy upwards, and 3½ ft of brown sandy clay which has weathered to loam at the top, and with, also near the top, occasional green patches infilling large borings similar to those described below, the whole being overlain by 1 ft of soil. The level of the borings is approximately 25 ft above the Chalk.

The next cutting of the abandoned railway (162560) eastwards of the above, 950 yd W. 18° N. of Milestone Farm, shows 9 ft of brown mottled light fawn fine sands below 1 ft of grey and buff mottled very glauconitic fine sand containing occasional green clayey streaks which is harder than the overlying 11 ft of greyish brown and buff mottled, glauconitic, shaly, sandy clay with some stiff clay seams ; 2 ft of weathered grey grauconitic clay above complete the section. In the upper part of the fine sands at the bottom of the exposure there are many borings which range up to 3 to 4 inches in diameter and extend to a depth of 2 ft and occasionally more, while one exceeds 4 ft in length ; they are filled by glauconitic sand similar to the overlying bed and their margins are usually sharp although some have a lining of green clay. A single dark-coloured partially worn flint pebble was also found at this horizon.

The base of the Thanet Beds was temporarily well exposed during widening of the Old Dover Road, 1050 to 1550 yd W. 40° N. of Milestone Farm. The Bullhead Bed was overlain by up to 4 ft of green glauconitic loam with brickearth above.

On the rising ground to the north-east of Canterbury numerous small springs are thrown out at the base of the sands occurring in the upper part of the Thanet Beds. These sands, fine in grade and golden brown to fawn in colour, are exposed for 10 ft, with gravel overlying (p. 270), in a disused pit (164585) 1000 yd N. 27° W. of the Mental Hospital adjacent to the road to Wingham. Whitaker (1872, pp. 90–1), states that nests of *Corbula regulbiensis* occurred in this pit and that a cone of *Pinites macrocephalus* had been found by G. Dowker.

The uppermost 3 to 4 ft of the Thanet Beds, consisting of fine brown and buff mottled glauconitic sands underlying the Woolwich Beds (p. 193), were noted in a small sand-pit (174590) on the golf course 1150 yd S. 42° W. of Fordwich church.

Some 42 ft of Thanet Beds, recorded as " mottled clay, sandy at base ", were passed through between the terrace deposits (p. 272) and the Chalk in a borehole for water (179563) 330 yd N. 13° W. of Hode Farm.

In the railway cuttings near Little Barton Farm and Bekesbourne the Thanet Beds are now only poorly exposed. Whitaker (1872, p. 91) recorded dark bluish grey marl, more sandy in parts, with fossils, overlain by light brownish grey marl, 10 ft, below sandy marl, with irregular beds of crumbling sandstone, light grey, somewhat greenish, with fossils, about 11 ft, in an exposure which was apparently near the road-bridge (169570) in the cutting south of Little Barton

Farm. Whitaker noted that the colour changes closely followed the configuration of the ground and considered the above 'beds' to be due to discolouration as a result of weathering, rather than divisions of the marl. In the cutting (188561) 150 yd W. 20° N. of the station (Bekesbourne Station) at Bekesbourne Hill 10 ft of light brown loamy sands were noted below 2 ft of well-bedded grey loamy clay. Large borings, such as those described above, infilled with highly glauconitic sands, were noted in the top of the sands. Ironstones were also seen at this level. The following fossils in the Geological Survey and private collections are recorded as having come from "Bekesbourne Cutting near Canterbury": *Schizaster* cf. *cuneatus* ; *Arctica morrisi, Cyrtodaria rutupiensis, Eutylus cuneatus* and *Pholadomya oblitterata*.

The Littlebourne Borehole (196576) probably penetrated nearly the whole of the Thanet Beds, but the overlying Woolwich Beds do not crop out nearby to provide a check on the horizon at the surface. The borehole passed through 44 ft of fine buff to brown and sometimes mottled fine sand, resting on 57½ ft of blue, brown or grey sandy clays with fine glauconitic clayey sands at the base, on Chalk. Only chip samples were taken, and the Bullhead Bed was not represented in the specimens.

According to Whitaker (1908, p. 205), a water borehole (175600) 590 yd W. 17° N. of Fordwich church encountered, below 15 ft of alluvium and gravel, 38 ft of Thanet Beds, of which the uppermost 25 ft consisted of blue clay, becoming green and more sandy at the base ; the remainder consisted of sandy strata, probably developed near the base of the formation.

The top 20 ft of Thanet Beds and the overlying Woolwich Beds (p. 193) are worked in the Trenleypark Wood Sand-pit (192592) (Plate VB), 850 yd W. 25° N. of Swanton Farm. The sand is fine, grey and glauconitic with, low in the face, ill-defined and irregular ferruginous buff-stained bands, which become more regular in the uppermost 5 ft. At this same level occur ironstone nodules resembling borings infilled with Thanet sand. The beds are irregularly ironshot.

A few feet of fine sand were noted in road cuttings near Fordwich and Elbridge House. Whitaker (1872, p. 92) found '*Cyprina*' in sandy marl on the north side of the stream at the latter locality.

Stodmarsh Borehole—Littlebourne—Wingham.

The Stodmarsh Borehole was sited upon gravel resting approximately at the top of the Thanet Beds, so that the thickness of 41 ft assigned to these beds and the gravel by Lamplugh, Kitchin and Pringle (1923, p. 175) is certainly incorrect. The Thanet Beds are approximately 100 ft thick hereabouts.

At the north-eastern buildings of Garrington Farms (206563) the Chalk (p. 160) is overlain by the Bullhead Bed below some 3 ft of greenish grey and buff mottled loamy sand ; the Bullhead Bed is gently undulating and the green-coated flints in it are mainly small, although some reach a length of 9 inches. The Thanet Beds in the outliers east of Littlebourne are much obscured by brickearth. The Bullhead Bed, according to Whitaker (1872, p. 94), was irregularly piped in the chalk-pit (250567) 130 yd E. 10° N. of Witherdens Hall. F. H. Edmunds noted on his map that 6 ft of red, buff and grey clayey loam were exposed in a pond bank (263559) 900 yd S. 13° E. of Twitham.

To the north of Wingham the Thanet Beds occupy the face of an escarpment which is capped by Woolwich Beds and Brickearth. On the slope of the river-cliffs (235595) 700 yd N. 37° W. of Wenderton Manor House soft shelly sandstone and grey and brown fine loam with moulds of fossils occur as debris.

Whitaker (1872, p. 92) listed "*Aporrhais? Astarte, Corbula Regulbiensis, Cyprina Morrisii,* and *Nucula*" from this locality. An old sand-pit (238599), 200 yd W. 35° N. of Deerson Farm, showed 4 ft of medium to coarse, firm, brown to yellow glauconitic sand and soft sandstone, in part ferruginous. At this farm a temporary section showed 6 ft of golden brown and grey glauconitic fine sand. The junction of the Thanet Beds with the Woolwich Beds which comprise a tiny outlier at Wenderton Manor House is exposed in a sand-pit (238592), and 150 yd to the west 5½ ft of greyish brown and buff mottled fine sand were noted. Whitaker (1872, p. 92) recorded "casts of *Cyprina*" from hereabouts.

The Bullhead Bed is seen in an old chalk-pit (242581) 620 yd N. of Wingham church and again, with up to 3 ft of greenish grey glauconitic very fine sand overlying, in a bank about 100 yd to the west. A drive cutting (245581) 650 yd N. 25° E. of this church shows nearly 3 ft of grey silty clay below 3 ft of pebbly drift, and about 100 yd northward of this locality another drive cutting exposes 2½ ft of light greyish brown very fine sand with moulds of fossils. This appears to be part of the "considerable thickness" in a "semi-indurated" condition, with impressions of shells, recorded by Prestwich (1852, p. 246). Referring to the sand-pit (246583) 600 yd S. 30° E. of Little Wenderton Farm, Whitaker (1872, p. 93) remarked that the sands of both the Thanet Beds and the Woolwich Beds were fine and sharp and of much the same colour, but those of the Woolwich Beds contained large dark grains while the Thanet Beds contained many small dark grains. He stated also that the Thanet Beds appeared to be more compact and were bored at the top, with sand of the Woolwich Beds filling the borings. In 1953 fine grey glauconitic sands of the Thanet Beds were seen for 1 ft beneath coarse brown clayey sand which, although in a weathered condition, was characteristic of the lowest Woolwich Beds; in the south-eastern corner of the workings 5 ft of Thanet Beds sands were exposed.

At a locality (247577) some 520 yd E. 27° N. of Wingham church 6 ft of grey streaked with buff, very fine, loamy sand with ghost fossils are exposed, at an horizon near the base of the beds. Some 400 yd to the east and 900 yd E. 11° N. of this church, Whitaker (1872, p. 93) described the following section (250576), showing the development of the lower part of the Thanet Beds in the eastern part of the district:

"
		FEET.
	d. Sandy marl, broken up (by weathering) ...	6
Thanet Beds.	*b.* Buff bedded clayey sand, the beds about a foot thick, dried hard and splitting into blocks ... ... about	8
	a. Base-bed. Marly sand with green grains, with about 3 inches of green-coated flints at the bottom and a few above (besides some small broken pieces) ... ... about	1½
Chalk.	Here and there a flint at top, and a few just below; cut into ... ... about	2½ "

At Shatterling, a pit (262583) south of the road 600 yd W. of the inn exposes over 4 ft of fine, khaki, glauconitic sand with ferruginous patches and streaks, underlying up to 4 ft of sandy drift with pebbles.   J.G.O.S.

WOOLWICH BEDS

The Mount—Boughton Street—Winterbourne—Upper Ensinge.

Woolwich Beds, estimated to be between 20 and 25 ft thick, with overlying Oldhaven Beds (p. 194) crop out around The Camp (Earthwork), in Perry Wood. They are now nowhere exposed, but Whitaker (1872, p. 162) stated that in the quarry (043556) 200 yd W. 25° S. of the cross-roads in Perry Wood they consisted of coarse sand, grey with dark grains, false-bedded, the lower part having the salmon tint common in this bed around Canterbury and showing some small faults, about 16 ft thick, overlain by a thin inconstant bed of soft sandstone with mollusc borings. This section was presumably that seen by Prestwich (1854a, p. 110). In another part of this quarry Whitaker stated that there was " pale yellowish sand, more than seven feet thick, between the pebble-bed and the Woolwich Sand ". From his description this bed is thought to be at the top of the Woolwich Beds, at the same level as the thin inconstant bed of soft sandstone mentioned above. A tiny outlier of coarse sand caps a mound 150 yd south-south-east of the above cross-roads.

In the north end of Conduit Wood the beds are probably overlain in part by Oldhaven Beds and both are much obscured by Head. On the western slopes of the hill (044550) behind a cottage 365 yd N. 10° W. of The Mount 3 ft of medium to fine sand of the Thanet Beds are overlain by 2 ft of medium to coarse clayey sands, which are ferruginous and hard in places, and are interpreted as basal Woolwich Beds. An occasional pebble is present near the junction, but with such a small exposure considerable doubt exists on whether the sands are strictly in place. Whitaker (1872, p. 162) stated that 'casts' of fossils were found in coarse iron sandstone at this locality.

On the western side of the Blean the Woolwich Beds form the steeper part of the escarpment slope. In Boughton Street (062593), immediately east of a chapel 580 yd E. 16° S. of the village church with spire, 6 ft of coarse greenish glauconitic sand with some ironstone were exposed in a trench. About 250 yd eastwards, on the north side of the road (064593) 820 yd E. 15° S. of the above church, coarse greyish brown loamy glauconitic sand is overlain successively by 7 ft of coarse pale greyish green glauconitic sand, 3 ft of medium to coarse greyish brown glauconitic sand with a loamy layer at the base and 2½ ft of Oldhaven Beds. In this pit Whitaker (1872, p. 165) recorded the following section of Woolwich Beds below about 1 ft of Oldhaven Beds:

" *g.* Thin irregular bed of purplish sand and stone. The pebbles from above have been splashed into the sand just below this.

h. Light-grey sand. The upper part almost white, being composed of grains of transparent colourless quartz and of pale buff quartz. The middle part false-bedded. The lower part bedded (by layers of clay), with here and there a small pebble. 5 or 6 feet, not clearly divided from the bed below.

i. Coarse pale green sand, about 5 feet.

k. Irregular alternations of brown and grey sand, the former chiefly at the upper part, passing down into:—

l. Light-grey sand, about 6 feet."

The index marks above are the same as Whitaker (1872, p. 267) gave to beds exposed in another pit on the opposite side of the road. His description of this pit was as follows:

Oldhaven Beds, nearly 2 ft, plus 10 ft nearby.

" *g.* Dark purple-grey sand, partly hardened into stone at top, not constant, passing down into the next.

   h. Light-grey fine sand, with a thin layer of light-grey clay in the middle
      and another at the bottom ; 3½ feet (together with the bed above), not
      clearly defined from—
   i. Coarse pale green-grey sand, with a thin layer of clay a foot down ;
      the lowest 18 inches greener and showing false-bedding ; 6 feet or more.
   k. Like sand, more or less ironshot ; 7 feet.
   l. Like sand, not so greenish as *i* ; false-bedded, 4 feet."

The two above sections are probably those described by Trimmer (1841) ; they
reveal the rather variable nature of the uppermost parts of the Woolwich Beds
on the western side of the Blean and should be compared with the section at
Winterbourne (below). The absence of the ironstone horizon is notable, but it is
not known whether this is due to failure of the bed or whether the sections
were not deep enough to reveal it.

A small pit at Horselees (064589), 1100 yd E. 33° S. of the church with spire
at Boughton Street, was also noted by Whitaker (1872, p. 165) : coarse grey sand,
partly pinkish, was overlain by very light grey rather fine sand with a few dark
grains and a layer of pipeclay at the bottom and another higher up, 4½ to 6½ ft
thick, below a thin irregular layer of purple-grey sand, hardened here and
there and almost cemented on to the overlying Oldhaven Beds, with a few
pebbles splashed into it.

In the disused sand-pit (067579) at Winterbourne, 1750 yd E. 25° N. of
Selling Station, some 25 ft of Woolwich Beds are exposed: a lower face shows
10 ft of current-bedded banded dark brown and khaki coarse sands with some
layers rich in glauconite, while on the north side of the pit a higher face shows
10 ft of coarse khaki-brown current-bedded glauconitic sand containing irregular
tubular semi-segmented nodules, with a crimson mottling particularly 3 ft
below the top, browner in the uppermost 2 to 3 ft and containing three
impersistent irregular bands of iron-pan, overlain in order by 1 in of pipeclay,
1 to 3 ft of light and dark brown and golden brown well-bedded fine sand with
small lenticles of ironstone, up to 2 ft of nodular ironstone, much shattered
by weathering (impersistent), and 4 ft of dumped sand. Elsewhere in the pit
the beds are affected by superficial trough faulting and are overlain by Head
(p. 226). Collins and Gill (1923, p. 306) found vivid red concretions, probably
from one of the above ironstones, in the woodland north of this pit, and
Whitaker (1872, p. 166) recorded green and brown coarse sand, clayey at the
base with a few flint pebbles, resting upon Thanet Beds in the road cutting
north-west of the pit.

A further abandoned pit (065576) at Winterbourne, 1450 yd E. 15° N. of
Selling Station is in Woolwich Beds and Oldhaven Beds (see Fig. 6, p. 196). A
temporary excavation in the pit originally exposed the upper surface of the
Thanet Beds, the Woolwich Beds section being as follows:

			Ft
	Oldhaven Beds:	8 ft to	18
10.	Grey glauconitic sandy clay with small lenses of leached sand with *Corbicula cuneiformis* (J. Sowerby)		1¼
9.	Gritty iron-pan	6 in to	1
8.	Greyish brown medium-grained sand with light grey gritty patches and darker streaks, and golden brown sand ; ironstone lumps		5½
7.	Hard concretionary ironstone with streaks and lenses of leached grey sands, combining northwards with bed 5 to become 2 ft of ferruginous sands ...		2¼

Ft

6. Coarse greyish brown sands ; dying out to north ...   1¼
5. Ferruginous coarse sands, occasionally hardened to
      ironstone with tubular nodules, lighter coloured
      markings, crimson stains and an occasional pebble   3½
4. Coarse   grey   speckled   current-bedded   sands.
      Ferruginous stains at top, with numerous tubular
      nodules and lumps   10
3. Lighter grey sands as bed 4, with bands of darker
      brown colour in places and a few small clayey
      streaks at bottom   4
2. Sands similar to beds 3 and 4, but more glauconitic,
      and clay seams and streaks more common ;
      irregular darker brown banding   4
1. Medium grained very glauconitic brown sands, often
      buff mottled, with clayey seams common, becoming
      coarser and with increase of clayey seams at base
      where occasional pebbles occur   17
      Thanet Beds. Just exposed.

Whitaker described the beds as follows : *a*. Loamy soil. *b*. Fine brown sand ;
beds above " on the eastern flank of the wooded hill, about a mile north
of the Ensinge Tunnel ". This cannot be precisely located but was probably
200 yd north of the above, on the side of the knoll known as Summerhouse
Hill. He erroneously regarded all the beds above and including the ironstone
as Oldhaven Beds. His diagram is copied in Fig. 5, with altered ornament.

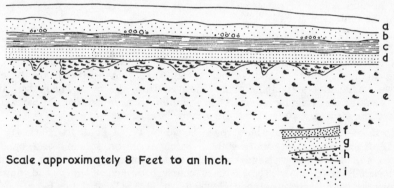

Scale, approximately 8 Feet to an Inch.

FIG. 5.   *Section in Woolwich Beds and Oldhaven Beds in Winterbourne Wood, near
Boughton Street.   After W. Whitaker*

Whitaker described the beds as follows: *a*. Loamy soil. *b*. Fine brown sand :
bottom part darker, clayey, and with a few small flint pebbles ; a foot or more.
c. Grey clay, a foot or more. *d*. Light grey rather coarse sand, a foot or
more. *e*. Brashy sandy brown ironstone, in parts haematitic, more than five
feet. With pipe-like fossiliferous masses (almost wholly made up of casts
of shells in ironstone) at top and 15 inches deep at most. *f*. Ironsand, partly
crimson at bottom. The iron probably derived by filtration from above, about
6 inches. *g*. Light grey sand. *h*. Half-inch layer of iron-sandstone, succeeded
by ironsand, 2 feet including bed *g*. *i*. Coarse sand, brown at top, of lighter
colour lower down. Beds *c* to *i* inclusive are now regarded as Woolwich Beds
and bed *b* as Oldhaven Beds.

Another old sand-pit (Scoggers Hill Pit) (064572) 1250 yd E. of Selling
Station is now much obscured and only the uppermost Woolwich Beds are
exposed ; about 4 ft of coarse brown gritty sands with irregular areas of lighter
colour, tubular concretions and streaks of pipeclay, are overlain by a thin seam
of pipeclay and then by 2 ft 10 in of laminated golden medium-grained sand
with an impersistent pipeclay seam and lenticular ferruginous sandstone in the
uppermost 2 ft, below 3 ft of ferruginous-stained grey medium-grained sand,
passing up into 1 ft 4 in of grey and buff sandy clay, beneath 1 ft 3 in of
disturbed Oldhaven Beds.

An old sand-pit (066564) 300 yd N. 25° W. of the western end of the Upper
Ensinge Tunnel reveals the following section in upward order : coarse sands,
hardened in places, 1 ft seen ; thin seam of pipeclay ; fine grey to buff sand
becoming coarse and buff upwards with a thin ferruginous seam in the upper
part, about 2 ft ; buff to greyish brown fine sand with a light grey clay seam
at the base, about 6 in ; grey clayey fine sand, $1\frac{1}{2}$ ft ; fine to medium-
grained brown and grey mottled sand, $1\frac{1}{2}$ ft.

Chartham Hatch—Upper Harbledown—Hales Place.

The Woolwich Beds are quarried (089575) 350 yd W. 12° S. of White Wall,
where some 12 ft of current-bedded coarse glauconitic sand, with irregular
brown-stained bands approximately 6 in apart, are exposed below Brickearth
(p. 237) and Head (p. 226) ; ill-defined darker patches, resulting from an
increase of glauconite in the sand, and tubular concretions were noted.

A pit (089578) behind cottages 440 yd N. 35° W. of White Wall was
described by Whitaker (1872, p. 166). He stated that 5 ft of coarse whitish
false-bedded sand rested on about 7 ft of pale greenish or brownish sand with a
pale salmon colour in places. The surface was covered by a drift of pebbles
from the Oldhaven Beds.

Up to 6 ft of sand from the Woolwich Beds are exposed below river gravel
(p. 266) and resting upon Thanet Beds (p. 185) in a pit (105570) 300 yd N. 20° E.
of the cross-roads at Chartham Hatch. The sands are coarse, hard and gritty,
with small clay patches and an occasional small flint pebble at the base. A
brown ferruginous staining, probably derived from the overlying gravel, affects
the beds and an irregular darker brown staining is also present. In parts of the
quarry the beds are cut out by the channelling of the gravel.

North of the main Canterbury road 700 yd N. W. of Poldhurst Farm (107586)
$7\frac{1}{2}$ ft of coarse brown sands with a patchy crimson staining and bands of
harder and darker brown sand and some streaks of clay, were noted underlying
Oldhaven Beds (p. 195). Some 3 ft of coarse brown glauconitic sand, hardened
in places, were seen in Upper Harbledown (118583), 830 yd E. 17° N. of
Poldhurst Farm.

A disused sand-pit (142593) in nursery grounds 400 yd E. 3° S. of Beverley
Farm exposed 14 ft of coarse grey glauconitic sands, gently current-bedded
in part, with an irregular reddish brown ferruginous staining, particularly near
tubular concretions and in the uppermost 6 ft, beneath channelled Oldhaven
Beds (p. 196). Some 50 yd N.E. of the last pit, near the entrance to the disused
railway tunnel, is a large old sand-pit which was recorded and figured by
Whitaker (1872, p. 167). He described the Woolwich Beds, which were completely exposed, as " Coarser sand, of a very pale yellowish or greenish tint,
the false-bedding shown by the frequent lines of green grains ; the lower part
darker and more yellow, and with a few pieces of iron-sandstone. The lowest
2 feet more clayey greenish and rusty, with a three-inch layer of clay . . . at
one part, and a very few small flint-pebbles at the bottom." On Whitaker's

scale-diagram of the section, the 3-in clay seam was shown as impersistent and at a level about 2 ft from the bottom of the formation, the total thickness of which, measured from the diagram, was 27 ft. An excavation for a boiler-house (144594) 120 yd E. 20° S. of the tunnel entrance proved Thanet Beds (p. 185) overlain by 8 to 10 ft of coarse glauconitic sand with an occasional pebble at the base and a few grey clay seams in the lowest 2½ ft.

A borehole at the derelict monastery (152599) 900 yd N. 29° E. of the church south of Hales Place encountered sandy strata between 49½ and 94 ft below surface. Of these strata the uppermost 10 ft are assigned to the Oldhaven Beds (p. 197). Below 94 ft, 4 ft of sandy clay were proved which are presumed to be basal Woolwich Beds, making a total of 38½ ft assigned to these beds.

Fordwich—Wingham.

The beds are magnificently seen in the large sand-pit (161600) at Shelford, 750 yd N.W. of the Sewage Works near Fordwich, where almost the complete thickness is exposed although the actual base is concealed. Some 27 ft of coarse khaki glauconitic sand were noted, mottled brown and gently current-bedded in many places, with a thin seam of clay locally just below the top of the beds. The tubular almost segmented nodules (p. 179) have their finest development in this quarry. A series of trough faults which affect all the Eocene strata exposed in this quarry is of superficial origin (p. 14).

The lower part of the Woolwich Beds was seen by Whitaker (1872, p. 172) in a pit (165585), now degraded, 870 yd N. 15° W. of the Mental Hospital near the Wingham road ; he recorded rather coarse greyish green and iron-shot sand 5 ft thick, resting upon ?Thanet Beds. An old sand-pit (176583) on the golf course 950 yd N.E. of this Mental Hospital exposes 3 ft of medium-grained grey glauconitic sand with ferruginous staining in parts, overlain by 5 ft of coarse glauconitic sand with bands and patches of ferruginous staining and some clayey streaks, below 0 to 1½ ft of sandy soil with pebbles ; tubular nodules occur throughout the section. Towards the northern end of the golf course (174590), 1150 yd S. 42° W. of Fordwich church, Thanet Beds are overlain by 6 ft of coarse glauconitic sand with iron-stained bands, and with impersistent clay seams up to 1 inch in thickness and patches of clay. The line of contact is tolerably even and small pebbles are scattered sporadically through the lowest 6 in of the beds.

In an old pit (181593) between the two roads 500 yd S. 6° W. of Fordwich church some 10 ft of coarse khaki and brown sands were noted below Oldhaven Beds (p. 197) and river gravel (p. 271).

The sands are worked together with the underlying Thanet Beds (p. 187) in the sand-pit in Trenleypark Wood (Plate VB). It is believed that the full thickness of the Woolwich Beds is exposed in this pit, although the recognition of the basal Oldhaven Beds is doubtful (p. 198). The thickness indicated for the Woolwich Beds is about 15 ft, which accords with the thickness estimated from mapping the outcrop. The sands are not as coarse as the typical Woolwich sand. They are current-bedded, grey and glauconitic, with a ferruginous staining in the uppermost 4 to 5 ft, where they are also appreciably harder. There is an irregular ill-defined green stain in the lowest 2 ft of the beds which passes upwards and downwards into grey and brown colours, while in the lowest 2 ft and associated with the green stain there are impersistent clay seams. A few small flint pebbles occur at the rather uneven contact with the Thanet Beds and also occasionally among and at the top of the clayey seams. To the north of the road (196597) 900 yd W. 4° S. of Elbridge House 3 ft of coarse brown and grey ferruginous sand underlie Oldhaven Beds (p. 198) and terrace gravel (p. 272).

Greenish clayey sands, similar to the basal beds exposed at Trenleypark Wood, crop out from below the Brickearth north of Wingham. The Woolwich Beds here do not seem to exceed 15 ft in thickness and are in a weathered condition.

A pit (238592) at Wenderton Manor House exposes 3 ft of medium to coarse glauconitic clayey sands. Near the road junction 500 yd N. of Little Wenderton Farm, Whitaker (1872, p. 173) recorded a pit (244591), 15 ft deep, showing false-bedded sand, partly salmon-pink in colour and partly hardened by iron, with green grains near the bottom. Whitaker (1872, p. 93) stated that the beds in an exposure (246583) 250 yd N. of Broomhill were light grey in colour, with large green grains, and were false-bedded. Only 4 ft of coarse ferruginous clayey sand resting upon Thanet Beds are now exposed here.   J.G.O.S.

Oldhaven Beds

Shottenden—Selling—Boughton Street—Upper Ensinge.

Whitaker (1872, p. 271) stated that the Oldhaven pebble beds in a pit at the top of Shottenden Hill contained "masses of fine light-coloured sand (chiefly at the lower part, where indeed there is often more of the sand than of the pebbles). The sand is almost white in parts, and has comparatively few dark grains." The total thickness there was 30 ft or more. He also recorded (1872, p. 162) that at Shottenden Hill the Woolwich Beds (p. 189) were overlain by $1\frac{1}{2}$ to 4 ft or more of fine light-coloured sand like that of the pebble bed, the bottom 2 in being rather ferruginous with here and there a pebble ; this was overlain by an unspecified thickness of sandy pebble bed with masses of sand. The above remarks referred to two pits close together, which are now probably combined as the mainly-obscured large old pit (043556) 200 yd S.W. of the cross-roads in Perry Wood, where ovoid black flint pebbles set in fine golden brown sand and containing a 1-ft bed of fine sand with a single line of pebbles in the centre, are now seen for $2\frac{1}{2}$ ft. The succession is therefore interpreted as comprising over 30 ft of pebble beds containing beds and derived irregular masses of fine sand, particularly in the lowest 4 ft or so. The derived masses of sand are presumed to be either from the Oldhaven Beds themselves or possibly from the Thanet Beds.

The Oldhaven Beds are also believed to occur on the ridge known as The Mount, but they are not as extensive as was formerly supposed. Excavation near The Mount itself proved that the pebbly material here was Head (p. 225), resting upon Thanet Beds. If they are present at all, the Oldhaven Beds must lie at the extreme northern end of the ridge and be completely covered by Head.

At Staplestreet, presumably near Mount Ephraim (064599), on the north-eastern side of the lane from Staplestreet to Dunkirk, Whitaker (1872, p. 266) recorded " Light-coloured Oldhaven Sand, about 10 feet, with thin and somewhat irregular layers of brown clay in the middle part (and also a little of the hard whitish marl, that occurs in a like position elsewhere) ; close to or at the bottom a thin layer of grey clay, sometimes splitting into two ; below which there is mostly an inch or two inches of almost white sand, with small flint-pebbles, and small pieces of the purple-grey sandstone from the top of the Woolwich Beds " resting upon Woolwich Beds.

North of the road in Boughton Street, Dr. J. R. Earp noted fine golden or pale yellow sand with numerous layers of loamy clay for 8 to 10 ft in a pit (064593) 820 yd E. 15° S. of the church with spire. A little westward in the same pit he noted that the Woolwich Beds (p. 189) were overlain by the pebble bed, 1 ft in thickness, below $1\frac{1}{2}$ ft of fine golden sand. Whitaker (1872, p. 267) described and figured two pits, close together on the south side of the roadway hereabouts, which are now obscured ; the following notes have been abstracted from his description: Light buff fine sand, at most 2¾ ft, resting on alternations of like sand with brown clay, resting slightly irregularly on the bed below ; the clay at the bottom of the hollows containing ironstone ; here and there a thin white clayey limestone at bottom, 1½ to 3 ft ; resting upon light buff fine sand, with some thin lenticular layers of clay. At and near the bottom, where the sand is partly of coarser grain, there are small soft nodules, sometimes 'casts' of shells, teeth and vertebrae of *Lamna,* and here and there a small flint pebble, 1¼ to 3¾ ft ; rests slightly unevenly on like sand, of a paler tint with 2 to 4 inches of light brown clay at the bottom, 4 ft. Resting on black flint pebbles in sand, thinning out to merely a few scattered pebbles at one part ; at another with a little fine yellow sand between the pebbles and the clay above ; and at another two layers of pebbles, separated by brown sand, 0 to 1¼ ft.

The Woolwich Beds ironstone caps several mounds near Winterbourne where it was originally mapped as Oldhaven Beds (p. 181). The section in a trackway ¼ mile east of Sugar Loaf Hill contained both Woolwich Beds and Oldhaven Beds according to Whitaker (1872, p. 165), but in fact was probably entirely in the former formation. Whitaker (1872, p. 268) also described two sections in Oldhaven Beds in the vicinity of the pit (067579) in Woolwich Beds (p. 190), in one of which 5 ft of ironstone, with here and there a pebble and with 'casts' of fossils, were overlain by " Sand, with a layer of clay and clayey ironstone, below which there are a few pebbles in the sand ; clayey and ferruginous at bottom, and passing into the bed below ". In the other pit sandy ironstone below was separated by a " Thin irregular layer of brown and whitish sharp sand ; not constant " from the pebble bed above.

The beds are well exposed in the overburden of the sand-pit (065576) 1450 yd E. 15° N. of Selling Station, as shown in Fig. 6. Since this sketch was made the face has been worked eastwards to expose the base of the London Clay (p. 198). Bed j in the diagram continues upwards to the London Clay and the total thickness of Oldhaven Beds is 8 to 10 ft, depending on the irregularities of the London Clay base.

The Oldhaven Beds have been dug together with the Woolwich Beds near the Upper Ensinge Tunnel, but are now nowhere well exposed ; their outcrop eastwards is much obscured by landslipping and clayey Head deposits. Whitaker (1872, p. 269) noted fine light-coloured sand near the cottages (089578) 400 yd N. 35° W. of White Wall.

Chartham Hatch—Upper Harbledown—Hales Place.

To the north of the main road (103586) 1100 yd N. 18° E. of Denstead Farm fine brown sand was seen at the bottom of a section below 2 ft of grey sandy clay with seams of stiff clay, which were overlain by 2 ft of clayey brown sand with a buff streaky mottling. The base of the Oldhaven Beds and the underlying Woolwich Beds (p. 192) were noted in another pit (107586) 700 yd of N.W. of Poldhurst Farm, where the basal pebble bed, here consisting of small flint pebbles and hard lumps of grey sandstone in a matrix of fine sand, was overlain by 1½ ft of fine buff-stained brown sand. A driveway nearby on the south side of the road 510 yd N. 32° W. of Poldhurst Farm shows 3 ft of fawn and golden fine sand beneath 1½ ft of fawn sandy clay with a thin iron-pan at the base. A section (111585) 400 yd N. of the farm was figured by Whitaker (1872, fig. 73, p. 269), and the following is based upon his description: the Woolwich Beds were overlain in upward sequence by sandy brown ironstone and ferruginous sand, with clusters of black flint pebbles here and there, sandy at the bottom, 2¼ to 3 ft ; light buff sand, 4 to 10 in ; ferruginous clay with concretionary ironstone in the thicker

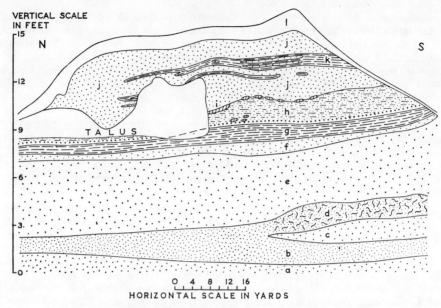

VERTICAL SCALE
IN FEET

N

S

TALUS

HORIZONTAL SCALE IN YARDS

0 4 8 12 16

Oldhaven Beds
- l. Surface soil, clayey.
- k. Greyish brown mottled clay.
- j. Greenish brown to brown, fine, sharp, glauconitic sand.
- i. Nodules of ironstone in a line dividing j and h.
- h. Khaki fine sand with many grey clay seams and occasional ironstone; line of pebbles at base.

Woolwich Beds
(pp. 190–1)
- g. Grey glauconitic sandy clay.
- f. Gritty iron-pan.
- e. Greyish brown medium-grained sand, occasionally gritty.
- d. Concretionary ironstone.
- c. Coarse greyish brown sand.
- b. Ferruginous coarse sand, crimson-stained and leached in places.
- a. Coarse, grey, glauconitic current-bedded sand.

FIG. 6.   *Section in Woolwich Beds and Oldhaven Beds near Winterbourne, 1450 yd E. 15° N. of Selling Station*

parts, 4 in to 1 ft 2 in; light brown loam and clayey sand, 2 in to 1 ft 3 in; grey clay, 3 to 4 in; fine light buff sand, which at one place occupied a hollow in the two underlying beds, 3 in to 2 ft; grey clay, 6 to 8 in; loamy soil, 2½ ft.

In the banks of a sunken trackway (117584) 690 yd E. 28° N. of Poldhurst Farm the Woolwich Beds, represented by 2 ft of coarse grey sands and a clay seam, are overlain by the basal pebble bed of the Oldhaven Beds, consisting of fine brown sand with rather sparsely distributed pebbles near the bottom. In a sand-pit (142593) at nurseries 400 yd E. 3° S. of Beverley Farm the Oldhaven Beds are channelled into the Woolwich Beds: the basal bed varies from 6 inches to 2 ft in thickness and in part of the channel exposed it contains a preponderance of small angular ironstone fragments, as well as rounded crimson sandstone and flint pebbles; this bed is overlain by up to 2 ft of massive khaki fine sand with an impersistent seam of pipeclay near the base, below 3 ft of well-bedded fine light brown sand, rather coarser and with iron-staining towards the bottom, below 0 to 2 ft of reddish sandy soil. Ironstone fragments also occur above the general level of the basal bed, where they are again associated with flint pebbles, and a pipeclay seam dies out abruptly against one of the lumps. These ironstones were evidently also derived, most likely from the Woolwich Beds. Some apparently nodular masses of red haematitic mudstone and brown limonitic mudstone which were noted in part of the face, may have had a similar origin.

Nearby, in a large old sand-pit at the entrance to the disused railway tunnel, blocks of fossiliferous ironstone (p. 181) up to 10 ft in length and 3 ft thick are both underlain and overlain by scattered black flint pebbles of the basal bed. Large borings up to about 2 inches in diameter in the blocks are also infilled by pebbly sand. Fine ferruginous sands, purple sandstone fragments and small pieces of ironstone are also present in the basal bed here. Whitaker (1872, p. 270) recorded an entirely different section on this site, reflecting the variable nature of the pebble bed: 18 ft of Woolwich Beds were overlain by " Small black flint-pebbles in sand, false-bedded, about 2 feet ", below " Pale buff fine sand, full of small green grains (disposed in lines and easily separable from their comparative lightness). About the middle a thin irregular bed of ferruginous clay, 10 or 12 feet." He also noted the presence of six small faults with throws of 2 inches to 1 ft to the south-east and north-west, probably similar to those in the Shelford Pit (p. 14). At the entrance to the tunnel the section seen by Whitaker (1872, p. 167) exposed the full thickness of the Oldhaven Beds: the Woolwich Beds (p. 192–3) were overlain by a thin pebble bed including lumps of purple-grey sandstone, beneath sandy brown ironstone with splintering flint pebbles in parts and a few moulds of ' Cyprina', 4 ft ; this was separated by a thin bed of black flint pebbles from false-bedded pale buff sand with occasional pieces of concretionary calcareous sandstone, poorly-preserved ' Cardium laytoni' and ' Fusus', and a thin seam of ferruginous clay near the bottom, 12 ft, beneath London Clay. A borehole at the derelict monastery (152599) 900 yd N. 29° E. of the church south of Hales Place penetrated a 1-ft pebble bed about 10 ft below the probable base of the London Clay (p. 193) and these 10 ft of strata are assigned to the Oldhaven Beds.

Fordwich—Wingham.

The large sand-pit at Shelford (p. 193) (161600) exposes the complete succession of the Oldhaven Beds, which attain a thickness of 17 to 22½ ft : at the base occur 0 to 2 ft of dark ovoid pebbles (up to 6 inches but mostly 2 inches long), purple sandstone blocks (up to 6 inches long with some longer) and large masses of bored ironstone enveloped by pebbles, sands and clays in places ; these are overlain by 14 ft of fine, gently current-bedded khaki sand, with streaks of glauconitic sand, moulds of fossils and occasional pebbles near the base, a bed of fragile shells in places about 3 ft above the base, and seams of laminated clay mainly in the uppermost 6 ft ; next above occur 5 to 6 ft of fine brown mottled sands with ochreous clayey sands and clay seams, below 1½ ft of khaki fine sands with clayey seams.

On the south side of the Great Stour in the vicinity of Trenleypark Wood there are outliers of Oldhaven Beds wedged below the 3rd Terrace. In a pit (181593) between the roads 500 yd S. 6° W. of Fordwich church a channel in Woolwich Beds (p. 193) is occupied by Oldhaven Beds which comprise: the basal bed, ½ to 1 ft thick, consisting of small, dark, rounded flint pebbles up to 1½ inches long, soft purple sandstone blocks and hard fine-grained grey sandstone in a matrix of coarse brown and buff sands, and containing a thin impersistent seam of laminated clay ; overlain by 5 ft of laminated khaki and golden fine sand, being very fine and finely laminated with clay seams in the lowest 1 ft ; below 3rd Terrace gravel (p. 271). The Oldhaven Beds are also exposed in places below the gravel (p. 271) (185591) 900 yd S. 28° E. of Fordwich church, and in Trenleypark Wood, in a tiny sand-pit (192593) 60 yd south of the main road near the entrance roadway to the sand-pit mentioned below, 2½ ft of Oldhaven Beds were noted with the pebble bed at the base.

In the working sand-pit (192592) in Trenleypark Wood (p. 187), just below the base of the gravel capping the mound (p. 240) there are pockets of red-stained brown and grey clay up to 1 ft in thickness containing small pebbles and greyish brown ironstone fragments. The pockets are cut off by the overlying gravel with which there has been some mixing; two of them are regarded as Oldhaven Beds *in situ,* but their relationships are obscure and they may possibly be part of the terrace.

The beds are seen again below gravel (p. 272) on the north side of the road (196597) 900 yd W. 4° S. of Elbridge House, where the pebble bed, varying from 3 in to 1 ft in thickness, is overlain by up to 3 ft of laminated fine brown glauconitic sand with a clay seam 1 ft above the bottom.

Soils containing pebbles from the Oldhaven pebble bed are widespread on the top of the Eocene escarpment to the north of Wingham, and are known locally as ' black pebble '. The pebbles are in fact derived from the basal beds of the local brickearth which was apparently formed by weathering of the Oldhaven Beds, and the latter must originally have had a wide outcrop. The Oldhaven Beds themselves occur in an ill-defined outlier at the extreme eastern margin of this district, 1900 yd E. 17° N. of Broomhill, where the pebble bed was found to exceed 4 ft in thickness.   J.G.O.S.

LONDON CLAY

Boughton Street—Dunkirk—Hales Place.

The heavy London Clay capping the Blean upland is intermingled with gravel at the surface (p. 244). Whitaker (1872, p. 294) recorded the basal bad, overlying Oldhaven Beds, just north of the high road between Boughton Street and Dunkirk, but the locality is now lost. Dr. J. R. Earp found septarian nodules in debris from pylon footings (075595) 600 yd N. of Dunkirk church.

In the top of the sand-pit (065576) (p. 195) the Oldhaven Beds are overlain sharply but unevenly by the basal bed of the London Clay, 1 ft thick, consisting of brown sandy clay with ferruginous streaks and occasional small pebbles, below 3 ft of grey and brown stiff clay.

Stiff light brown clay 3 ft thick was seen (087593) near the lane to Bossenden Farm, 1370 yd E. 13° N. of Dunkirk church, and 5 ft of blue clay, becoming loamy at the surface, were observed in a stream-bank (118598) 1400 yd W. 5° N. of Hothe Court.

A borehole in school buildings (St. Edmund's School) (134592) encountered 9 ft of yellow clay with reddish brown nodules, overlying 47 ft of blue clay with fossil wood, pyrite and septaria. The lowest beds of the London Clay, however, were variously described as " Imperfect claystone " (Whitaker 1908, p. 96) and " Mottled clay " (Brown in MS.), 7 ft in thickness. The underlying beds cannot be classified; they comprise: greenish sand, 74 ft; loamy sand, 2 ft; and greenish sand, 2 ft; *" Cyprina Morrisii "* was recorded from the lowest 4 ft.

Whitaker (1872, pp. 270 and 167) saw the base of the London Clay in pits near the entrance to the railway tunnel on the defunct Canterbury–Whitstable line, and at the tunnel entrance the lowest 1½ ft of the formation were exposed. At Hales Place he described (1872, p. 166) grey shaly London Clay with a boss of Oldhaven Beds protruding up into it.

London Clay is exposed in the overburden of the Shelford Sand-pit (p. 193), where the basal bed is about 1 ft thick and consists of grey glauconitic sandy clay with rare, black, rather imperfectly rounded pebbles along the irregular base and with buff mottling and ferruginous nodules near the top; this is overlain by 20 ft of greyish blue shaly clay with a patchy ferruginous staining, occasional flattened harder nodules, pyritized woody remains and selenite, and commonly with the bedding disturbed. When this section was measured the London Clay was overlain by 10 ft of disturbed clay and gravel, but much of this was later removed by excavation and its place taken by weathered and disturbed clays (p. 269).

In Trenleypark Wood an outlier of London Clay is preserved due to folding.

J.G.O.S.

Chapter IX

PLEISTOCENE AND RECENT

GENERAL ACCOUNT

INTRODUCTION

ACCUMULATIONS of clay, loam, sand, gravel and calcareous matter of Pleistocene and Recent age are collectively termed Drift. In this district such deposits were formed by three main agencies and they are grouped accordingly, as Solifluxion, Fluviatile and Marine-Estuarine Deposits.

Solifluxion Deposits (Head *sensu lato*)**.** The solifluxion deposits cap mounds and the surfaces of solid strata, mantle valley slopes and fill the bottoms of minor valleys and depressions. They either remain as disturbed relics of solid strata or they have flowed to their present positions under the influence of gravity, and bedding is rare in them. Their constituent materials are of local origin and the deposits differ in composition according to the nature of their source, ranging from chalky boulders and powdery chalk, through clay, loam, sand, and gravel. They include Clay-with-flints, Sand in Clay-with-flints, Coombe Deposits, Head Brickearth, Head Gravel and Head (*sensu stricto*). There is every gradation between the different solifluxion deposits, and between the solifluxion and the fluviatile deposits. The gradation in the latter case is dual, for some of the solifluxion deposits have been derived wholly or in part from river terraces, while some of the younger spreads pass into Dry Valley and Nailbourne Deposits and Alluvium. The association of Head Brickearth with river terraces is notable and it has been shown that this relationship is not always transitional (p. 240). There is also a gradational relationship with landslips.

It is uncertain how much of the solifluxion deposits resulted from simple downwash and how much was moved by the process of freezing and thawing. There is palaeontological evidence, mainly from neighbouring areas, that some solifluxion spreads were formed by surface washing in a temperate but damp climate (Dines and others 1954, pp. 132–4 ; White 1928a, pp. 67–9 ; Kerney 1963) : however, signs of frost action are unmistakably present within this district. The bases of many solifluxion deposits, where they rest on Chalk, are so folded and contorted that their formation purely by slope-washing is unlikely. Festooning is found within the deposits and the upper parts of the river terraces show the effects of frost-churning with disruption of the bedding ; there is also evidence of the effects of ground-ice to a depth of 50 ft below the surface (p. 14).

It is apparent that the solifluxion deposits cannot all be of the same age and in the absence of fossils only their relationship to river deposits can give any suggestion as to when they were formed ; their altitude can offer no certain indication.

In the majority of the solifluxion deposits the constituents have been transported laterally for a short distance, but in a few little or no such transport has taken place. The Clay-with-flints probably belongs to the latter type of deposit, while the capping of The Mount, near Selling, which was formed by the disruption, more or less *in situ,* of the Oldhaven Beds, is a particularly good example; others are provided by some of the Head Brickearth spreads (p. 213).

Fluviatile Deposits. The fluviatile deposits comprise terraces and fans of river gravel, alluvium, and a generally older alluvium shown on the maps as Dry Valley and Nailbourne Deposits. Peat, which has formed where vegetable debris has accumulated, is also included within this group.

Marine-Estuarine Deposits. The marine-estuarine deposits consist of shingle banks thrown up by waves (Storm Gravel Beach Deposits) and a mixed series of sands, silts and clays formed on saltings or tidal flats. Associated dunes of wind-blown sand are also shown on the maps. J.G.O.S.

CLAY-WITH-FLINTS

The most widely distributed deposit on the Chalk plateau is composed of clay with scattered flints, aptly termed the Clay-with-flints. The deposit is much dissected, but stretches northwards from the crest of the Chalk escarpment to abut against the main outcrop of Thanet Beds in the extreme north-west of the district described here and against outliers of these beds. It extends from above 600 ft O.D. to below 200 ft; it is exceptionally low in the area around Brogdale (006593), where it is only about 125 ft above O.D., but it is too thin to be shown on the geological map, as is commonly the case in the vicinity of the Thanet Beds.

The Clay-with-flints comprises unbedded stiff clay, silty clay or sandy clay; it is generally reddish brown or brown in colour, but also shows darker streaks and patches of manganiferous appearance and some, rarely seen, streaks of crimson, grey or purple where the deposit has not been so well oxidized because of poor drainage. Near the contact with the Chalk, and usually for a few inches only, the clay is stiffer, unctuous and dark grey in colour. Flints of all shapes and sizes are set in the clay, for the most part without any semblance of order or bedding, the proportion of flints to clay being extremely variable; the flints range from the smooth Tertiary-rounded pebbles, which are more commonly brown than black in colour, through battered, subangular flints from river gravel, to unworn flints derived directly from the Chalk; angular flint flakes and chips are of frequent occurrence; green-coated flints from the Bullhead Bed at the base of the Eocene, rotten flints, probably from Pliocene deposits, and fragments of silicified sandstone or 'sarsen-stone', probably derived from the Thanet Beds, are all found locally. Ironstones of Tertiary, and perhaps Lower Greensand, age are also common. Sandy deposits, presumably of Pliocene origin, are intermingled with the Clay-with-flints (p. 204).

Reddish brown loam, containing flints which are mostly angular chips, possibly formed as a soil from the Clay-with-flints, rests irregularly on that deposit to a thickness of about 3 to 4 ft. Generally this loam is absent on the west sides of spreads of Clay-with-flints on north–south trending ridges, where the boundary of the Clay-with-flints and Chalk is usually abrupt and only a few piped relics lie below the marked slope-change associated with the boundary. Towards the middle of the ridges pockets and spreads of the flinty loam are common and sometimes extensive. These tend to increase eastwards and may pass gradually into the Head on the valley sides without any sudden change of slope. The Clay-with-flints, although itself classed as a solifluxion deposit, is one of the most important sources of material for other solifluxion deposits found within the Chalk country.

A threefold subdivision of the Clay-with-flints in France was made by Brajnikov and Furon (1934), accompanied by an X-ray analysis by Brajnikov and Urbain (1934). The three divisions are briefly as follows: (i) That resting directly upon Chalk, deep reddish brown in colour, with large unworn flints and a few metres in thickness. (ii) That overlying and often extending beyond (i), underlying Tertiary sands and composed of fawn sandy clay with slightly bruised and rounded flints. (iii) A post-Oligocene formation, with fragments of Cretaceous flints and rounded debris of Tertiary age, resting upon the Tertiary, upon (ii), upon (i), or upon the Chalk. Both (i) and (iii) appear to be represented within this district.

Accounts of the mode of origin of the Clay-with-flints are varied. Whitaker, who named the deposit (in Hull and Whitaker 1861, p. 54), thought it to be the residue of solution of Chalk with addition of debris from Tertiary rocks (Whitaker 1889, pp. 281–2). Reid (1899, p. 37) found that the proportion of clay to flint was such that the deposit could not have been formed by solution of the Chalk alone. Sherlock (1912) considered that the Clay-with-flints of Buckinghamshire had resulted from glacial action, while Edmunds (1948, pp. 56–7) thought it was likely " to have been assembled under sub-glacial conditions, as a *mélange* of local materials disturbed and partly re-sorted by local ice-caps or snowfields ", supplemented by solution of the Chalk. Wooldridge and Linton (1955, p. 55) suggested that the proportion of residue from solution of the Chalk varied according to whether the deposit rested upon the Pliocene land-surface, the sub-Pliocene bench or the sub-Eocene bench. Loveday (1962) restricted the term Clay-with-flints to " reddish coloured, highly tenacious clay containing flints found immediately overlying Chalk " ; he considered that clay illuviation and possibly clay synthesis from the weathering products of overlying pervious deposits were involved in the formation of Clay-with-flints, in addition to Chalk solution which provided the flints.

West of the Stour gap the Clay-with-flints rests upon Upper Chalk, but east of the gap it spreads on to Middle Chalk where the base of the Upper Chalk retreats from the escarpment, and on to Lower Chalk at several places on the escarpment edge.

The base of the Clay-with-flints is irregular because of intense piping into the Chalk, and the thickness of the deposit is therefore variable. Between Challock Lees and Molash, and at least as far north as Badlesmere, it is believed to average at least 10 ft, although there is no evidence from exposures. East of the Stour the deposit seems generally to be thicker:

near Stelling Minnis cesspits have been dug to 20 ft without encountering Chalk, but the uppermost few feet of these pits are in Head Brickearth. The Clay-with-flints is rarely and poorly developed on Chartham Downs and north-east of the Canterbury–Dover road.

Anomalous patches of Clay-with-flints occupy spurs within the Stour valley between Godmersham and Chartham, and also to the north of Petham. They are exceptional only in that they are apparently younger than some of the river deposits and that they rest on bench features or on gentle slopes, rather than on a plateau. Apart from the Petham developments, they occur about 100 ft lower than the local plateau with which the normal deposit is associated, and at about 200 ft O.D., in some cases falling below this level. They differ markedly from nearby Head deposits by being more clayey in composition. Their contained flints appear to have been transported either from adjacent river gravels, Chalk or perhaps from true Clay-with-flints. They are similar in position and possible origin to the patches of Head at Eastwell Park (p. 208), which are perhaps older, and to decalcified Coombe Deposits near West Brabourne (p. 231). They occupy the remnants of an early land surface which has been largely destroyed by the present lateral drainage system, but which was formed in the time intervals between 5th Terrace and 4th Terrace and between 4th Terrace and 3rd Terrace.   J.G.O.S.

Petrology of the Clay-with-flints. The matrix of the Clay-with-flints specimens (MR 24820–4) is mainly a brown silty clay. Particle-size analysis by Mr. J. Dangerfield shows it to have a very variable grain-size distribution (Table 4 ; Fig. 9). To prepare the analyses, the field samples were broken up by hand and each one split until a fraction of suitable size was obtained. The fraction to be used was broken down further by hand and by using a rubber pestle, and when necessary by gentle pressure with a ceramic pestle. It was then dry-sieved through a 300-mesh B.S. sieve. Material retained on the 300-mesh sieve was placed in a very weak solution of Calgon (sodium hexametaphosphate), warmed and gently stirred to complete dispersal. Next it was washed through the B.S. sieves (from No. 7 to No. 300) listed in Table 4, using distilled water. A sub-sieve particle-size analysis by the pipette method was carried out on a weighed portion of the <300 mesh fraction. The apparatus used and method adopted were in principle those of British Standard No. 1377: 1961 (British Standards Institution 1961). The clay grade in the result includes dispersed iron oxide.

The results contrast with those of Loveday (1962) who found the matrix of the Clay-with-flints from localities in the Chiltern Hills to be predominantly clayey deposits, with little silt or sand. Histograms (Fig. 9) reflect the figures of Table 4 (p. 215), which show the very poor sorting.

Among the coarsest fragments, flint is dominant. The sand grades are largely composed of quartz, together with sandstone cemented by buff hydrous iron oxides, angular flint, a little glauconite, white siltstone and a trace of muscovite, biotite and microcline. In some sandstone fragments, the subangular to rounded individual sand grains are coated with very lustrous dark brown limonite. Many of the quartz fragments are very flaky.

There are rare carbonized woody fragments, whose identity was confirmed by Dr. F. W. Anderson. Black to dark brown manganiferous grains are abundant in the sand grades of a specimen from Capel Street (Table 4, No. 14). A spectrographic analysis by Mr. K. L. H. Murray of a not very pure sample confirmed the presence of plentiful manganese and silicon, together with aluminium and iron, and smaller quantities of calcium, magnesium, titanium and barium. As commonly happens with some manganese minerals, only a very poor X-ray pattern was obtained. No structure was found in the manganiferous grains when examined in thin section (E 30670), but a few enclosed quartz.

X-ray examination by Mr. B. R. Young of the clay fraction showed it to contain substantial amounts of a montmorillonite mineral, or smectite, which is presumably derived directly from the Chalk, where this is also a common clay mineral (see also Sabine, Young and Dangerfield 1963). The manganese (now mainly in the sand grades) is perhaps derived also from the Chalk, where it occurs in small quantity. Knowledge of the processes of accumulation of manganese nodules is however at present insufficient to allow their presence in the Clay-with-flints to shed light on the origin of the formation.

P.A.S.

SAND IN CLAY-WITH-FLINTS

Along the crest of the North Downs escarpment there occur scattered patches of sand with subordinate beds of ferruginous sandstone. These are thought to be equivalent to the Lenham Beds of the Maidstone area (Reid 1890, p. 49 ; Worssam 1963, pp. 85–99), although palaeontological confirmation is lacking. West of the Great Stour valley the outliers are few and small but eastward to Cape Street they are more numerous and of greater extent, resting upon a bench or platform between about 550 and 600 ft above sea level at the escarpment, and tilted towards the north-east.

The sands are generally red or brown in colour, due probably to the oxidation of abundant glauconite, but orange, yellow, crimson, pink, grey and white sands have been recorded. They are commonly fine grained, but coarse sand has been found in places, and beds with pebbles of quartz and flint up to $\frac{1}{4}$ in and more in diameter occur locally. Wooldridge (1927, p. 53) stated that the lowest lithological unit throughout the East Kent area was a fine sand with flint pebbles of small and uniform size ; above were red loamy sands, which passed up into yellow sands ; the combined thickness of these three divisions did not exceed 10 to 15 ft ; they corresponded to the beds noted by Prestwich (1858, pp. 326–7) in a pipe at Harrietsham (Maidstone Sheet 288). Above the last-mentioned yellow sands came a " thick mass of bright red sands with ' slabby ' ironstone, . . attaining a maximum thickness of 40 ft. or more." No evidence was found during the present survey to substantiate this succession.

Chatwin (*in* Worssam 1963) concluded that the Lenham Beds should be referred to the Pliocene, and that they are very little, if at all, older than the Coralline Crag. However, in the present district it is seldom possible to draw a definite line between the sands and the Clay-with-flints. The sands are underlain by mottled grey and black unctuous clay, with large unbroken flint nodules, which is evidently the residue left after dissolution of chalk ; in places they are overlain by stiff reddish brown or brown clay, with flints in all stages of destruction. This composite deposit has been lowered into solution pipes in the Chalk, with varying degrees of disturbance, and the whole has suffered erosion ; piping is generally most advanced on the flanks of the hills, so that the remaining deposit is usually thicker there than on the hill-tops, where peaks of the lower clay, and even of chalk, locally penetrate to the ground surface. In these circumstances it is thought proper to regard the deposits as drift, and as a sandy facies of the Clay-with-flints.

The sands seem to be distributed on a surface some way below that on which they were originally deposited, but it is not clear how far they have been removed. Very commonly festooned sandstone beds can be seen in pipes, the limited disturbance suggesting that the distance travelled has been small and probably almost entirely vertically downward. This impression is supported by the parity of the levels at which the several masses occur, since greater descent must have produced increased disparity. Wooldridge (1927) considered that the sands were deposited on a pre-existing peneplaned surface, whose gentle undulations had not been completely obliterated by marine abrasion. Of this surface only a narrow ridge remains west of the Stour gap, but it is represented by a bench some 10 miles wide east of the River Stour, although sands are preserved over a width of only about 5 miles. Wooldridge regarded the north-eastward slope of the surface as being in part original and in part due to a slight post-Diestian uplift of the Wealden anticline. The coastal limit of the plane in the present district was, according to Wooldridge, some distance south of the existing Chalk escarpment.   G.B.

Petrology of the Sand in Clay-with-flints. Wooldridge (1927) attempted to correlate and date these sands by studying the suite of heavy detrital minerals contained in them and concluded that they were of Diestian age. In order to test the consistency of the method and to add to the data available, heavy residues from samples from four localities in the present district were separated and examined ; some 12 000 grains between 0·05 and 0·45 mm in size were counted and the results are set out in Table 3, together with corresponding figures by Elliot, Wooldridge and Edelman for comparison. Wooldridge's figures have been recalculated as percentages, ignoring the error due to his not having listed minerals present in small quantity.

The suites that Wooldridge obtained from localities along the North Downs and in the Chiltern Hills were essentially mixtures of ilmenite (listed as an opaque mineral), limonite after glauconite (dissolved in the separation process and not listed), and zircon, with subordinate amounts of tourmaline, staurolite, rutile, kyanite and other minerals, in decreasing order of abundance. He noticed the " constant presence of andalusite, sillimanite, speckly kyanite, monazite and spinel, and the paucity or virtual absence of garnet." Staurolite, andalusite and locally kyanite occurred as large grains, tourmaline as small grains. He also recorded the occurrence of muscovite, biotite, anatase, hornblende, calcite, chlorite, green pyroxene and corundum. In East Kent green pyroxene was everywhere present, garnet was locally concentrated, while andalusite, tourmaline and kyanite seemed to be less abundant than elsewhere.

TABLE 3

Heavy mineral constituents of Sand in Clay-with-flints and related deposits

	1	2	3	4	5	6	7	8	9	10	11	12	13	14
Opaque minerals	43·0	61·2	64·8	66·6	56·3	52·8	54·7	45·9	47·3	66·0	56·5	62·5	25·0	50·0
Zircon	31·7	22·0	18·5	13·7	24·1	25·2	24·0	29·0	32·0	23·0	31·0	9·0	6·0	10·0
Rutile	7·8	5·6	3·9	2·9	8·1	11·7	10·6	12·1	10·6	2·5	4·0	4·0	1·5	7·5
Tourmaline	5·8	4·3	3·0	7·1	4·0	4·6	4·1	6·9	4·2	3·0	3·0	8·0	1·5	15·0
Staurolite	3·7	3·5	3·4	4·1	2·5	1·8	2·3	2·5	1·3	4·0	5·0	10·5	1·5	7·5
Anatase	3·5	1·2	1·0	0·8	1·4	1·7	1·5	0·7	3·3	—	—	—	—	0·5
Andalusite	1·8	0·7	4·0	2·5	0·3	0·3	0·4	0·3	0·3	—	—	2·0	—	3·0
Kyanite	1·6	0·9	0·8	1·3	2·2	1·0	1·2	1·7	0·6	1·5	0·5	4·0	0·8	4·5
Brookite	0·4	0·1	0·1	0·3	0·3	0·3	0·4	0·3	0·1	—	—	—	—	0·5
Sillimanite	0·3	0·2	0·1	0·3	0·3	0·2	0·3	0·3	0·2	—	—	—	—	0·5
Sphene	0·2	0·1	0·1	0·1	—	—	—	—	—	—	—	—	0·8	0·5
Monazite	0·1	0·1	0·1	0·3	0·4	0·3	0·4	0·3	0·1	—	—	—	—	—
Amphibole	0·1	0·1	0·2	0·1	0·1	0·1	0·1	—	0·1	—	—	—	18·0	—
Spinel	—	√	√	√	√	—	—	0·3	√	—	—	—	—	—
Chloritoid	—	—	—	—	√	0·1	0·1	—	√	—	—	—	—	—
Garnet	—	—	—	—	—	0·1	—	—	√	—	—	—	23·1	—
Epidote	—	—	—	—	—	—	0·1	—	√	—	—	—	20·2	—
Saussurite	—	—	—	—	—	—	—	—	—	—	—	—	0·8	—
Augite	—	—	—	—	—	—	—	—	—	—	—	—	0·8	—

The figures are percentages of the heavy mineral fractions

Analyses by Miss H. A. H. Macdonald.

1. Sand-pit (138443) on west side of road, 370 yd E. 5° S. of Sixmile Cottages. (E 28811, A–F).
2. Sand-pit (145441) in Park Wood, 580 yd W. 42° N. of the Gate Inn. (E 28812, A–H).
3. Side of lane (200397,) 450 yd E. 20° S. of the inn at Paddlesworth. (E 28813, A–D).
4. Cliff top (244379), 1650 yd N. 15° E. of Copt Point. (E 28814, A–G).

Analyses by S. W. Wooldridge (1927, p. 100). Figures recalculated as percentages.

10. Average for Folkestone area.
11. Paddlesworth, Kent.

Analyses by R. W. Elliot (*in* Worssam 1963, table 2, p. 89).

5. Old quarry on edge of scarp at Deans Hill, 1¼ miles N. of Harrietsham. (E 25273, A–C).
6. Between Eagles Wood and Reynolds Wood, about ½ mile N.E. of Harrow Inn, Warren Street, N.E. of Lenham (E 25274, A–E).
7. Small quarry on scarp, 700 yd W. by N. of Pivington Farms, Lenham (E 25275, A–B).
8. Lenham " Old South Pit ", 1000 yd N.N.E. of Lenham church. (E 25277, A–B).
9. Old quarry on edge of scarp, 300 yd S. of Coldblow, Thurnham. (E 25276, A–D).

Analyses by C. H. Edelman (1933, p. 62). Figures recalculated as percentages of total residue.

13. Average figures for A-Group (Quaternary of the Netherlands).
14. Average figures for B-Limburg Group (Quaternary of Lim...

In samples of Lenham Beds from the present district Groves (1931, p. 86 and table III) found an assemblage in which zircon, rutile, tourmaline, staurolite and kyanite were common; he also recorded the local occurrence of muscovite, andalusite and chiastolite, and of rare monazite and sphene; garnet was found at Folkestone only, where it was "very common".

The present results conform broadly to those of Wooldridge. Rather less kyanite was found, although this mineral was always present; also always present were anatase and brookite, agreeing with Elliot's findings, and sphene, which neither Elliot nor Wooldridge recorded. Spinel was found in only one sample, and garnet, pyroxene and corundum were absent. Tourmaline, staurolite, andalusite and kyanite were usually found amongst the larger grains, anatase, brookite, amphibole and monazite amongst the smaller.

Edelman and Doeglas (1933) grouped those Tertiary rocks of the Netherlands which are younger than the Asschien according to their content of heavy minerals. The A-Group was characterized by the predominance of epidote, amphibole and garnet, and was known only as marine deposits; the B-Group was characterized by the lack of "epidote in real quantities", while garnet and amphibole occurred in some samples: this group was known only as continental sediments; the intermediate Transition-Group, in which epidote and rutile were predominant, was mainly of marine origin. Edelman (1933) extended the classification to the Quaternary and tabulated, among others, the average heavy mineral content of the A-Group and B-Limburg Group (a subdivision of the B-Group). These figures (recalculated as percentages of the total heavy mineral residue) are reproduced in Table 3. The resemblance of the heavy mineral content of the Kent sands to the B-Limburg assemblage is apparent, and Edelman (1933, fig. 1) included the Lenham Beds in the B-Group. As Elliot (*in* Worssam 1963, p. 90) has pointed out, however, the latter is known only as continental deposits, while the Lenham Beds are of marine origin.   H.A.H.M.

HEAD

The deposits shown as Head on the map are lithologically variable; those on or near the Eocene beds generally comprise clays, silts, sands or sandstone, usually mixed with battered flints from river gravel and also angular flints. Tertiary-rolled flints and green-coated flint nodules from the Bullhead Bed are commonly present. On the Chalk outcrop the Head is composed of clays and loams with many flints and is for the most part derived from the Clay-with-flints. It shares with Head Brickearth the tendency to rest on slopes facing between north-west, north-east and south-east. On or near the Gault outcrop the deposits comprise brown to blue tenacious clays or brown loams with white-patinated angular flints, mostly at the base, and chalk fragments. The deposits are rarely found on the Lower Greensand and Wealden strata.

The Head spans several periods of formation, during any one or all of which material may have contributed to a particular spread. Locally it is possible to detect two periods of formation, where one deposit is found to be derived from another (pp. 225–6), but there can be no certainty that the individual phases at different localities were coeval. Some of the deposits of Head have accumulated as the result of recent downwash, particularly where the relief is relatively high, as in the Chalk valleys. Indeed Dines and others (1954, pp. 124–32) described part of the Head deposits of the Chatham district under the title 'Hillwash Head'. This lead is not followed here because of the uncertainty in separating the different types of Head. The spreads are extremely variable in thickness. Those which could be termed 'Hillwash' are usually thin, up to about 5 ft thick, although they may greatly exceed this figure in pipes in the Chalk and where occurring in hollows and valley bottoms. The 20-ft deposit at Bagham (p. 226) is one such example.

J.G.O.S.

Spreads of reddish brown or brown clay with flints between Westwell and Boughton Aluph (Fig. 7) are shown as Head on the geological map. They resemble Clay-with-flints except that they lie below the Chalk escarpment. Similar deposits occur near Lenham (Worssam 1963, pp. 101, 110), while other comparable deposits, mapped as Clay-with-flints, occur in the Stour and Petham valleys (pp. 203, 218).

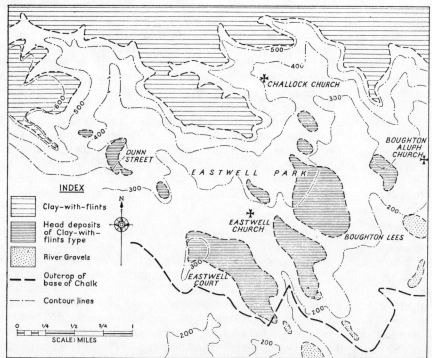

Fig. 7. *Sketch-map to show occurrences of Head deposits of Clay-with-flints type near Ashford, and their relation to Clay-with-flints and River Gravels.*
Drift deposits other than these three are omitted

The highest spread, west of Dunn Street, is about 450 ft above O.D., while the extensive spreads in the eastern part of Eastwell Park, which appear to lie on the dissected remains of a gently sloping plain, decline from 350 ft to about 250 ft O.D. to the east and south-east, where the deposits pass into 'Hillwash Head' (not shown separately on the 1-inch map) or into Head Brickearth. The deposits were noted by Topley (1875, p. 168), who wrote of this ground, north-west of the gravelly area of the Stour valley, as being " in places covered with a subaërial deposit, containing angular flints ".

In no exposure of the clay has bedding been noted ; in some the clay descends in solution pockets into the Chalk and it has in general the appearance of a residual deposit. However, the many unworn and little-broken flint nodules which it contains must have been carried down from the escarpment, as flints do not occur in the Middle and Lower Chalk beds on which the deposit rests. It seems likely that it developed in part from the weathering during Pleistocene interglacial phases (cf. Loveday 1962, p. 96) of a chalky scarp-foot drift such as the Coombe Deposits.

About ⅜ mile north-west of Goat Lees a spread appears to be in contact with a patch of 4th Terrace gravel. In the main, however, the deposit must be older than the 4th Terrace in the Ashford area. Towards the Chalk escarpment the various spreads all end abruptly at least 200 yd before the steepest part of its rise, giving some measure of the amount by which the escarpment has receded since the deposits were laid down.   B.C.W.

HEAD GRAVEL

Certain gravels in the north-west corner of the district included in the Canterbury (289) Sheet are classed as Head Gravel. They form spreads on hill-tops, with a slight northward slope that suggests derivation from the south, and were probably deposited where the large dry valley west of Hanslett's Forstal, at an early stage in its development, opened out on reaching the Thanet Beds outcrop. In the Chatham district (Dines and others 1954, p. 106) such gravels were described as an older group of Head deposits. However, the gravels near Hanslett's Forstal have a notable correspondence in altitude with 3rd and 4th Terrace gravels of the Lampen Stream and neighbouring valleys east of Canterbury (p. 270), which are similarly situated towards the foot of the Chalk dip-slope and it is possible that they are of similar age to these terrace gravels, and that their mode of origin may have been fluviatile.   B.C.W.

COOMBE DEPOSITS

In the past the term Coombe Deposits has been applied to a wide range of deposits found in the Chalk valleys, such as valley gravel and brickearth, but it is here restricted to those Head deposits which consist mainly of derived chalk. Mantell (1822, pp. 277–8) described the ' Calcareous Bed '. in the cliffs at Brighton as " formed of the ruin of the chalk strata, with an intermixture of clay " ; later (1833, p. 31), he added " the harder masses are provincially termed *Coombe rock* ". This name was used by Bromehead (*in* Dewey and others 1924, p. 106), White (1924, p. 79) and latterly by other authors, for all the chalky deposits, whether indurated or not, and was included within the broader division of Coombe Deposits. However, for descriptive purposes it seems best to confine this expression to the cemented parts of the deposit, and to use the term Coombe Deposits for the uncemented material.

Coombe Deposits mantle slopes and fill the bottoms of coombes cut in the Chalk escarpment ; they also lie in hollows or upon the escarpment slope, thus belying their name. In places they spread to the Gault outcrop where they form a light workable soil in contrast to the adjacent clays. They consist of slipped and washed chalk debris in the form of pebbles, pellets or powder, or a mixture of these ; angular flints are usually present. The deposits are white, with a light brown irregular ferruginous banding in many places. Parts of the surface layers are commonly decalcified to form a brown flinty clay.

Reid (1887) considered that similar deposits in Sussex were formed by solifluxion under subarctic conditions, with the Chalk frozen hard so that water could not percolate. Such conditions are here envisaged for the formation of the coombes themselves, but most of the deposits of this district seem to be later than the main period of coombe formation. Spreads near Brook and at Folkestone have yielded molluscan faunas which indicate decreasing severity of the climate in Late-glacial times and such depositional conditions may be applicable to the Coombe Deposits generally.　　　J.G.O.S.

HEAD BRICKEARTH

Superficial loams shown on the geological map as Head Brickearth, are so called because they are suitable for making bricks. However, the once extensive use of brickearth for this purpose has dwindled and within the present district is now (1960) confined to one brickworks near Hawkinge (pp. 236, 298).

The loams are dark or light reddish brown to buff in colour and are unbedded, apart from rare lines of flints and where interbedded with sand and gravel. Normally they are friable and sandy, but every gradation exists to a heavy clay loam. On weathered faces a marked vertical jointing is commonly present.

Thickness of the deposit varies greatly within each spread ; 22 ft were proved at Molash (p. 234), where the base was not reached, but generally 15 ft would be regarded as greater than average. In the thicker and more extensive spreads the unweathered brickearth is a light brown or khaki colour and highly calcareous. The calcareous matter occurs as tiny fossils and chalk fragments, as an infilling of tubes up to about 5 mm in diameter and also as small ' race ' nodules. The uppermost 5 ft or so of these large spreads, and the whole thickness of the smaller spreads, are of a deeper colour and contain numerous cavities and tubes from which the calcareous matter has been dissolved. The base of this decalcified layer is sharp and tolerably even but some cavities and tubes are also present in the calcareous levels below. Pitcher and others (1954, p. 309) found that in brickearth at Pegwell Bay the content of acid-soluble carbonate fell suddenly from 25 per cent in the lower layer to 5 per cent in the upper.

At the base of the brickearth there is always a gravelly layer which may be composed of battered subangular or angular flints. Tertiary-rolled flints, flints from the Bullhead Bed or fragments of Lower Greensand, and in many places it is indistinguishable from Head. In areas where there was a dearth of material this layer is represented only by an occasional Bullhead or Tertiary-rolled flint. Some brickearths overlie river terraces while others interfinger and pass laterally into the river gravel.

The loams are found at any elevation, resting on valley slopes or on gently inclined surfaces such as that of the Clay-with-flints or the Eocene plateau. Many of the scattered small patches on the Clay-with-flints may be piped relics of very much larger spreads and it is possible that some of the larger patches may also be so preserved (p. 235).

Measurements made from the 6-inch geological maps reveal that the total area of brickearth outcrop in this district is about 34·6 square miles ; of this only 6·6 per cent rest on slopes which face in directions between north-west and south (Fig. 8), and 9·9 per cent on slopes facing between south and south-east. The relative absence of brickearth on slopes which face between north-east and east (11·6 per cent) is almost certainly due to the paucity of slopes of this orientation, since the dominant drainage pattern trends north-east, rather than to some depositional phenomenon.

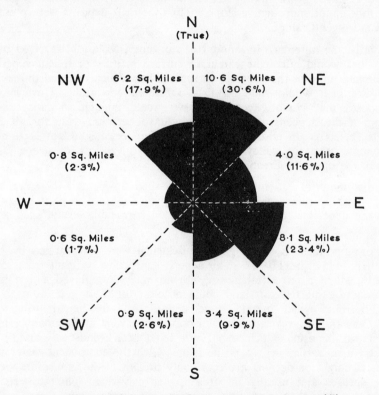

Approximate Total Area of Brickearth 34·6 Square Miles

FIG. 8.   *Diagram showing the areas of Brickearth present on slopes which face within the compass points indicated*

The preference for certain slopes, which is similar to that exhibited by other Head deposits, is a conspicuous feature in many valleys, particularly in the Chalk, and is most probably the result of different erosion of the deposits, rather than of non-deposition. Such erosion may be accounted for in several ways as described below, and it is likely that the whole explanation lies in a combination of these processes.

Firstly, the distribution of the deposits appears to be related to the asymmetry of some of the valleys, for the west-facing and south-facing slopes, on which deposits are lacking, are usually the steeper. The "tilting with an appreciable easterly component" which, according to Coleman (1952, p. 75) caused the lower reaches of the Great Stour to migrate east-wards, may have had a more widespread effect than has hitherto been realized. Such a tilt, if in a direction oblique to the trends of the valleys, would have caused the Chalk streams to cut their valleys into an asymmetrical form, eroding more on their eastern and northern banks and removing the drift accumulations thereon in the process. Ollier and Thomasson (1957) and Thomasson (1961) maintained that asymmetrical valleys developed at the same time as down-cutting. Erosion affected the steeper slope more than the less steep, causing the former to undergo parallel retreat, while in the valley bottom the down-cutting stream moved laterally to maintain a position at the foot of the more active slope. Thus the drift deposits were preserved on the more gentle slope.

Secondly, the difference in amount of sunshine (insolation) received by the valley sides would affect the extent of surface melting of frozen superficial deposits and more of the deposits on those slopes which received the greatest amount of heat would slide off in a saturated condition. However, this presupposes that climatic conditions were cold enough to cause regular freezing of the deposits, which may not have been the case for all Head deposits (p. 200). In addition insolation differences would not account for the asymmetry of the valleys, but would have given rise to steeper slopes facing in the opposite directions.

Thirdly, material can more easily be washed down, and therefore off, a steeper slope by rainwash. The author believes this to be particularly true where the deposit is permeable and rests on permeable strata, such as the Chalk.

Various opinions on the origin of brickearth were summarized by Dines and others (1954, pp. 110–2). The deposit can probably result from several processes, all of a terrestrial nature. Wind may have played a part in the formation of some deposits, and could account for those spreads near Wye and Canterbury which mantle gravel deposits and the neighbouring valley sides. Generally, however, brickearth occurs in a very similar manner to the stony Head deposits, which could not have been wind-transported; also solid strata or earlier drift deposits from which the component materials of brickearth might be derived are commonly present in the immediate vicinity of the spreads and usually lie at a higher elevation. This suggests that solifluxion, rather than wind, was the main transportive agent. Many spreads seem to be simple downwash deposits from a loamy parent-rock such as the Thanet Beds or the Sandgate Beds, assisted in some cases by frost-churning. Some of the solifluxion sludges were carried away by streams and are inter-bedded with and pass laterally into gravels. These deposits are particularly noticeable in, but are not restricted to, those valleys or parts of valleys where no supply of sand was at hand, as in the Chalk tributary valleys and in the Little Stour valley. Some of the brickearth is found to be interbedded with sand and was presumably deposited in water. In an old brickworks near Canterbury (p. 239) the brickearth exposed in 1952 had graded bedding, with sand streaks and lenses, and may have been formed in a manner similar to a varved clay.

Brickearth deposits which mantle or replace the deposit from which they were derived were probably formed as a frost-soil, as described by Pitcher, Shearman and Pugh (1954). Some of the spreads described here, in particular those near Wingham, which are derived from Thanet Beds and are associated with shattered chalk, probably originated in this way. It seems that even so clayey a formation as the London Clay can weather down under suitable conditions to produce a heavy brickearth. Within this district such brickearth is associated with landslipped London Clay, and is perhaps formed by intense frost action upon the landslipped material. The most suitable source rocks for more normal brickearth spreads, however, are those which are themselves loamy and break down readily. The Thanet Beds, with which brickearth is constantly associated, are a good example.   J.G.O.S.

Petrology of the Brickearth. The samples of brickearths examined are essentially composed of fine sand and silt (63 to 89 per cent) but show considerable variation in the frequency of occurrence of the different grades (gravel and sand 8 to 61 per cent, silt 23 to 79 per cent, clay 9 to 30 per cent ; Table 4). The samples are mainly well sorted but some show normal to poor sorting. Only one sample (MR 24807) came from an unweathered section. It is a calcareous clayey silt containing about 15 to 20 per cent calcite. The loss of weight on extracting with cold dilute hydrochloric acid was 21 per cent, but this includes a small proportion, probably a few per cent, of other acid soluble matter, principally limonite. The remaining ten brickearth samples examined are practically free from calcite. Acid loss in nine specimens amounted to 2·3 to 5·4 per cent, but a CO_2 determination by Mr. G. A. Sergeant on the sample which showed greatest loss (5·4 per cent) was only 0·02 per cent.

Particle-size analysis by Mr. J. Dangerfield (summarized in Table 4) followed the method described on p. 203. As in the Clay-with-flints, the clay fraction includes dispersed iron oxide, and calcite (some extremely fine) is present in the unweathered brickearth sample. Typical cumulative frequency curves and a histogram are shown in Fig. 9. The ' coefficient of sorting' ranges from 1·3 to over 4, but small changes in the amount of clay grade make considerable differences to the figure and render it of doubtful value. Figures in Table 4 are expressed to two decimal places to allow accurate summation.

The two high level brickearths (on Clay-with-flints, Table 4, Nos. 6, 7), are well sorted, which may indicate a less composite origin than for some of the other samples. The brickearth on Woolwich Beds (No. 1) has a greater proportion of coarse and medium sand, a grain-size distribution also shown by the Woolwich Beds themselves. A brickearth related to Head (No. 10) is poorly sorted with 3 per cent of fine gravel present. The cumulative frequency curves are generally smooth.

A thin section (E 30564) of the very friable calcareous rock (Table 4, No. 1) shows it to be a calcareous silt, composed of angular quartz in a ground of finer quartz, calcite, glauconite, plentiful small fossils, including shell fragments and chambered organisms, limonite ore and clay. Tubes (generally ascribed to rootlets) have central orifices of about 0·3 mm with rims of

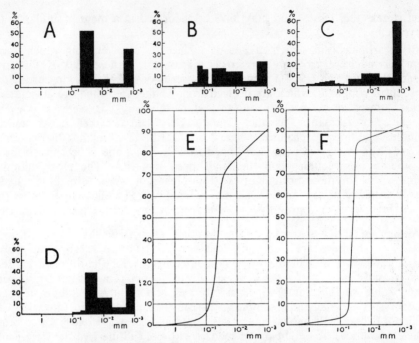

A–C.   Block diagrams ('histograms') showing variable grain-size distribution of
        Clay-with-flints.   A is No. 16 of Table 4; B is No. 15; C is No. 14.
D.   Block diagram ('histogram') of typical Brickearth specimen. No. 9 of
        Table 4.
E.   Cumulative frequency curve (weight percentage). Brickearth No. 3 of Table 4.
F.   Cumulative frequency curve (weight percentage). Brickearth No. 7 of Table 4.
        Very well sorted.
        The diagrams are cut off at 10⁻³mm.

FIG. 9.   *Diagrams showing the grain-size distribution o
           Clay-with-flints and Brickearth*

calcite of very fine grain among which there is only a little quartz silt. The
diameter of orifice plus rim is about 0·6 mm. The carbonate is largely
calcite ($\omega=1\cdot658$) but a little carbonate has refractive index ($?\omega)=1\cdot680$.

Examination of the separate rock fractions and the thin section of the
unweathered brickearth shows the maximum tube length (which survived the
separation) to be about 4 mm. The coarse sand fraction is mainly composed
of the rootlet tubes. Quartz is dominant in the medium and fine sand grades.
Plentiful muscovite and a little pale-brown biotite occur among the fine sand ;
glauconite is common in the medium and fine sand, some as casts of
chambered organisms, and from X-ray evidence glauconite may also occur
in the clay fraction. Chambered fossils and derived *Inoceramus* fragments
are common and there is plentiful limonitic iron ore, and flint. A little alkali-
feldspar including microcline is present. Heavy minerals include zircon,
rutile, staurolite, hornblende, epidote, brown and green tourmaline, garnet,
kyanite, sapphire and zoisite. Many foraminifera and other small organisms
were reported by Dr. F. W. Anderson to be derived Chalk types.

In other samples, quartz is dominant. It is generally subangular but some
is well-rounded and some contains magnetic inclusions. Flint, glauconite,

TABLE 4

Particle-size analysis of Brickearths and Clay-with-flints

	Fine Gravel	Coarse Sand	Medium Sand	Fine Sand	Silt	Clay	Fraction 0·01 to 0·05 mm
B.S.							
Sieve No.	>7	>25	>72	>300	Sub-sieve		(approx.
Mm	>2·411	>0·599	>0·211	>0·053	>0·002	<0·002	per cent)
Brickearths							
1	0·39	2·65	3·66	6·05	63·50	23·71	53
2	0·06	0·15	2·02	46·80	32·14	18·73	22
3	Nil	0·87	1·40	22·26	62·93	12·52	50
4	1·25	0·80	4·02	12·96	49·85	30·14	25
5	0·19	0·15	0·19	60·81	22·76	15·53	17
6	Nil	0·10	0·42	15·99	67·25	15·73	55
7	Nil	0·77	1·39	10·03	78·70	8·52	71
8	Nil	Nil	Nil	7·95	66·65	25·37	58
9	1·94	0·68	0·64	4·27	63·34	28·93	50
10	3·12	1·36	2·16	6·11	61·16	25·40	48
11	0·67	0·14	3·68	23·77	45·27	26·42	36
Clay-with-flints							
12	Nil	0·94	0·14	21·29	72·35	5·28	—
13	Nil	0·19	6·87	40·66	10·80	41·47	—
14	1·22	1·79	8·50	6·48	23·23	58·74	—
15	1·61	1·97	2·39	38·85	32·82	22·34	—
16	0·38	0·55	1·18	1·96	61·13	34·84	—

1. Brickearth on Woolwich Beds. 300 yd N. of White Wall (089575). MR 24813.
2. Brickearth on Thanet Beds (near Chalk boundary). 450 yd W. 16° N. of Staple church (265568). MR 24817.
3. Calcareous brickearth on Thanet Beds (near Chalk boundary). 500 yd S. 25° E. of Wingham church, Kent (245570). MR 24807.
4. Brickearth on Thanet Beds. 400 yd W. 40° S. of Denstead Farm (099574). MR 24809.
5. Brickearth on Thanet Beds. Road cutting 300 yd E. 20° S. of Preston Farm (025603). MR 24816.
6. Brickearth on Chalk and Clay-with-flints. 100 yd E. of Inn at Stelling Minnis (143470). MR 24811.
7. Brickearth on Chalk and Clay-with-flints. Hawkinge brickworks (223409). MR 24810.
8. Brickearth on Chalk. 1620 yd S. 42° E. of Canterbury (East) Station (157562). MR 24812.
9. Brickearth on Chalk. 500 yd N. 31° W. of Petham church (129515). MR 24814.
10. Brickearth on Gault. 800 yd W. 40° S. of Horton Park (120396). MR 24819.
11. Brickearth on Folkestone Beds. 400 yd E. 40° S. of church at Brabourne Lees (084401). MR 24818.
12. Clay-with-flints. Trench 180 yd S.W. of Wytherling Court (035532). MR 24820.
13. Clay-with-flints. 200 yd S.W. of the Inn at Stelling Minnis (141469). MR 24821.
14. Clay-with-flints. Sewer trench at Capel Street, S.E. of Cauldham (246384). MR 24822.
15. Clay-with-flints. 300 yd S.W. of the Inn at Stelling Minnis (140468). MR 24823.
16. Clay-with-flints. 500 yd N. 31° W. of Petham church (129515). MR 24824.

ironstone, siltstone and mica flakes are present; among heavy minerals, zircon, staurolite, rutile, hypersthene, tourmaline and rare alkali-amphibole (probably riebeckite) were found.

The clay fractions of six samples were found by Mr. B. R. Young to contain an abundant clay mineral with 10Å basal spacing, probably illite or glauconite, and kaolin, with appreciable amounts of quartz. Several specimens yielded a weak line at 14Å, which was still present on films of material which had been heated to 550°C for two hours and is, presumably, due to chlorite.

Some British brickearths have previously been compared with loess (Pitcher, Shearman and Pugh 1954). Russell (1944), who studied the extensive Lower Mississippi deposits, postulated a number of main criteria, including a characteristic grain-size distribution of at least 50 per cent of the particles between 0·05 mm and 0·01 mm, and the presence of calcareous tubular concretions commonly enclosing rootlets. Half the samples fail these criteria on grain size (see also Sabine, Young and Dangerfield 1963). Only the calcareous brickearth, where there are rootlet tubes remaining, bears close comparison with loess, although the origin of loess itself is disputed and may vary in different parts of the world. The other brickearths now examined may be weathered types, formerly calcareous, and comparable with loess, but the evidence is no longer available to allow decisive argument.   P.A.S.

DETAILS
Clay-with-flints

Hoath Wood—Frittenfield—Molash—Shottenden.

Around the Eocene outlier south of Hoath Wood much of the Clay-with-flints has been derived from the Thanet Beds, the sediments of which grade into the sandy clay of the Clay-with-flints; the latter is brown'at the surface but in the mass it also contains reddish brown and grey clays of Thanet Beds type. Brown flinty clay, 2 ft thick with deeper pockets, was exposed at localities (974579 and 984581) respectively 950 yd W. 16° N. and 500 yd N. 35° E. of Scooks. A similar thickness of stiff brown clay with white-patinated flints was noted (975559) 700 yd N. 40° W. of Hockley, and 420 yd E. 34° S. of this place 0 to 3 ft of stiff chocolate-brown flinty clay overlie Chalk. Thanet Beds sand occurs as debris and is believed to be intermixed with the Clay-with-flints thereabouts.

Near Belmont the loam commonly found at the surface of the Clay-with-flints is absent and the deposit, which is perhaps only 4 ft in average thickness, gives rise to a heavy flinty clay soil. To the west of Throwley the deposit becomes more loamy at the surface and contains many flint nodules from the Bullhead Bed. The underlying Chalk reaches the surface and occupies a slight rise above the plateau north of Parsonage Farm (998562). Such occurrences are common where the Clay-with-flints is relatively thin, but this patch, 500 yd by 200 yd, is unique on account of its size. At Wilgate Green, 650 yd E. 30° S. of North Wilderton, a degraded chalk-pit (999568) indicates that the deposit is about 6 ft thick.

The spread south of Tong Green is loamy at the surface and the majority of the flints scattered at the surface are under 3 inches long; a few larger flints and nodules from the Bullhead Bed are also present. The section in a trench (984517) some 850 yd E. 44° N. of Monkery Farm was recorded by the land-owner. It displayed the contents of a large pipe for a distance of 60 ft and to a depth of 8 ft. The pipe consisted of a core of buff loam (brickearth) with grey streaks and patches passing upwards into the subsoil; this was surrounded successively by dark brown clay with large flints and by red and grey mottled clay, and a pinnacle of brown clay also rose into the core of loam. J.G.O.S.

Northward of the main road to the west of Paddock a surface layer of reddish or yellowish brown silty loam with angular flints, 1 to 2 ft thick, overlies dark red or reddish brown clay with flints.

Much yellow to orange or red sand and clayey sand, with flint pebbles in places, occurs in an area of woodland (975492) ½ mile west-north-west of Frittenfield. An old pit 100 yd S.W. of Frittenfield showed Chalk at ground surface at one place and 6 ft of clay with flints at another some 10 yd distant. B.C.W.

Sandy and stiff clays with white chips of flint were seen for 2 to 6 ft in an exposure (998533) 600 yd S. 40° E. of Bell's Forstal, 'hillwash' making up a large proportion of the section. Two pipes of Thanet Beds are also exposed here and loams of these beds occur in the fields to the north. Brown clay with large flints was seen for 4 ft in a section (002541) 1000 yd E. 28° N. of Bell's Forstal. J.G.O.S.

Around Pested (005515) stiff red clay was seen in many shallow excavations, and in the soil are numerous angular flints, small green-coated flint nodules, and fragments of dark brown ferruginous sandstone. A thin section of this sandstone (E 28169) from a block found loose on the surface near Tower Farm (020514) is described by Dr. R. Dearnley as showing a ferruginous conglomerate with subangular quartz grains (0·25–0·4 mm) and grains of sutured quartz cemented by ferruginous (limonite–haematite) matrix.

A temporary excavation (011505) ¼ mile S. 11° W. of the inn at Challock Lees cross-roads, was 6 ft deep and showed, resting on an irregular chalk surface, ½ to 1 ft of dark brown clay with large flint nodules, overlain by 2 to 5 ft of pale brown silty clay. Excavations to 4 ft depth at a school 250 yd S. 22° W. of here showed stiff dark red clay with flints and several blocks of fine-grained white quartzite ('sarsen-stone').

On the highest part of the plateau south of Challock Lees stiff red or reddish brown clay with flints is widespread. Between Challock Lees and Molash, below the 500 ft contour, green-coated flint nodules and fragments of dark brown ferruginous sandstone, as well as flints, are abundant in the soil. B.C.W.

An old pit (011526) 300 yd N. 17° E. of Hegdale Farm exposed 1 to 5 ft of stiff chocolate-coloured clay with darker streaks and flints resting unevenly on chalk. Fine golden and greenish brown sandy clay is present in the deposit which also contains Tertiary-rolled flints, ironstones and green-coated flints. J.G.O.S.

A roadside section (027517) ¼ mile S. 22° E. of Molash church showed 3 to 4 ft of red and pale brown mottled stiff plastic clay, with patches of reddish brown sandy clay, the whole being full of flints, either broken nodules or small angular chips, and with occasional small green-coated flint nodules and blocks of coarse-grained ferruginous sandstone. A similar roadside section was seen 350 yd to the north-east. The thickness of the Clay-with-flints nearby is about 16 ft. B.C.W.

Adjacent to the outlier of Thanet Beds at Lees Court there is a considerable amount of loam within the Clay-with-flints; at a locality (016560) 450 yd W. 18° S. of this place stiff flinty clay underlies 3 ft of reddish brown loam containing large angular flint nodules, small unworn flint nodules and green-coated flint nodules.

A section (017536) 530 yd N. 2° E. of Dryland Farm shows dark chocolate-brown flinty clays unevenly overlain and enveloped by 6 ft of greyish brown and red speckled clays with pockets of flints, below 1 ft of flinty loam soil. In a hedge-bank (022538) 850 yd N. 38° E. of this farm 3 to 6 ft of brown flinty clays were noted, while similar clays were seen to a depth of 5 ft at a locality (032542) 875 yd E. 40° S. of Woods Court, and many Tertiary-rolled flints were scattered in the neighbouring fields. In a pit (034536) 480 yd N. 25° W. of Wytherling Court the Chalk is overlain by 4 to 6 ft of chocolate-brown flinty clay.

In an isolated patch of Clay-with-flints which is benched into a slope and surrounded by Head, a pit (053545) 1050 yd E. 17° N. of the cross-roads at Shottenden shows chalk beneath 3 ft. of brown clay with unworn chips of flint.

Godmersham—Chartham—Petham—Hastingleigh—Stelling Minnis.

The anomalous patch of Clay-with-flints ½ mile east-south-east of Godmersham occupies a wooded spur and consists of stiff chocolate-brown and red flinty clays ; the spread ¼ mile north of Godmersham comprises stiff clays and brown gritty clays with water-worn flints, Tertiary-rolled flint pebbles and also some reddish brown sands. Spurs south-east of Old Wives Lees, at a level between 4th Terrace and 3rd Terrace, are covered by staff flinty clays. Three other occurrences ⅓ mile north of Chartham lie below the level of 4th Terrace ; the lowest of these, originally noted by Whitaker (1872, p. 29), is exposed in the railway cutting (106558) 750 yd N. 5° W. of Chartham church, where pipes up to 3 ft deep contain white angular flint chips, a few small water-worn flints, battered subangular flints, Tertiary-rolled flints and small rounded siltstones, in a matrix of reddish brown gritty clay and clayey sand. In one pipe a layer with a clayey matrix overlies a mass with similar constituents in a sandy matrix, suggesting that the deposit has been derived in part from an old deeply piped river gravel.

Near Petham anomalous deposits of Clay-with-flints are benched into the valley side at about 200 ft O.D. One patch is a continuation of the main development on the plateau, which itself contains water-worn flints. A temporary section at Wootton Farm (123518) exposed 3 ft of abraded flints, some of which had been subsequently split, in a matrix of stiff red and brown loamy clays. Water-worn flints scattered on the surface at the northern tip of the dissected spread of Clay-with-flints between Godmersham and Petham were probably derived from the river gravel of the 5th Terrace.

At Solestreet (096493) the deposit was reported to be 25 ft thick and it was 6 ft thick 1250 yd W. of Pett Street (073471). A deep auger hole (094461) 540 yd S. 30° E. of Hassell Street showed 4 ft of stiff brown flinty clay passing down into sandy clay, 4 ft ; this in turn passed down into brown slightly clayey coarse sand, 3 ft ; resting on stiff brown flinty clay, seen for 3½ ft. The total thickness of the deposit was said to be about 16 ft thereabouts.

A chalk-pit (119466) 150 yd S. of Ittinge Farm shows a pipe which is below the level of the nearby spread of Clay-with-flints and sand. This pipe is 5 ft wide, is exposed to a depth of 5 ft and has a lining of stiff brown flinty clay, which is succeeded inwards by reddish brown loam containing flints and iron-stones ; this grades inwards into reddish brown and grey loam with sandy streaks ; in the core of the pipe is reddish brown medium-grained sand.   J.G.O.S.

Near Hastingleigh Tertiary-rolled flint pebbles, decayed flints, cherts and quartz pebbles are common in the soil. Between Hastlingleigh and Stowting Common the residual ridge above the Chalk escarpment is capped by stiff reddish brown and grey clay with many flint nodules and pebbles, and near the trackway (100430) 1200 yd south-south-east of South Hill the deposit includes much sand.

At Stowting Common (123435) an auger hole proved 14½ ft of loamy brown clay, which was increasingly sandy towards the base ; chalk was not reached. Green-coated flints were found on the surface (146451) 1000 yd S. 30° W. of Wheelbarrow Town. G.B.

Brown flinty clay about 3 ft in thickness is exposed (130488) 350 yd N. 35° E. of Wadden Hall, where the deposit is probably more than 20 ft thick. At North Leigh (132472) a well record indicates that the Chalk is only 7 ft below the surface, that is considerably less than the locally reported thickness of the Clay-with-flints. At Stelling Minnis cesspits have been dug through Head Brick-earth and Clay-with-flints, to 20 ft without encountering chalk, and the deposit has been worked (probably for brick-clays) to 7 ft depth in a pit (143481) 1200 yd N. 4° E. of the inn at Stelling Minnis. At Broxhall Farm (158513) 3 to 5 ft of stiff red flinty clay are exposed.

Milestone Farm—Kingston—Hemsted—Brockman's Bushes.

Much loam from the Thanet Beds and water-worn flints from river gravel are incorporated in the small spread near Milestone Farm ; 2 to 4 ft of loam and clay with angular and battered flints and pipes of Thanet Beds rest on chalk in a pit (170558) 100 yd N. of the farm.

Water-worn flints are scattered on the surface of the deposit around the river gravels west of Kingston. Brown flinty clay 2 to 5 ft thick was noted in a section (177509) 550 yd N. 10° W. of Reed's Farm. Large flints are imbedded in the 2 to 5 ft of stiff chocolate-brown clay exposed (165488) 200 yd E. of Lynsore Court and 6 ft of similar clay with split unworn flints were seen (168479) 250 yd E. of Palmstead. J.G.O.S.

In an old pit (193499) 280 yd W. of Heart's Delight 4 ft of reddish brown flinty clay were seen resting on chalk. The soil on the spur north of this pit is gravelly, while to the south patches of brickearth, locally exceeding 4 ft in thickness, rest on the clay ; in a temporary excavation (192495) made in a depression 600 yd W. 40° S. of Heart's Delight, 4 ft of clay were overlain by 8 ft of brickearth.

Chocolate-brown clay 3 to 4 ft thick, with unworn flint nodules and a few flint chips, overlies chalk in pits (170462 and 179463) 480 yd W. and 600 yd E. of Jacques Court.

In a chalk-pit (162449) just inside Clavertye Wood dark brown to black clay with flints fills pipes in the chalk. A grassed-over pit (164445) ¼ mile W. 5° N. of Exted is 9 ft deep and was presumably dug for brick-clay ; brickearth is present at the surface here, but for the most part the pit is in stiff brown clay. Flints on the surface (163440) 600 yd N. 40° E. of Mount include some with brown and white patinas and many derived from Tertiary deposits.

Near Hemsted, where the depth to chalk is about 8 ft, debris from a trench (141418) dug to 5 ft suggested that sand underlies the surface clays. On the crest of the escarpment north-east of Horton Park, Tertiary-rolled flints, decayed flints and ironstones are commonly scattered over the surface.

Small outliers of Clay-with-flints are present on hillocks (151405 and 158403) respectively 1100 yd W. 20° S. and 700 yd S. 35° W. of Lyminge church ; at the latter place the deposit is gravelly in part and 50 or 60 ft lower than that in Postling Wood.

The Clay-with-flints on the summit of the Middle Chalk outlier at Brockman's Bushes varies from stiff reddish brown clay to reddish brown loamy or sandy clay, with flints. It includes patches of sands with small flint pebbles.

G.B., J.G.O.S.. B.C.W.

Walderchain Wood—Paddlesworth—Aylesham—Chalksole—Abbot's Cliff.

The Clay-with-flints on the ridges south-south-west of Walderchain Wood is overlain by patches of brickearth which locally attain mappable proportions, as north of Dreal's Farm. Clay was formerly dug, presumably to make bricks, from a pit (186437) 700 yd W. 4° S. of Standardhill Farm.

In the road cutting (192398) 350 yd west of Paddlesworth 1 to 2 ft of grey flinty clay intervenes between Chalk and Sand in Clay-with-flints. In the Paddlesworth–Hawkinge area patches of sand and loam are common on and in the Clay-with-flints; in an excavation for a cesspit (209416) 550 yd W. 20° S. of Densole Farm, stiff brown flinty clay irregularly underlies and overlies red, brown and white sand, the composite deposit exceeding 8 ft. in thickness.

Reddish brown and brown clay with many flints is seen in pipes in a chalk-pit (211449) 1050 yd N. 20° E. of Brandred Farm, where the pipes are up to 5 ft deep, and in another (220466) 350 yd E. 15° S. of Denton Court, while pipes in a pit (226459) 150 yd N. 30° W. of Hill House Farm are lined with dark brown clay and filled with brickearth. An auger hole only 10 yd east of a sand-pit (223451) in the wood 600 yd E. 35° N. of the inn at Selstead proved 1 ft of brickearth on 6 ft of pale grey, light brown and red unctuous clay, with scattered broken flints, without reaching either sand or chalk.

In the railway cutting (242522) 300 yd S. 10° E. of Aylesham Halt 2 to 3 ft of brown clay rest on chalk of the *Uintacrinus* Zone; farther east the clay exceeds 4 ft in thickness. West of Aylesham patchy loam and clay, with flints, many of which are green-coated, are too impersistent to be shown on the map.

The thickness of the Clay-with-flints in a hole (260433) 7 ft in diameter, 900 yd E. 21° S. of Chalksole, was 12 ft on one side and 24 ft on the other, due to piping. East of Little London the deposit is about 15 ft thick.

On the crest of Creteway Down sand and sandstone are abundant in the clay, while at the top of the cliffs south of Capel Court and at Abbot's Cliff, Clay-with-flints can be seen in pipes, in some places with associated sand and brickearth. South of Satmore the clay surface is strewn with broken nodules of brown and grey flint, and pebbles of quartz and flint.   G.B.

SAND IN CLAY-WITH-FLINTS

Frittenfield—Godmersham—Solestreet—Hastingleigh—Stowting Common.

In the wood (978486) on the summit of the escarpment, 600 yd south-west of Frittenfield, small diggings, now overgrown, were apparently for sand. Fine-grained orange to pink sand overlain by 1½ ft of sandy loam with flints was proved by auger at the edge of a field just north of here. Sand was also noted around the wood 300 yd south-south-east of Frittenfield. In the banks of a track (022502), 150 yd east of Rattle Hall, is dark red clayey sand, and between 50 and 100 yd south-east of the track, yellow sand was seen. Yellow sand was formerly dug in a pit about 5 ft deep at the northern end of the sand patch ⅝ mile E. 23° S. of Rattle Hall. In this patch, as in others ⅜ mile E. 30° S. and ⅝ mile S.E. of Rattle Hall, the ground surface is at or slightly above the level of the surrounding Clay-with-flints. Masses of sand in the dry valley north-north-west of Warren Farm are probably piped into the Chalk. On the 500-ft contour here (030495), and also 150 yd to the west-north-west, are small quarries in yellow to orange or pink sand, with massive ironstone beds.

Beside the road up White Hill (037489), 500 yd E. 10° S. of Warren Farm, is a group of pipes at about 300 to 350 ft O.D. ; the sand in them has thus descended more than 200 ft below the Clay-with-flints plateau. On the largest pipe, west of the road, are some small partly overgrown pits ; one of these is 10 ft deep and shows 4 ft of orange to dark red unbedded sand. The wall of this pipe is exposed in a cutting beside the road, and from surface indications the pipe appears to measure 120 yd from north to south and up to 70 yd east–west. In a quarry to the south (p. 138 and Fig. 4), the chalk at the south-west end of the face, in line with this pipe, is down-faulted, presumably as a result of solution. Towards the north-east end of the quarry face the chalk is similarly down-faulted ; the edge of another pipe of sand is there seen in section. B.C.W.

There are two small patches of sand, probably resting in pipes, on the spur west-south-west of Godmersham ; a pit (045497) in the western patch, 2050 yd W. 21° S. of the church, shows 8 ft of reddish brown clayey sand with slightly abraded flints, resting on chalk.

The most northerly occurrence of Sand in Clay-with-flints within this district lies about ½ mile north of Solestreet (097502), at about 450 ft O.D. This deposit has been extensively dug, and ironstone fragments, flints, loamy sands, brickearth and iron-slag litter the outcrop.

Piping probably accounts for some of the masses of sand on the upland east and north-east of Wye ; however the spread 400 yd south-west of Marriage Farm, which is associated with light and dark grey streaky loam, appears to be lenticular within the Clay-with-flints. The largest patch, immediately east of the Crown on Wye Downs, has been extensively dug.

West of Bodsham Green indications of red gritty clays are commonly found within the outcrop of sand, which in places is sharply defined from the neighbouring Clay-with-flints, while elsewhere their juncture appears gradational. Augering proved Clay-with-flints overlying this sand (p. 218). In a bomb crater (104461) 950 yd W. 6° S. of Holt Farms 4 ft of coarse reddish brown ferruginous sand were exposed.

In a pipe exposed in a chalk-pit (126469) 500 yd N. 19° E. of Dean Farm the sands are fine-grained, ferruginous and a rich golden brown in colour, with occasional ironstone fragments, corroded angular flints and Tertiary-rolled flint pebbles. Clay-with-flints to a thickness of 1½ ft interposes between the sand and the Chalk. J.G.O.S.

An excavation for a reservoir 18 yd by 40 yd (091454), 700 yd N. 43° W. of the inn at Hastingleigh, displayed sand in shades of red, brown and grey, with slabs of deep red ferruginous sandstone in one part ; into this sand peaks of stiff brown clay with flints protruded from below. East of Hastingleigh many water-worn flint pebbles, large unbroken flint nodules and ferruginous sandstone fragments are present in the sandy soil. Coarse brown sand containing pebbles of flint and quartz was seen in the roots of a fallen tree (094439) 200 yd W. 16° S. of South Hill ; the sand at the south-western end of this mass appears to be banked against the Clay-with-flints which caps the ridge above the escarpment.

Red and brown sand is seen in pipes lined with mottled grey and black clay in chalk-pits 1000 yd S. 10° E. of Dundas Farm (113428), 450 yd S.W. of Whatsole Street (112440) and in the roadside bank (113447) at the hairpin bend 450 yd N. 15° W. of Whatsole Street ; the last is roughly 50 ft below the local base of the Clay-with-flints. The sand south of Whatsole Street is mixed with the Clay-with-flints: 100 yd S. 25° W. of the farm fine red ferruginous sand with some sandstone fragments was poorly exposed at the lane side (114442) ; 150 yd farther south-south-west an auger hole penetrated 4 ft of stiff brown clay resting on 3 ft of fine red clayey sand ; 110 yd to the east-south-east another hole revealed over 8 ft of fine red sand, but between the two holes chalk was proved at 4 ft, overlain by mottled brown, black and grey clay.

In the wood (115451) 250 yd north-west of the church near Elmsted Court small unmapped patches of fine brown sand have been dug ; sand also occurs as patches or lenses in the Clay-with-flints in an arc north and east of the church. Between Lymbridge Green and Stowting Common (124436) sand has been dug in small pits, and sand is also present in the Clay-with-flints south of the road from Stowting Common to Pett Bottom: an auger hole between these two occurrences proved 14½ ft of brown loamy clay, which became increasingly sandy downwards.

Sand is present on high ground in the wooded area south and east of Sixmile Cottages (135443): 4 ft of fairly fine soft red sand were exposed in a small pit (138443) at the roadside 370 yd E. 5° S. of Sixmile Cottages and sand was also formerly dug in pits in Park Wood (145441). At its north-eastern end this mass of sand appears to be overlain by Clay-with-flints and brickearth ; some confirmation of this was obtained from an auger hole (147442) on the west side of the road 500 yd north of Mockbeggar cross-roads which penetrated the following beds in downward sequence : brown brickearth, clayey towards the base, 4 ft 3 in ; pink, white and pale grey clay and silty clay, 9 in ; mixed multicoloured clay and brown sand with sandstone fragments, 1 ft 2 in ; buff brickearth, 7 in ; loose orange-brown fine sand, 6 in ; pink and white silty clay, 4 in ; white and brown fine sand, 5 in.

Only traces of sand and sandstone were found on the ridge (160430) 500 yd south of Mount, where sand is shown on the Old Series Geological Map (Sheet 3).

Horton Park—Tolsford Hill—Elham—Paddlesworth—Hawkinge—Cauldham.

The patches of sand on the escarpment near Horton Park and Stowting appear to lie mostly in pipes ; however Clay-with-flints apparently overlies the sand 850 yd N. 20° W. of Hemsted, while sand is believed to be present below Clay-with-flints 250 yd E. 30° S. of this place (p. 219). There are extensive diggings in the spread 1200 yd E. 20° N. of Horton Park : Whitaker (1872, p. 341) recorded that " The sand above Monk's Horton is buff and yellowish, but red where there are concretions of ironstone ; hardened masses contain flints and flint-pebbles."

J.G.O.S.

On Tolsford Hill fine-grained yellow and orange sand, with flint pebbles at one place, was seen in molehills and small excavations. Similar sands were noted in two patches at Brockman's Bushes. At the head of a dry valley (161385) a partly overgrown excavation showed what appeared to be a downwashed deposit of sand with closely packed large, slightly worn flints, flint pebbles and occasional blocks of fine-grained dark brown ferruginous sandstone. The sandstone, which resembled that typical of the Lenham Beds, yielded some indeterminable moulds of fossils.

B.C.W.

Traces of fine brown and red sand were found along both flanks of the ridge north-west of Standardhill Farm, and this sand appears to form an irregular lenticular mass within the Clay-with-flints ; this impression was supported by an auger hole (190438), 210 yd W. 15° N. of the farm, where 4½ ft of light grey and reddish brown clay rested on 3 ft of sandy clay, on 4 ft of red clayey fine-grained sand with thin seams of sandy clay, on 2¾ ft of brown clay with small flint chips. Some sand occurs in the superincumbent clay, however, and occasional quartz pebbles are found on the surface.

A further body of sand forms the high ground west-south-west of Mountscourt Farm ; in a pit (187422), now deteriorated, near the road 200 yd S. 33° W. of the farm, Topley (in MS.) recorded 10 ft of moderately coarse sand with beds of clay 1 to 2 inches thick at one place and irregular beds of ferruginous sandstone elsewhere. The sand was yellowish to bright red, with patches of white where the sand was most friable. In this vicinity a loose fragment of coarse sandstone contained pebbles, up to ¼ in diameter, of clear and milky quartz, and rarely of flint.

The sand deposit south-west of Little Shuttlesfield (182405) seems to be thin and patchy. At Great Shuttlesfield Farm (177413) sand was at one time dug around the farm buildings ; in a temporary hole 2 ft or more of red and brown sand were irregularly overlain by 3 ft of brown flinty clay. Sand is also present in the Clay-with-flints 400 yd to the east, but the deposit is probably not continuous.

A small deteriorated pit (223451) in the wood 600 yd E. 35° N. of the inn at Selstead is 15 ft deep and displays fine red and brown clayey sand, overlain by pale grey, red and brown clay with brickearth above. Here Topley noted (in MS.) 5 ft of yellow and whitish sand and brickearth (see also Whitaker 1872, p. 342). The sand can be traced for some 250 yd along the hill-side, but probably does not extend far to the east under the Clay-with-flints (p. 220).

Red sand, clayey in places, forms high ground south and east of Winterage Farm ; Topley (in Whitaker 1872, p. 342) noted that sand had been dug to depths of 12 ft and 8 ft respectively, in two pits to the south of the cross-roads east of Winterage Farm. At its northern end this mass of sand seems to lie within the Clay-with-flints, and pockets of sand also occur in the surrounding Clay-with-flints.

Sands also form the high ground at Paddlesworth ; of these Prestwich (1858, p. 324) wrote that they were 30 to 40 ft thick and consisted of ochreous and ferruginous sands, more or less argillaceous, with subordinate fine quartzose grits and seams of iron-sandstone, some of which formed blocks 3 to 4 ft wide by 1 ft thick, or even more. These blocks were common in and about the hamlet, and some contained flint pebbles and unrolled flints. In fragments of ironstone Prestwich found some pieces of fossil wood pierced by *Teredo*, together with what appeared to be the mould of a bivalve shell. Red sand with sandstone fragments is exposed in the lane-side (200397) 450 yd E. 20° S. of the inn at Paddlesworth.

At the northern end of the mass of sand at Hawkinge, Topley (in Whitaker 1872, p. 342) saw " a pipe of red and yellow sand, about 4 feet thick, with an inch layer of white pipeclay ", the red sand weathering to pink ; in another chalk-pit he recorded " 4 or 5 feet of sand and iron-sandstone with 3 or 4 inches of white and mottled pipeclay, passing into mottled clayey sand."

At the western end of the body of sand south-west of Capel Street fine red and brown sand with hard sandstone fragments was formerly dug to a depth of about 8 ft in small pits (238378) beside the road 800 yd S. 22° E. of Hope Farm. In the chalk-pit at the top of Dover Hill (238377) sand in pipes up to 10 ft deep is everywhere underlain by from 2 to 12 inches of mottled grey and black clay containing a great accumulation of unbroken flints. At the top of the cliffs above Folkestone Warren sand is seen in pipes in the Middle Chalk: one pipe (244379), 1650 yd N. 15° E. of Copt Point, is about 20 ft deep, containing brown to crimson sand with disordered masses of ferruginous sandstone ; where the sandstone bedding can be discerned the beds are seen to follow roughly the contours of the chalk surface. Some 50 yd to the south-west other pipes are lined with up to 1 ft of unctuous mottled grey and black clay with many flints, which is succeeded inwards by about 1 ft of brown sandy clay with corroded flints, and by disturbed red and brown sand and ferruginous sandstone filling the pipes.

The deposit was well displayed in a sewer trench along part of the main road from Folkestone to Dover, south-west of the road to Cauldham (Cauldham Lane): a very uneven piped chalk surface was overlain by up to 4 ft of reddish brown or chocolate-brown flinty clay, in places mottled with grey and in places manganiferous, beneath coarse red and brown sand ; peaks of clay commonly protruded through the sand to ground level, and at places 50 yd and 140 yd S.W. of Cauldham Lane chalk was touched at depths of 8 ft and 5 ft respectively. At one place clay was seen to overlie sand, possibly due to overturning at the edge of a pipe. A trench along Cauldham Lane for 300 yd north-west of the main road showed 5 to 6 ft of fine red clayey sand with impersistent beds of hard red ferruginous sandstone. In this area Topley (*in* Whitaker 1872, p. 342) put the thickness of the sand at 50 or 60 ft, but this was probably about twice the true maximum figure.   G.B.

HEAD

Westwell—Boughton Lees—Kennaways.

The banks of the Pilgrims' Way 300 yd E. of Dunn Street (995479) expose 4 ft of brown and reddish brown clay with angular flints, on chalk. The Head north of Eastwell Court slopes gently north-eastwards towards the valley containing Eastwell Lake, the northern boundary of the spread being indefinite due to downwash.

A temporary pit (017473) ½ mile E. 5° S. of Eastwell church showed 3 ft of reddish fawn loam with flints, including some large nodules, and fragments of sandy ironstone and sarsen-stone '. Just west of the triangle of roads at Boughton Lees (021471) temporary excavations up to 4 ft deep showed stiff reddish brown to grey mottled clay and reddish brown loamy clay with white-patinated angular flints, mostly broken nodules, crowded in patches. At one place was a pinnacle of rubbly chalk.

The Head on valley slopes around Challock is generally reddish brown or dark brown flinty clay, derived directly from the Clay-with-flints. On the south side of the valley (978497) ½ mile north-north-west of Frittenfield are some old trenches up to 5 ft deep in reddish brown or orange loamy clay with angular flints and flint pebbles.

A temporary excavation for a soakaway (004506) 800 yd N.E. of Beech Court exposed the deposits of the dry valley bottom, without however reaching their base. It showed 2 ft of brickearth, containing festooned lenticles of fine gravel of angular flint chips in the top 1 ft, overlain by 1 to 1½ ft of brown loam with sharp angular flints in places close-packed ; capped by 1½ ft of soil (partly recent hillwash).   B.C.W.

An old chalk-pit (984524) 600 yd E. 10° N. of Rushmere shows about 5 ft of reddish brown sandy clay with flints. The spread of Head near Bethel Row (999543) is very gravelly in parts ; southwards it passes imperceptibly into Clay-with-flints, from which formation it must have been derived. A collapsed dene-hole (008534) 300 yd N. 20° W. of the cross-roads ⅔ mile south of Leaveland exposed 4 ft of brown sandy loam with flints. In an old pit (012526) 350 yd N. 32° E. of Hegdale Farm there are large scattered flints in reddish brown loam seen for 3 ft.

Selling—Shottenden—Bagham—White Wall—Fordwich.

A large patch at Selling, derived mostly from Thanet Beds, is composed of brown clay with a few flints, but changes to sandy clay with many more flints near Hogben's Hill. On the upper valley-side south-east of this place is a patch of somewhat older Head, continuous with the above and not differentiated from it on the map. It is exposed (035566) 700 yd W. 8° N. of the inn at Selling, where Whitaker (in MS.) noted 5 ft of roughly-bedded sandy gravel composed mostly of Tertiary-rolled flints, but also with rare subangular flints. This pebbly drift is derived from the local Oldhaven Beds ; it is the oldest drift deposit near Selling and may have spread originally over a wide area. Tiny patches cap the Thanet Beds outliers at Owens Court and Gushmere and rest on Thanet Beds 200 yd S. 20° W. and 400 yd S. of the inn at Selling.

Very similar pebbly drift which caps The Mount, north of Shottenden, has hitherto been regarded as Oldhaven Beds. An excavation (043548) 40 yd W. of the highest point showed Thanet Beds overlain by 3 ft of compact gravel consisting of pebbles up to 2 in long in a matrix of brown and orange fine sand, below 1 ft of humus. Patches of grey sand with pebbles appeared to be included masses of Oldhaven Beds. Another excavation at the summit of The Mount exposed 6 ft of the pebbly drift in a black to grey (at surface) matrix of fine sand with patchy ferruginous cement, beneath 9 in of humus. The pebbles were partially and irregularly graded, with the smaller ones unconsolidated and set in very little matrix. This capping probably resulted from the almost complete destruction of the Oldhaven Beds, more or less *in situ*. The deposit, however, need not necessarily be as old as the patches of pebbly drift noted above, because of its proximity to the source rock. Pebbly downwash from this cap mantles much of the local Thanet Beds and Chalk. Similar Head crowns the hill in Perry Wood, to the north, but is not thick enough to be shown on the geological map.

There are at least two generations of Head in the large spreads east of Shottenden and at Old Wives Lees. Firstly a mass of sandy clays and clays with flint nodules, battered flints from the nearby river terraces, Tertiary-rolled flint pebbles, ironstones, flints from the Bullhead Bed, and patches of glauconitic loam from the Thanet Beds, occupies the long even gentle slopes. Secondly a downwash from this deposit drapes the steeper valley slopes.

A lane cutting (058552) 100 yd S. of Stonestile Farm shows 5 ft of stiff clay and sandy clay, tightly packed with Tertiary-rolled flint pebbles and some broken unabraded flints, resting irregularly on chalk which is itself disturbed and contains pebbles near the contact. At Old Wives Lees the Head mantling the gentle slopes passes indistinctly into the river gravels, with which it may be contemporaneous.

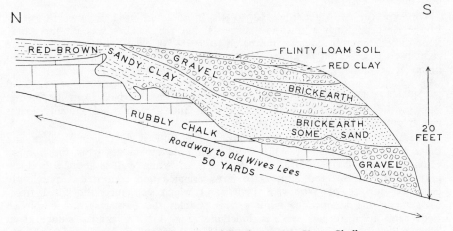

FIG. 10. *Sketch-section of Head resting on Upper Chalk
in a road cutting at Bagham*

Head deposits on the steeper slopes are derived partly from the 3rd Terrace and therefore post-date its formation.

Road-widening at Bagham (075537), 300 yd W. of the station, showed Head, composed of gravel interbedded with brickearth, up to 20 ft in thickness, resting on a highly contorted surface of rubbly chalk (Fig. 10).

In the roadside (051529) 250 yd N. 40° W. of Young Manor Farm 3 ft of flinty brown loam, passing into brickearth westwards, overlie 2 ft of hard flinty disturbed chalk. At Winterbourne, in workings (065576) for Tertiary sand (p. 190) 1450 yd E. 15° N. of Selling Station, an 11-ft section at the north end of the pit is cut slightly obliquely to the slope into a minor valley. The Head exposed consists of bedded lenses of gravel, grey clay, brickearth and medium-grained sand, but towards the eastern end of the section, where Woolwich Beds occur, it contains much less gravel and becomes mostly a coarse grey and brown mottled loam with interspersed flints (mainly chips) and Tertiary-rolled flint pebbles, and the bedding, which is parallel to the base of the deposit, becomes very indistinct. There are also indefinite masses of coarse sand present and a seam of grey clay passes directly into the clay seam in the top of the Woolwich Beds.

The same Head deposit is well exposed in the disused sand-pit (067579) 1750 yd E. 25° N. of Selling Station. The section facing north here is parallel to the valley slope and shows Woolwich Beds with a fairly even surface overlain by 3 to 6 ft of brown and grey silty clay with Tertiary-rolled flint pebbles and small chips of flint, below $1\frac{1}{2}$ ft of clayey reddish brown loam with flints. The west-facing section consists mostly of gravel made up of flints, Tertiary-rolled flint pebbles, and fragments of ironstone and purple sandstone (from the Oldhaven Beds) up to 9 in long, passing downslope into sands with gravel seams. A vertical channelled contact with disturbed Woolwich Beds sand was also noted. The disturbed nature of this sand is revealed only by the presence of two seams of ironstone with an occasional flint. In the east-facing section 7 to 8 ft of loamy and gravelly rubble contain a variable amount of sand from the Woolwich Beds below. Masses of tightly-packed gravel in the deposit pass laterally and suddenly into seams of gravel in either sand or loam and the whole mixture passes north-wards into brickearth with a gravelly base. The uppermost 4 ft of this brick-earth are cut by a channel about 5 ft wide, infilled by coarse sand with streaks of ironstones and flints. The brickearth and gravel fall into three sets of superficial trough faults which run parallel to the valley side in a similar manner to those near Shelford (p. 14).

Whitaker (in MS.) recorded 12 to 13 ft of loam in a brickworks pit at Lower Ensden. Such a thickness suggests that the locality was on the valley floor thereabouts. In contrast the patch (079561) 700 yd E. 40° N. of this place consists of coarse sand from the Woolwich Beds with water-worn and unabraded flints. It is, perhaps, a disturbed relic of a river terrace but the marked bench with which it is associated is largely artificial. An exposure (083564) 1250 yd E. 33° N. of Lower Ensden showed 5 ft of fine sands derived from the Thanet Beds, with some coarser sands and orange and grey mottled gravelly loams mostly confined to the base.

The patches of Head on the escarpment north-east of Upper Ensinge and those in the valley west of White Wall are all clayey sludges from the London Clay. The deposit (089575) 350 yd W. 12° S. of White Wall is about 6 ft thick, gravelly and sandy, and appears to have dragged and contorted a previously disturbed surface of Woolwich Beds sand which contained flint-filled frost-wedges. In the absence of tell-tale flint pebbles the passage from Woolwich Beds sand to disturbed sand is indefinite (Fig. 11). A similar sludge from Thanet Beds occurs in the road cutting (181594) 460 yd S. of Fordwich church, where 3 ft of gravelly loam are overlain by 5 ft of buff loam.

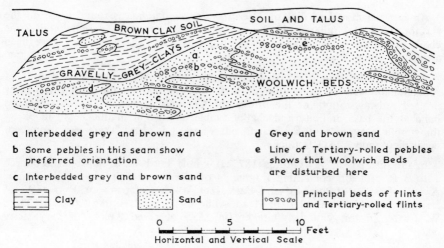

a Interbedded grey and brown sand

b Some pebbles in this seam show
   preferred orientation

c Interbedded grey and brown sand

d Grey and brown sand

e Line of Tertiary-rolled pebbles
   shows that Woolwich Beds
   are disturbed here

----- Clay

∙∙∙∙∙ Sand

∘∘∘∘∘∘ Principal beds of flints
        and Tertiary-rolled flints

```
0        5        10
├────┼────┼────┼────┤ Feet
Horizontal and Vertical Scale
```

FIG. 11. *Sketch of a section 350 yd W. 12° S. of White Wall,
showing Head resting on Woolwich Beds*

Wye—Petham—Hastingleigh—Exted—Bridge—Barham.

At Wye a patch partly resting upon the 3rd Terrace is probably the decalcified remnant of Coombe Deposits. A temporary exposure (059471) 550 yd E. 14° N. of Wye church showed 2 ft of light grey chalky loam overlain with a sharp but uneven junction by up to 4 ft of dark brown loam with angular flint chips and occasional Tertiary-rolled flint pebbles. In a second exposure the brown flinty loam was resting on coarse brown clayey loam with small angular flints and roughly circular patches of light grey loam with chalk pellets seen for 2 ft.

In the roadside (056467) 250 yd S. 36° E. of Wye church 3 to 4 ft of coarse gritty brown loam with water-worn chalk fragments, subangular flints and Tertiary-rolled flint pebbles were noted. In a trench (052466) 350 yd S. 35° W. of this church 8 ft of coarse rough angular flint gravel in a matrix of coarse brown sand rested with an irregular and channelled (to 3 ft) base on fine laminated green sand of the 3rd Terrace (p. 269). At the bottom of the angular gravel was a discontinuous seam, a few inches thick, of derived chalk pellets and pebbles.

A borehole at Wye College (055469), 150 yd E. 15° S. of the church, proved 12 ft of drift deposits before reaching the Chalk. This, however, must include some river terrace.

In a pipe (091493) 20 ft wide and 8 ft deep in Upper Chalk (p. 144), 650 yd E. 10° S. of Winchcombe Farm, flinty loam, becoming clayey near the margins, was noted. In the core of the pipe Clay-with-flints was strung out downslope below the soil. Downslope from and ending abruptly against this core were 2 to 3 ft of loam with very few flints, and an isolated mass of similar loam was trapped within the core. The gravelly material of the Clay-with-flints was composed of large angular flints, water-worn ironstones, quartz pebbles and rare cherts and phosphatic nodules.

The spreads north-west of Petham comprise a gravelly soil with angular flint chips, water-worn ironstones and Tertiary-rolled flint pebbles. At Debden Court (135522) the Head is about 12 ft in thickness. J.G.O.S.

The Head on the north-north-east-facing slope north-east of Hastingleigh consists of brickearth, sand, clay and flints; it has been cut by a later tributary valley (097452), which is floored by a younger Head.

An internal mould in flint of *Micraster coranguinum* was found loose on the surface (168440) ¼ mile S. 13° W. of Exted ; the nearest *M. coranguinum* Zone chalk was some 2½ miles away to the north-east, but the fossil may have been artificially transported.   G.B.

At the road-bridge (176535) 1000 yd W. 35° S. of Bridge church 6 ft of rotten chalk with an admixture of loam are exposed on the west side of the disused railway. South of the bridge the Chalk is capped by a 3-in tufaceous band at the base of flinty and chalky rubble up to 8 ft in thickness ; in places the rubble is overlain by brickearth, but where it is absent there is a prominent decalcified zone 2 to 3 ft thick.   J.G.O.S.

In the disused railway cutting (187522) ¼ mile south of Bishopsbourne church 1 to 3 ft of loam containing fragments of chalk and flint rest on the Chalk. A well (199509) of the Mid and East Kent Water Company 1450 yd N.W. of Barham church proved 12 ft of Head, while in the road cutting (205500) 500 yd W. 6° N. of this church Whitaker (in MS.) observed " pipes of very stony drift ".

Goodnestone—South Barham—Shepherdswell—South Alkham—Dover Harbour.

In 1961 part of an associated skeleton of a mammoth was found by workmen in an excavation for a soakaway (259552) at the road junction at Claypits, 800 yd E. 33° N. of Goodnestone church. The remains, which included a femur, humerus, scapula, jaw and tusk, lay 5½ to 7 ft below ground level, in a colluvial deposit of brown silt containing chalk fragments and flint nodules. They were identified by Dr. A. J. Sutcliffe of the British Museum (Natural History), who recorded that in the excavation the bedding was locally fairly well defined and showed the following sequence :

		Ft	in
Humus and brickearth (buff silt)	1 ft 9 in to	1	11
Chalk pellets and silt ; with some pebbles ...	... 8 in to	–	11
Stratified brown silt ; small chalk pebbles			
in basal 4 inches	... 9 in to	1	0
Chalk pellets and flints	... ...	–	5
Brown silt	... ...	–	6
Silt, chalk fragments and small flints	... 7 in to	1	6
Brown silt with large flints	10 in to	1	1
Brown silt	... 0 to	–	11

The chalk surface was subsequently reached at 15 ft below ground level.

Just west of South Barham, in a silage pit (200485), Head rests on an uneven piped Chalk surface. In the lower part the deposit consists of brickearth containing a few large flint nodules, but in the upper part it is gravelly, with a majority of angular flints and a minority of rounded flint pebbles and unbroken nodules.

In a chalk-pit (228465) 330 yd E. 15° S. of Wootton church, the Chalk is overlain by 1 ft of brown flinty clay beneath 2 to 3 ft of flinty loam ; in another pit (221468) 450 yd N. 42° W. of the church the Chalk is capped by 1 ft of chocolate-brown clay below 2½ ft of brickearth, and in another 125 yd to the east 4 ft of flinty brickearth are exposed. The north bank of the lane (226503) 300 yd south of Womenswold church, displays 2 to 4 ft of flinty loam resting on an irregular piped chalk surface. ' Bullhead Bed ' flints are among those scattered on the surface of the Head 250 yd south-west of Frogham (255502), and also 450 yd S. 30° W. of Coxhill. In silage pits (251472) just south-west of Coxhill, up to 6 ft of brickearth with flint chips were exposed.

The railway cuttings north-north-west of Shepherdswell Station show Head resting on a very irregular chalk surface. Normally 2 or 3 ft thick, with pipes up to about 10 ft deep, the deposit is thicker, with few flints, beneath a minor valley (256487); elsewhere flints are common in the brickearth matrix, being mostly small and broken. In general the brickearth grades down into reddish brown clay up to 1 ft thick, usually with a thin layer of dark brown, almost black clay, at its base; big undamaged flint nodules are normally concentrated in the clay, but such nodules also occur in places in the brickearth above. In some pipes layers of flint nodules are arranged in festoons, suggesting that the deposit has sagged into the developing cavities.

The spread of Head at South Alkham is gravelly in parts, and appears to have been dug for gravel in old pits (252420) beside the road 550 yd W. 43° S. of Alkham church. This deposit does not adjoin the Dry Valley and Nailbourne Deposits along the whole of its length, suggesting that deepening of the valley, which is here entrenched in Melbourn Rock, has proceeded since the Head was deposited.

In the bottom of a truncated minor valley (248382) 1000 yd W. 22° S. of Capel Court, Head is exposed at the cliff top: it consists of up to 5 ft of brown loam containing flints and chalk rubble. Similar material is seen at the top of the low cliff (315402) north of the railway sidings at Dover Harbour and probably represents the remains of Head on the side of a valley now removed by the sea. In foundation excavations (303399) 850 yd E. 36° S. of Farthingloe 4 ft of brickearth with occasional flint chips were exposed, and were included as part of the Head in the vicinity.   G.B.

Stanford—Sandling Junction—Brabourne.

The north face of the Swan Lane sand-pit, Stanford (119392) (p. 99), exposes 3 to 4 ft of ochreous-mottled grey to buff vertically jointed loam, with a thin discontinuous layer of angular flint gravel at its base. The base is festooned where it rests on Gault clay, but not where it rests on Folkestone Beds.

North of Stanford, on the slope west of the East Stour River, a clay wash, in places with flints, passes south-westwards into brickearth. The slopes of the valley south of Perry Wood carry an orange-mottled, grey silty clay wash which merges into brickearth in two patches adjacent to the East Stour flood-plain (p. 233). North of the wood the slope down to Shrine Farm bears a wash of re-sorted glauconitic clay. Similar clay occurs on a dip-slope of Folkestone Beds north of the Gault outlier by Sandling Junction.   B.C.W.

Where the spreads of Head at Brabourne and Monks Horton rest on Gault they comprise brown clays and loams with chalk fragments and clean white chips of flint.   J.G.O.S.

HEAD GRAVEL

Hanslett's Forstal area.

The patch of Gravel on a hill-top at Elverland (978583), just south of Hoath Wood, consists mostly of flint pebbles. The best-developed spread, ½ mile north-east of Hanslett's Forstal, has a flat surface declining gently north-eastwards from about 170 ft above O.D. to 150 ft at Judd's Hill in the Faversham (273) Sheet. A temporary excavation (985600) showed 3 ft of gravel of large sub-rounded and battered flints in a loamy matrix, on Thanet Beds sand.   B.C.W.

COOMBE DEPOSITS

Digges Court—Eastwell—Bilting—Wye—Brook.

The Coombe Deposits at Digges Court (973477) consist of chalky wash containing abundant chalk pebbles and angular flints; the ground surface is flat and slopes gently south-westwards. Similar material, more than 4 ft thick, floors the valley west of Eastwell church. The chalk pebbles are about the same size as the nodules in the Melbourn Rock, from which they may well have been mainly derived. B.C.W.

Resting on the valley slope (058495) ¼ mile north-east of Bilting a small patch of loam with numerous chalk pellets passes into Head, while south-westwards it is concealed by Head Brickearth.

The large spread ½ mile north of Wye gives rise to a gentle and notably even slope within the uneven topography of the valley side. It backs against the gently rising ground capped by the Melbourn Rock and thins out laterally against rising ground at Wye and Olantigh Towers. The deposit consists of a mixture of powdery and pellety grey chalk which in places is decalcified at the surface to a heavy flinty brown loam. Numerous white-patinated chips of flint up to 3 in long and occasional Tertiary-rounded flints are scattered over the surface.

Chalky detritus forms markedly even floors to the deep coombes in the Chalk escarpment near Brook. In the coombe (Devil's Kneadingtrough; Plate VIB) on Broad Downs (076452) a deposit consisting of chalk pebbles, up to 2 inches in long diameter but usually smaller, in a matrix of dirty brown loam with occasional flints, was exposed in butts for a rifle range which were dug to a depth of 6 ft.

The deposit near Brook church (066443) has an apparently flat surface which ends abruptly against a rise in the ground in the west, but slopes gently in the north-west to the alluvium of a tributary of the Great Stour. A stream at the southern margin is incised into the Gault, leaving the Coombe Deposits perched on the north bank. The clayey alluvium of the same stream grades into the extreme southern corner of the deposit. In the north-east, feeders fill the bottoms of coombes in the Chalk escarpment and the stream valleys, and also mantle gentle slopes. Springs which break out at or near the base of the Chalk are deeply incised into these valley fillings.

The Brook church deposit is exposed in a ditch (066441), 200 yd S. 5° W. of the church, where 6 ft of fine grey marl with ill-defined bands of ferruginous staining were noted. Tightly packed abraded chalk pellets, with some semblance of bedding and with an occasional angular flint, were seen for 3 ft, 120 yd S. 20° E. of the church. According to Kerney, Brown and Chandler (1964) the deposits on the Gault outcrop near Brook span the period of the Allerød interstadial. They recognized the eight pollen zones of Godwin (1961) in the Coombe Deposits here and in the nearby coombe, of which the lowest three zones in a borehole (064445) 280 yd N. 35° W. of Brook church consisted of (I) gravel of subangular to rounded pieces of chalk with a few flints in a ferruginous-stained heterogeneous matrix containing much sand and clay, about 3 ft 9 in thick, passing up into about 3 in of calcareous muds and silts with a fauna of molluscs; (II) the Allerød horizon: about 7 in of mainly organic silts and muds from which a radiocarbon date of 9950 ± 160 B.C. was obtained; at the summit of this zone there was a thin soil horizon; (III) chalky muds and silts passing up into the surface soil, together about 9 ft 1 in thick. Kerney, Brown and Chandler considered that the material of this third zone had been transported from the coombes in the chalk escarpment by solifluxion, by release of water from snowfields and by the scarp springs during times of damp cold climate,

(*A* 10045)

A.   Folkestone Warren

B.   The Devil's Kneadingtrough, on Broad Downs, near Wye

(*A* 9140)

with intervals of local dryness during which the silts were laid down, mainly as a loess. They stated that the coombes were cut mainly during this period, between the years 8800 to 8300 B.C. In the Devil's Kneadingtrough the same authors proved a thickness of up to about 30 ft of Coombe Deposits, composed of a lower series of white or pale-coloured variable chalk rubble and mud overlain by brown humic chalk muds. Locally near the springs intervening deposits of chalky mud and tufa were encountered. The lower series is stratigraphically continuous with the zone III deposits near Brook, while the remainder, in which they recognized five zones, has yielded Neolithic to Bronze Age, Iron Age and Modern relics, and a molluscan fauna indicating first a wooded and later a cleared environment. Kerney, Brown and Chandler considered the flatness of the floor of the coombe to be due to ploughing.

Patches of heavy loam with chalk pellets at Hampton and Beddlestone Farm and the capping of Spelders Hill (shown as Head) are probably partially or completely decalcified Coombe Deposits, which formed on a plateau sloping gently southwards to the 4th Terrace deposits at Mersham.   J.G.O.S.

Postling—Etchinghill—Cheriton—Dover Hill—Dover Harbour.

Powdery chalk wash occurs at The Pent (141393) and also floors a broad hollow at Postling (146390). In some fields on this latter spread the soil is flinty. At 200 yd S. and 500 yd S. 11° E. of Douglas Farm ditches showed 3 to 3½ ft of fine-grained chalky wash, with angular flints in places, on Gault clay. A borehole (158379) 100 yd north of Bluehouse (Whitaker 1908, pp. 187-8) proved 8 ft of wash, resembling Chalk Marl, on the Gault. At Beachborough (169381) a fan of chalky detritus extends eastward from the foot of Summerhouse Hill over a gentle slope of Gault clay. Small exposures here showed chalky wash with chalk pellets and a few angular flints. Between Etchinghill and Pean fine-grained chalky detritus occurs, partly on the Gault clay outcrop, on the lower slopes of coombes.   B.C.W.

North of Danton Farm (191374) a spread of light brown and grey marl extends some way across the Gault from the base of the Chalk. McDakin (1893 ; 1894, p. 135) described how a house at the foot of the escarpment in this vicinity was destroyed in 1891 by an 'avalanche' caused by the sudden thawing of a mass of snow which had accumulated on the frozen chalk on the top of the Downs ; the water so released carried down " the loamy beds on the top of the chalk, and the rubble of a much older fall ". This material was presumably added to the local spread of Coombe Deposits. In the gentle depression north of Cheriton (198378) the Gault is patchily covered by a deposit of yellow and white powdery marl, rather similar to that near the head of the valley north of Morehall (207375), which is in part tufaceous. Farther east the Gault outcrop is partly obscured by spreads of white and brown marl. A section at the northern end of the disused (Broadmead) brickpit (212375), 400 yd S. 34° W. of Castle Hill, was cleared and recorded by Kerney (1963, pp. 209–10) : Gault clay was overlain successively by (a) flint and chalk gravel, 7 cm, (b) greenish grey calcareous mud with scattered fragments of chalk, 35 cm, (c) grey chalk mud with charcoal fragments (fossil soil), 15 cm, (d) white chalk mud with seams of fine chalk rubble, 70 cm, (e) chalky hillwash with *Helix aspersa* Müller (probably of Roman or post-Roman age) and (f) modern soil ; (e) and (f) together were 90 cm thick. Kerney (1963, pp. 204–8) also described the deposits which fill a depression in the Chalk escarpment west of Dover Hill. These were exposed to a maximum depth of about 20 ft in a cutting (235376) 950 yd S. of Hope Farm : frost-shattered Lower Chalk was overlain by (a) chalk rubble, which passed upwards into (b) chalk mud and finer rubble ; (b) was divided into two parts by (c) a band of grey chalk mud with charcoal fragments (fossil soil),

Q

which was about 25 cm in thickness. The whole was overlain by (d) chalky hillwash and (e) soil, together 90 cm thick, which Kerney concluded were of Early Iron Age date or later. The fossil soil (c) was strongly folded and was foreshortened by thrusting and imbrication ; these disturbances Kerney attributed to mass sliding downhill of the saturated chalk muds, possibly as a result of repeated freezing and thawing of the ground. Radiocarbon dating of the charcoal in the fossil soil gave the date 9984 ± 210 B.C. (Godwin and Willis 1961, pp. 73–4), which is within zone II of the Late-glacial Period.

In beds (a) to (c) at Dover Hill Kerney found a fauna comprising *Cochlicopa spp.*, *Pupilla muscorum* (Linné), *Abida secale* (Draparnaud), *Vallonia costata* (Müller), *V. pulchella* (Müller), *Arianta arbustorum* (Linné), *Hygromia hispida* (Linné), *Helicella itala* (Linné), *Punctum pygmaeum* (Draparnaud), *Euconulus fulvus* (Müller), *Retinella radiatula* (Alder), *Vitrina pellucida* (Müller) and *Agriolimax sp.* This fauna was repeated in beds (a) to (d) at Castle Hill, with *Catinella arenaria* (Bouchard-Chantereaux), *Succinea pfeifferi* Rossmässler, *Columella columella* (Martens) and *Vertigo pygmaea* (Draparnaud) in addition. Changes in the abundance of the species present at different levels in the deposits indicated that the fossil soil at both localities coincided with the close of a relatively warm phase which Kerney correlated with the Allerød interstadial.

The low cliff (241371) north-east of the Roman Villa north of Copt Point shows Gault clay and Chloritic Marl overlain by 4 to 5 ft of pale buff and white loam containing fragments of chalk and flint and glauconite. The loam grades up into some 5 ft of light brown loam and grey soil, with flint, chalk and other debris. The landward extent of the deposit is probably small.

Murchison (1851, p. 384) recorded that the section on the north side of the railway siding (315402) ½ mile north-east of Shakespeare Cliff showed a chalk rubble, which resembled the lower portion of one part of the Brighton breccia east of Kemp Town in being composed of broken chalk almost *in situ*. The breccia had a maximum thickness of 50 to 60 ft, but when followed westwards from the station towards the Shakespeare Tunnel it was seen to thin out and to be supplanted in its lower part by solid chalk. In its upper portion Murchison remarked that it merged into a band of " flint-drift ", the ' Head ' of the present classification. The section now shows Melbourn Rock at the Shakespeare Tunnel end, but eastwards the face is progressively obscured by grime, so that it cannot be certainly said whether the chalk is solid or not. However, in places bedding can be discerned near the bottom of the face, and it is possible that the Coombe Deposits do not extend to the base of the cliff, even at the east end.   G.B.

HEAD BRICKEARTH

Head Brickearth associated with Wealden, Lower Greensand and Gault.

A spread of brickearth (976396) on the Weald Clay outcrop ¼ mile west of Bartlett Farm rests on a flat surface and appears to be a terrace of the tributary of the East Stour River that flows east from the alluvial tract south of Bartlett Farm. The soil is a brown loam and small exposures in hedge-banks showed loam mixed with coarse sand or gravel of Wealden pebbles.

East of Dignash brickearth mantles gentle slopes west of the stream that rises at Westwell, and near its south end the spread extends across an interfluve into the valley of a parallel stream to the west. Up to 5 ft of brown, stoneless silty clay loam are exposed in the banks of the lane (990463) which leads southwards from Nash Court, and about ¼ mile south-east of Nash Court brickearth more than 3 ft thick lies on the interfluve between the Westwell stream and that to the east. A spread of brickearth extends mainly on the western side of the Stour valley, almost continuously from one mile north-east of Ashford to Bilting. Tongues of brickearth in the tributary valleys are in all cases best developed on north-facing or north-east-facing slopes. Along its north-western margin the brickearth adjoins spreads of Head deposits or of river gravel, against which it shows a well-defined boundary in most places. The main spread may locally attain a thickness of about 15 ft ; it appears to override terraced river gravels, and its surface declines gently to the Stour alluvial tract, which it adjoins with no great break of slope.

At a large old brickpit (022444) 300 yd west of Spearpoint Corner, some 6 to 8 ft of brickearth were formerly dug and in 1952 a temporary excavation in the floor of the pit showed 2 ft of brickearth with some angular flints, with a sharp base slightly festooned into sandy gravel (p. 268). In another brickworks (032447) about $\frac{1}{8}$ mile farther east, 6 ft of brickearth were dug in 1954. Three disused brickworks, with pits 5 to 6 ft deep, occur respectively $\frac{1}{2}$ mile W.N.W. of Kempe's Corner, north of the road to Boughton Lees ; $\frac{1}{4}$ mile E.S.E. of Kempe's Corner, south of the road to Wye ; and $\frac{1}{2}$ mile S. 35° E. of Kempe's Corner. Topley (1875 p. 168) noted that the last-mentioned pit showed the brickearth to be a good clean loam, with slight signs of stratification.   B.C.W.

About Bilting and Wye the brickearth is closely associated with Head, the latter apparently underlying the spreads of loam. An excavation (043469) 1250 yd W. 4° N. of Wye church showed 3 ft of brickearth, with an occasional angular flint or split Tertiary-rolled flint in the uppermost 1 ft where it was contaminated by surface washing. The loam which overlies 3rd Terrace gravel at Wye race course exceeds 8 ft in thickness. A trench (053466), 380 yd S. 32° W. of Wye church, showed 2 ft of Head, consisting of coarse brown sandy loam with angular flints up to 3 inches long, overlain by 4 ft of brickearth under 1 ft of brown loamy soil. This patch of brickearth oversteps the Head on to the sands of the 3rd Terrace (p. 269). The two spreads south-east of Wye are unusual in that they rest on slopes which face southwards ; they exceed 4 ft in thickness and are underlain partly by a chalky loam.

At Brabourne Lees brickearth is restricted to valleys cut in the Folkestone Beds. It has been dug at a brickworks south of the road junction 150 yd south-east of the church, and in a sand-pit (084401), 400 yd E. 40° S. of the church, 2 to 3 ft of coarse angular gravel, made up of flints, ironstones and Tertiary-rolled flint pebbles, are overlain by some 6 ft of brickearth.   J.G.O.S.

The brickearth between Sellindge and Stanford occurs on gentle slopes and passes across interfluves between one tributary of the East Stour and the next, leaving small tracts of the underlying solid strata showing through. Old brickpits 4 to 5 ft deep, were noted south of Horton Priory (106392) ; 200 yd east of Somerfield Court (106378) ; and inside the race course south of Westenhanger (123368). The pit (130373) at a disused brick and tile works 200 yd east of Westenhanger Station exposed in 1955 some 5 ft of rather clayey brickearth ; Topley (1875, p. 171) noted 8 ft of " good clean loam without gravel " here.

On the Hythe Beds dip-slope a spread of brickearth extends across the northern part of Lympne Airport. Around the southern fringe of this spread the brickearth overlies, at a depth of 2 to 3 ft, glauconitic clay derived from the Sandgate Beds.   B.C.W.

In Folkestone the loam mantling part of the dip-slope of Folkestone Beds passes on to the outlier of Gault. It was dug to 7 ft depth at a site (218357) 600 yd S. 30° W. of the Central Station, without proving the base. The spread was formerly dug in what is now a housing estate, in Radnor Park, and in the Kingsnorth Gardens, respectively 350 yd S. 35° W., 200 yd N., and 150 yd

S. of this station. The third of these localities may be that north-east of Inglis Farm, where Topley (1875 p. 164) recorded 20 ft of brickearth, with a few flints and bones near the bottom. Topley listed the following fossils from the brickearth of the Folkestone district, revised from those published by Mackie: *Bos primigenius, Bos urus, Cervus elaphus, Equus* and *Sus scrofa.* Loam with occasional flints is seen for 7 ft in the West Cliff (230358) 370 yd W. 10° N. of the station at Folkestone Harbour; near the base are boulders of sandstone from the Folkestone Beds, up to 1 ft in diameter, and a 3-inch seam of angular flints and phosphatic nodules. According to Mackie (1851, p. 259), this deposit overlies the terrace deposits exposed to the east (p. 276).

Mackie (1851, pp. 260–2) recorded a core of a horn of *Bos urus* found in an excavation in the tiny patch of brickearth at Darlington Place, approximately 950 yd W. 36° N. of the station at Folkestone Harbour. He noted the presence of microfossils from the chalk in the Folkestone brickearths and also recorded about 8 ft of brickearth containing *Helix concinna, Pupa* and *Succinea oblonga* at London Street (232364), 600 yd N. 10° W. of the station at Folkestone Harbour. No brickearth was detected during the recent survey either here or in the area ¾ mile west-north-west of Copt Point, where a spread is shown on the Old Series 1-inch map.

Head Brickearth associated with the Chalk. The large spread of brickearth south-west of Sheldwich church rests in shallow valleys in the Clay-with-flints from which formation it is apparently derived.   J.G.O.S.

An area of loamy soil with flints extends between Beech Court and the cross-roads by the inn at Challock Lees. In parts of this tract, shown on the 1-inch geological map as patches of brickearth, the loam attains a thickness of 3 ft.

B.C.W.

A well in a patch of brickearth lying at the head of a dry valley (024521) 120 yd S. 10° E. of Molash church was dug to 22 ft without bottoming the deposit, which is probably contained in a pipe in the Chalk. Many of the smaller patches on the Clay-with-flints are perhaps preserved in a similar manner, for solution depressions, which are without a brickearth filling but affect the surface of the Clay-with-flints, have been noted at the heads of dry valleys at several localities.

There are many flints scattered on the surface of the patch of brickearth at Godmersham Park, where much of the material seems to have been carried down the valley draining part of King's Wood. In the lane cutting (067517) 80 yd S. 20° E. of Hurst Farm 3 ft of the brickearth rested on about 1 ft of angular flint gravel.

Near Dane Court (056530) and Chilham the loams, without exception, occupy east-facing and north-facing slopes and are intimately associated with Head deposits which themselves are no more than a gravelly facies of the brickearth. The large spread which mantles the spur at Chilham and the side of the Stour valley southwards has been worked to a depth of 6 ft at Mountain Street.

Brickearth lying on the lower slopes of valleys near Crundale is again associated with Head. At Crundale House (082487) the 3 ft of loam exposed in a trackway cutting contain a few white flint chips less than ½ inch long.

Brickearth flanking the Alluvium and Dry Valley and Nailbourne Deposits from Bagham (078534) to Swarling Farm (130528) exceeds 8 ft in thickness near Thruxted and was seen by F. H. Edmunds for 7 ft in a lane cutting (117533) 500 yd N. 35° E. of Sappington Court. In the railway cutting at Shalmsford Street (098549) 12 ft of brickearth overlie 2 to 3 ft of gravel. In the old river-cliff on the north side of the road (092554) 800 yd N. 22° W. of the road–rail bridge at Shalmsford Street 6 ft of flinty and chalky loams were seen.

Two tiny patches at Petham, marked by a slight bench feature on the valley side, appear to have formed as a delta of the tributary valley to the west. In the northern patch (128516), 550 yd N. 30° W. of Petham church, the following section was seen:

	Ft	in
Made ground, consisting of flint wash, chalk rubble and brown loam, in two places filling diggings to 4 ft ... usually	1	3
Brickearth, with occasional small flint chips in poorly defined seams	5	2
Angular unbedded flint gravel tightly packed in a matrix of brickearth, passing southwards into a seam within the brickearth. The latter is grey and reddish brown in colour below the gravel seam ... ... 6 in to	3	6

The top surface of the gravel is sharp but irregular and rises locally in a hump at the maximum thickness. The gravel also contains a hollow in the upper surface where the flints are lightly scattered in the loam. Small spreads, also associated with bench features and believed to have been formed in a similar manner, are present in the Petham valley at Sheep Court and Dean Farm; in the Lower Hardres valley at Catt's Farm and Upper Hardres Wood; at Pett Bottom, Bursted Wood, Covet Wood and Duskins.

Spreads capping the Clay-with-flints and probably relics of a much larger deposit are present near Bavinge Farm, and between Little Waddenhall (136487) and Jacques Court (174463). At Bavinge Farm (103465), where the loam has been dug to a depth of 7 ft, the almost circular shape of the outcrop suggests preservation within a large pipe. Workings to about the same depth are seen at Stelling Minnis where the total thickness of loam exceeds 9 ft. In the side of the lane (150531) 350 yd W. 6° S. of Lower Hardres church 3 ft of brickearth overlie river gravel of the 4th Terrace. In the disused railway cutting (171548) 300 yd S. 20° W. of Renville loam averaging 4 to 6 ft in thickness, and resting on chalk (p. 156), was noted on the map by F. H. Edmunds. The large spread on the west bank of the Little Stour River northwards from Bourne Park to Bridge is also allied to Head (p. 228). The loams have been dug at brickworks ⅓ mile south-west of the latter place to a depth of 8 ft.   J.G.O.S.

On the Clay-with-flints plateau south of Misling Farm (131446) brickearth is present on a north-west-facing slope, but west of Stone Hall it faces east, and between Stone Hall and Eastleigh Court (138455) it lies banked against the north-east-facing slope.

On the plateau in the Park Wood–Elhampark Wood area, and south-west of Bladbean, brickearth generally occurs on the ridges, but there is a tendency for it to extend down the south-east-facing slopes of shallow valleys: this is illustrated 1 mile south-south-west and ¼ mile south-east of Wheelbarrow Town (145446 and 157456) and ½ mile east of the Gate Inn (156438). A deep auger hole (147442) beside the road in Park Wood, 500 yd north of Mockbeggar cross-roads, proved 4¼ ft of brickearth resting on clay and sand (p. 222).

The brickearth on the interfluve (167420) just south of Ottinge, dominantly on north-facing and east-facing slopes, appears to be relatively free from flints. Brickearth was formerly dug for brickmaking from the spread on the north-facing side of the tributary valley east of Derringstone, where the excavation (208492) is now some 6 ft deep.

At Kingston brickearth rests on north-facing and east-facing slopes, associated with bench-features on the valley sides: 200 yd N. 13° E. of Kingston church, an excavation for a cesspit (198514) was in 6 ft of brickearth with very few flints. A similar pit (194510), 520 yd. W. 30° S. of the church, was close to the south-west margin of the brickearth spread, but showed 8 ft of brickearth with roughly-bedded angular flints, suggesting either that the brickearth occupied a pipe, or that it rested against a low cliff.

The brickearth on east-facing and north-east-facing slopes (220518) to the east of Ileden Wood is littered with flint chips and occasional flint pebbles, some of which are green coated. At Nonington (254522) Head Brickearth faces north-west, north and north-east, and a cesspit 200 yd S. 4° E. of the church was reported to have shown 7 ft of brickearth on chalk ; this thickness probably included some basal gravel. In the road bank 150 yd S. 22° E. of the church, about 5 ft of brickearth, including some flinty gravel at the base and hillwash debris at the top, were seen to rest on chalk. The spread of loam at Fredville (257510) is on slopes facing east-south-east and east ; the surface east of Fredville house is irregular, and may have been dug over for brickmaking.

In the valley (259485) ¼ mile N. 17° E. of Shepherdswell Station there is a disused brickyard, and in the yard of Longlane Farm (259489) 4 ft of light brown structureless loam, with small flint chips, were exposed in a bank. This deposit merges into a valley-bottom loam which is exposed in the railway cutting 650 yd N. 18° W. of the station.

Two small patches of brickearth (199455) east of Oxroad Farm show a preference for east-facing slopes, but a larger spread (195449) north-west of Dreal's Farm blankets an undulating area indiscriminately. At Mountscourt Farm (187424) a mass of brickearth is banked against, and probably partly overlies, the Sand in Clay-with-flints (p. 223).

Brickearth on the ridge at Acrise (194423) rarely exceeds 4 ft in thickness, and rests on Clay-with-flints. North-west of Pay Street (205412) brickearth again seems to overlie Sand in Clay-with-flints, usually with an intervening layer of flinty clay. It occupies east-facing and south-east-facing slopes ½ mile east of Paddlesworth (205399), and at Hawkinge aerodrome (215397). At Swingfield Minnis (217426) and the plateau to the north-east, brickearth caps high ground, but where the surface is broken up by shallow valleys trending north, north-east and north-west it shows a marked preference for slopes facing east, south-east and north-east, to the virtual exclusion of any brickearth deposit on opposing slopes. The deposit was evidently formerly exploited at Foxholt (217426) and 400 yd E. 33° N. of Hoad Farm (214432): at the latter place the existing excavation is 5 ft deep, but brickearth was found to extend downwards at least a further 4 ft.

The ridge between Hawkinge and Ellinge is capped by brickearth, and at Fern-field Brickworks (223409), 1700 yd W. 42° S. of Great Everden Farm, the deposit is light brown in colour, containing rare chips of flint, ironstone fragments and well-rounded quartz pebbles up to ½ inch in diameter.

Brickearth has been dug to depths of 6 ft and 4 ft in excavations (266403 and 269405) respectively 250 yd W. 11° N. and 250 yd N. 41° E. of the mill at West Hougham. West of Hougham Court (270394) brickearth is more extensive on the north-west side of the ridge, with occasional flint nodules, quartz pebbles and flint pebbles in the soil.   G.B.

Head Brickearth associated with the Tertiary rocks. Westward of the Blean the majority of the brickearth spreads rest on the Eocene-cut plateau ; they envelop outcrops and pipes of Thanet Beds and seem to be composed of redistributed loams from these beds.

In a well at the Faversham Waterworks (013596), on the crest of the ridge 850 yd W. 16° N. of Westwood Court, 4½ ft of brickearth rested on Thanet Beds (p. 182). In an old brickpit (016598) 600 yd N. 40° W. of the same place Whitaker (in MS.) recorded 12 ft of loam overlying gravel. In part of the pit he also described a section—presumably on or near the valley floor, since it included beds of fluviatile origin—which showed 5 ft of brickearth overlain by an irregular bed of clayey sand and clay with pockets up to 3 ft deep, below 1 ft of light clayey brickearth and 3 ft of washed loam with flints, at the surface.

The loam 300 yd S. of Copton (018588) exceeds 4 ft in thickness ; this spread extends across the Chalk to the Thanet Beds outliers at Owens Court and Lees Court, and spills off the Eocene-cut platform, down minor valleys, to join up with neighbouring spreads.

Between Colkins (034593) and the railway Whitaker (in MS.) noted the abrupt incidence of loam on the slope of the hill, with a line of pebbles, subangular flints and 'sarsen-stones' where it rested on Thanet Beds. In the railway cutting south-east of the level-crossing he saw the same brickearth, 10 to 12 ft thick, with pebbly gravel and 'sarsen-stones' at the base ; in places the gravel had the appearance of being false-bedded and it was sporadically distributed in lines and patches in the lowest 6 ft. In a pit (037601) 400 yd W. 30° N. of Brenley Corner 2 to 4 ft of brickearth containing a few flints rest very irregularly on chalk.

Near the Blean, where the outcrop of the Thanet Beds is more extensive and continuous, the brickearth is found mantling valley sides, with a marked preference for north-facing and east-facing slopes. Brickearth was encountered to a depth of 6 ft 9 in in the water borehole (053594) at Boughton Street, 860 yd. E. 44° S. of Nash Court ; the lowest 9 in were gravelly and rested on Thanet Beds (p. 183). The loam was seen by Whitaker (in M.S.) for 6 to 7 ft in the brickpit (053593) 1050 yd S. 37° E. of Nash Court.

At Winterbourne, in the sand-pit (065576) 1450 yd E. 15° N. of Selling Station, 5 to 6 ft of brickearth are exposed at the south end of the pit. The loam contains broken flints and pebbles in the lowest 2 ft and in scattered masses. This deposit is too small to be shown on the 1-inch map.

At a brickworks (076562) near the railway cutting north of Lower Ensden Whitaker recorded (in MS.) that the brickearth was 18 to 20 ft thick. In the entrance-cutting to the sand-pit (089575) 350 yd W. 12° S. of White Wall 6 ft of reddish brown loam are exposed ; at the base is a well-defined seam of small Tertiary-rolled flint pebbles and angular white-patinated chips of flint, and similar flints occur sporadically in the mass of loam. An excavation (100571) 550 yd S. 18° W. of Denstead Farm exposed 3½ ft of brickearth overlying 2 ft of gravel in a matrix of coarse brown sand. Whitaker (in MS.) saw the gravel here to be 4 ft or more in thickness and composed of " subangular flints, flint pebbles, lumps of Tertiary iron-sandstone and silicified wood (from the Thanet or Woolwich Beds) ". The brickearth fills the bottom of this valley throughout: much of the spread seems to be a stoneless loam, but locally there are gravelly patches, and, ¼ mile south-west of Hall Place, there is sand at the surface, perhaps the result of rain-wash.

The large brickearth spread near Howfield Farm and Tonford Manor is closely allied to river gravel and Dry Valley and Nailbourne Deposits ; a temporary section (116562) 250 yd W. 11° N. of the former exposed 5½ ft of brickearth, with an irregular band of subangular flints near the base, resting on gravel of the 2nd Terrace. In the river-cliff (117563) 300 yd N. of this farm 5 ft of rather clayey brickearth with chalk pellets and some flints were noted.

At Thanington the loam is continuous with that in Canterbury, and conceals and envelops the river terraces in a similar manner. A temporary section (130566) 220 yd W. 30° S. of Thanington church showed 5 ft of clean reddish brown loam. Similar loam was seen to 4 ft in the road bank (134563) 500 yd E. 33° N. of Cockering Farm. In the banks of Hollow Lane (142560) 1250 yd E. 7° S. of this farm the Chalk is weakly silicified near its contact with the gravel which unevenly overlies it ; this gravel has a matrix of gritty loam and is 2 to 2½ ft thick, being overlain by 3½ ft of brickearth including a seam of angular flint gravel 0 to 6 inches thick. The brickearth also has a sharp, near-vertical, apparently erosional contact with the Chalk, against which the latter is also silicified for about 1 inch.

In Canterbury the loams have been extensively dug for brickmaking. The limits of working are mostly shown by the outcrops of gravel which have been uncovered, but some excavations in the centre of the city may be concealed by made ground, which is up to 14 ft thick. The following localities which lie north-west of the Great Stour are described in relation to the northern railway station (Canterbury West Station). An excavation for a school boiler-house (138577) 1200 yd W. 38° S. of this station showed the passage of brickearth into river gravel (Fig. 12). Resting upon disturbed clays of the Thanet Beds at a locality (137576) 1350 yd W. 38° S. of this station were 5 ft of brickearth, and in a trench (141581) 500 yd W. 24° S. of this station sand and gravel of the river terrace were overlain by 1 ft of gravel consisting of angular flints and Tertiary-rolled pebbles in a brickearth matrix, below 4½ ft of brickearth ; some yards to the west this brickearth expanded to a thickness of over 7½ ft. At the western edge of the cemetery (136583) 1050 yd W. of this station 4 ft of brickearth with occasional embedded flints were noted ; 5½ ft were seen 350 yd E. 40° N of this station, where the total thickness of loam is probably about 7 ft. In a borehole at nurseries (144587) 400 yd N. 20° W. of this station the combined thickness of brickearth and gravel was 24 ft.

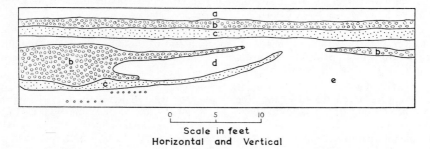

Scale in feet
Horizontal and Vertical

a. Loamy soil.   d. Brickearth with sandy seams and patches.

b. Unbedded gravel.   e. Brickearth with dark brown streaks and
                          with sand seams at base.

c. Loamy sand.

FIG. 12.   *Section* 1200 *yd W.* 38° *S. of Canterbury* (*West*) *Station*

In an old brickpit (146589) immediately east of the disused railway, 600 yd N. of this station, 8 ft of brickearth rested on gravel ; a few angular flints were scattered in the lowest 3 ft of loam and in part of the face they were concentrated in two beds respectively 1 and 2 ft above the base. Dewey and others (1925, p. 282) recorded an average thickness of about 4 ft of reddish yellow brickearth on gravel (p. 274) in the main pit (St. Stephen's Pit) on the opposite side of the disused railway. A trench (145593) at right angles to the road 1060 yd N. 3° E. of this station, cut down the side of a gentle valley, exposed above Thanet Beds, 7 ft of brickearth with two tongues of gravel ; the latter did not reach the ground surface, but expanded rapidly into the bank. At Hales Place (148594), 4 ft of loam, with angular flints and Tertiary-rolled flint pebbles at the base, rested on Woolwich Beds. In the old brickpit (Edwards Brickworks) (152590) 1200 yd N. 10° E. of Canterbury Cathedral Dewey (*in* Dewey and others 1925, p. 282) recorded, above the gravel : " 6 feet of stoneless brick-earth with a bed of ' race ' nodules at its base which separates this material from 11 feet of more loamy and plastic brick-earth ". A borehole on the south side of the railway 1020 yd E. 19° N. of the church south of Hales Place proved 17 ft of drift, including an unspecified thickness of gravel.

South-east of the Great Stour and south of the principal railway station (Canterbury East Station) the removal of surface brickearth has revealed the underlying gravel. Reid (1891, p. 52) remarked that the brickearth capping here was variable and thickest in the east.

In the railway cutting (150572) 450 yd E. 18° S. of Canterbury East Station 8 ft of the loam were exposed, and in a chalk-pit (154571) 900 yd E. 10° S. of this station it was 3 to 6 ft thick, with angular flints and flints from the Bullhead Bed at the very irregular base, resting on chalk. In the lane bank (Puckle Lane) (153570) 720 yd E. 30° S. of the same station, 4 ft of loam with a few flints at the base rested on Thanet Beds. A laundry borehole (157566) 1270 yd E. 34° S. of the station reached chalk at a depth of 20 ft, but such a thickness of brickearth would only occur in a pipe and this figure may include some Thanet Beds. Reddish brown loam, $5\frac{1}{2}$ ft thick, with at its base a thin bed of flints from the Bullhead Bed, subangular flints and angular fragments of a tabular flint course, was seen in the road cutting (158564) 1560 yd E. 41° S. of the station. Excavations for a tank at garage premises in St. George's Place (153575), 500 yd S. 20° E. of Canterbury Cathedral, revealed river gravel (p. 274) overlain by 6 ft of brickearth with a patchy dark brown to black staining, beneath 1 ft of soil; a few flints were scattered towards the base of the loam. On the north side of the roadway another digging showed 4 ft of loam underlying 3 ft of made ground. Foundation excavations in St. George's Street (151577), 230 yd S. of the Cathedral, exposed 3 ft of brickearth, with the base not seen, underlying $7\frac{1}{2}$ to $8\frac{1}{2}$ ft of made ground. Three trial borings at the site of new bank buildings 100 yd N.W. of the last excavation proved the brickearth to be 9 to $11\frac{1}{2}$ ft in thickness, overlying gravel, penetrated only for 4 ft; at the surface there were 11 to $13\frac{1}{2}$ ft of made ground. A record of "stoney and clayey mixture" for 10 ft, below 14 ft of made ground, in a borehole (152578) sited in what is now a car park 220 yd E. 40° S. of the Cathedral, is regarded as referring to brickearth; below this, and resting on chalk, were 2 ft of "gravel with Bullhead flints".

Dewey and others (1925, pp. 281–2) recorded a thickness of 6 to 9 ft of stoneless brickearth overlying gravel (p. 274) in the Vauxhall Brickworks (Gaskins Pit) (161589), 1500 yd N.E. of Canterbury Cathedral, and on the basis of implements found in this pit Dewey (1926, p. 1432) regarded the brickearth as Late Mousterian in age. The small section remaining in 1952 (167591), 2200 yd E. 36° N. of the Cathedral, showed greyish brickearth with flecks of manganese, alternating with impersistent seams of medium-grained sands, 3 ft, overlain by brickearth with occasional very faint dark banding and embedded tiny chips of flint, 5 ft, below soil, 6 inches. The upward passage from sand to loam is gradational and the lower part of the loam succession suggests that it was formed in a manner similar to a varved clay.

A temporary section (162587) near the southern margin of the spread 1500 yd E. 31° N. of the Cathedral showed 4 ft of loam with patches of flints. In the housing estate (Barton Estate) $\frac{1}{2}$ mile south-west of Little Barton Farm, sections showed about 6 ft of loam, with flints and chalk fragments in the lowest 1 to 2 ft.

The railway cutting at Bekesbourne Hill (182565) 600 yd N. 20° E. of Hode Farm, shows 6 ft of loam resting on gravel. This brickearth spread caps the Chalk and Thanet Beds, and was probably formed by the redistribution of material from the latter formation; it is, however, continuous with the loam on the left bank of the Little Stour River, which extends from Bishopsbourne to the northern margin of the 1-inch map, a distance of some 6 miles, and is thought to be of fluviatile origin, since it is intimately associated with river gravel throughout its length.

Soakaways (189556) 540 yd N. 5° W. of Patrixbourne church, proved 10 to 12 ft of loam resting on gravel. South of the roadway (195562) 500 yd S. 20° W. of Howletts, 4 ft of clean loam were noted, while Whitaker (in MS.) stated that at Howletts (196568) the loam exceeded 15 ft.

Brickearth in a minor valley and river gravel capping a knoll form the overburden of the Trenleypark Wood Sand-pit (192592), 850 yd W. 25° N. of Swanton Farm (Plate VB). The gravel on the knoll thins and continues as a bed about 1 ft in thickness down the gentle slope of the valley. It is overlain by 5 ft of impure brickearth with scattered angular and split flints, which are crudely bedded and streaked by dark brown sandy masses and seams of sand up to about 4 inches in thickness. The bedding roughly follows the slope of the surface of the Tertiary sands below. At the surface the brickearth becomes very flinty and impure. Interposing suddenly between the basal gravel and the brickearth is a seam, which expands to a maximum thickness of 3 ft, of lighter coloured calcareous brickearth in which the sandy seams are also present. These deposits are thought to have been formed in still water pounded up within the valley. The differentiation of calcareous and non-calcareous loams here appears to be original, in contrast to the brickearths in the Wingham area.

In the disused sand-pit (196597) north of the road 900 yd W. 4° S. of Elbridge House 3 to 4 ft of pebbly loam rest on Woolwich Beds (p. 193), and 6 ft of loam were seen in the road cutting (203597) 150 yd W. 20° N. of this place.

Brickearth and its associated gravel were exposed in sewer trenches at Little-bourne (Fig. 13). A section (205561) showing 5 ft of brickearth, of which the lowest 2 ft were gravelly, was noted 100 yd S. 40° E. of the western farm of Garrington Farms.

There are several old workings for loam and gravel between Littlebourne and Wingham. Whitaker (in MS.) recorded that at a locality (220576) 650 yd S. 25° W. of Ickham church the loam was 1 to 3 ft thick and rested on gravel. In the brickyard (224571), now obscured, 600 yd N.W. of the cross-roads at Bramling, he noted that the section was about 15 ft in depth, comprising brown brickearth with a bed of scattered Tertiary-rolled flints and a few pieces of flint, resting on grey and yellowish loam with an irregular bed of scattered Tertiary-rolled flint pebbles and a few flints from the Bullhead Bed in a clayey matrix; at the bottom the loam was more sandy and yellowish, not unlike the sand of the Thanet Beds. In the Trapham Borehole (234571), Burr (1913, p. 732) recorded 23 ft of drift, which presumably included some Thanet Beds.

On the east side of the Little Stour, a temporary excavation (236598) 500 yd W. of Deerson Farm showed 8 ft of loam resting on gravel. The railway cutting (235575) 650 yd W. of Wingham church is cut 12 ft into loam.

In a cottage garden (235564) on the south side of the lane 870 yd E. 20° S. of the cross-roads at Bramling the piped surface of the Chalk is overlain by 1 to 3 ft of gravel consisting of flints from the Bullhead Bed, Tertiary-rolled flint pebbles and white-patinated angular flints, and enclosing a lens of brick-earth 6 inches thick; the gravel is overlain by 4 ft of brickearth, the even base of which indicates that it was deposited later than the piping which affects the surface of the Chalk; the gravel here is therefore not a true basal bed of the brickearth.

In the overburden of a sand-pit (246583), 250 yd N. of Broomhill, Whitaker (1872, p. 93) saw 5 ft of buff, bedded, calcareous sandy loam with some pebbles, overlain by 5 ft of brown loam with pebbles here and there.

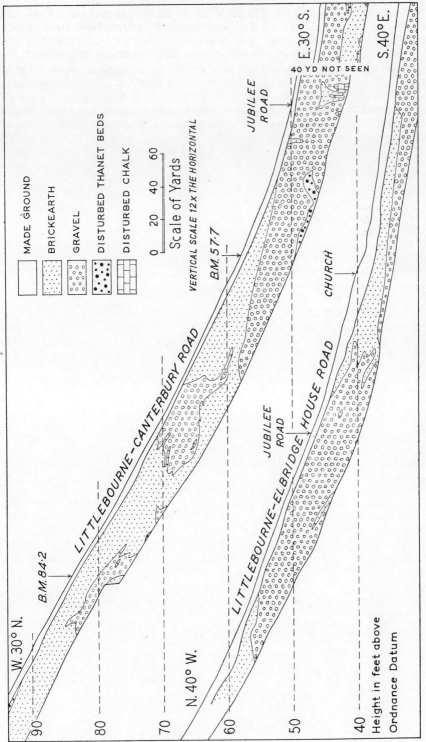

Fig. 13.   *Trench sections at Littlebourne*

The railway cutting (244570) 450 yd S. 20° E. of Wingham church shows light khaki-brown highly calcareous loam, seen for 2½ ft, underlying 5 ft of reddish brown loam which have been decalcified apart from an occasional fragment of chalk. The lower limit of decalcification is fairly even and well defined.

In the roadside (249540) 370 yd W. 40° N. of Bonnington Farm 6 ft of loam were noted. An excavation (255569) 640 yd E. 32° S. of Dambridge Farm, exposed Thanet Beds overlain by 4 ft of calcareous loam with a seam of Tertiary-rolled flint pebbles at the base and a seam of small angular chips and pebbles of flint in the lower part; the section was capped by 6 inches of pebbly wash.   J.G.O.S.

Chapter X

PLEISTOCENE AND RECENT (*Continued*)

GENERAL ACCOUNT

RIVER GRAVELS

SANDS AND GRAVELS which were laid down during the formation of the present drainage system rest upon benches, cap high ground or mantle valley slopes adjacent to their parent streams. These deposits, together with benches which were shaped by streams but carry no deposit (p. 4), form an incomplete record of the development of the present topography and drainage pattern.

The main constituent of the gravels is usually flint, varying in every degree from almost unworn nodules and angular chips formed by frost action or by impact, down to well-worn cobbles, which have been in many cases derived from earlier terraces. Rounded black flint pebbles, water-worn during the Tertiary period, are commonly present and are not restricted to terraces near Tertiary outcrops. Siltstones and ironstones from the Wealden, cherts from the Lower Greensand and ironstones from the Lower Greensand and Tertiary formations are widespread, but are subordinate to flints except in terraces of the Great Stour west of Ashford and those near Kingsnorth ; in these areas Wealden fragments are common and are, in some places, the main constituent. Fragments of Hythe Beds are rare in spreads far from the outcrop. Fragments of chalk are uncommon in terraces except beneath the water-table (see below). Less important constituents include ' sarsen-stones ', some of which may be of considerable size, clear quartz pebbles, dark siliceous pebbles (lydites) from the Lower Greensand, silicified wood from the Tertiary formations and phosphatic nodules from the Gault.

A large proportion of the gravels in all the terraces described herein is bedded only in a rudimentary manner, while some is unbedded. Such gravels, which may form the entire deposit seen or may be channelled into underlying bedded deposits, have been called " torrent-bedded " (Wooldridge *in* Dewey and others 1925, p. 277), an apt description of their appearance today, although much of the derangement is the result of other causes.

Most of the disturbances affect only the uppermost few feet of the gravels, where buckling and dragging of the bedding into columns is commonplace and the bedding is frequently destroyed. In some localities, particularly where the terraces rest on clay, the underlying strata are also affected (pp. 269 and 274). These disturbances are considered to be due to ground ice, and it is probable that the present area has experienced more than one period of ice formation, because the older and higher terraces, most of which rest on London Clay, have been more strongly affected than the younger gravels. So much so, that Coleman (1952, p. 78) assigned the former to a separate group, despite recognizing their similarities to other gravel deposits. Although the relics of gravel present on the London Clay are in places associated with flat surfaces (cf. Coleman 1952, fig. 1), gravel is not restricted to such flats, nor are the flats covered wholly by gravel and sand. In the writer's opinion, the surface of the London Clay is covered by a mélange of disturbed gravel, sand and clay formed by cryoturbation during successive cold periods. This mélange is exposed in the overburden of the Shelford Sand-pit (p. 269). The ' polygons ' at Harbledown (p. 274), which lie at a lower altitude, are thought to be the result of a similar process affecting gravels resting on clay of the Thanet Beds.

Frost-churning, however, cannot account for all the crudely-bedded gravel in the terraces, because in places overlying undisturbed sands are present. In these cases the gravel may have been deposited under fluvio-glacial conditions and the disturbances may have been caused by the melting of derived ice. Indeed Coleman (1952, p. 77) considered the gravel at Chartham Hatch to have been deposited at a maximum distance of 25 miles from an ice-front.

Another factor in the formation of disturbances in the gravel must have been the decalcification of included chalk fragments. Chalk is widely distributed along the river valleys and large amounts must have been carried away and deposited among the terrace gravels, as is now found mainly below the water-table. The effects of the removal of this chalk by solution in percolating rain-water, may, therefore, have been considerable. The unusual 2nd Terrace gravel at Littlebourne (Fig. 13), in which the gravel is wholly unbedded, may be an example of this process.

Previous authors (Dewey and others 1925, pp. 277, 279 ; Dewey 1926, p. 1431) have suggested that the terraces were deposited during rigorous climatic conditions. Their views are supported in this area and the picture of conditions during the formation of the majority of the terraces is thought to be one of large quantities of rapidly flowing water moving, besides their load of chalk, sand and gravel, blocks of floating and rolling ice charged with debris. In most cases the material must have choked the water courses and become piled into aggradations of debris, with the higher levels channelling into the lower and overstepping them to rest upon solid strata (cf. Dewey and others 1925, p. 280). However a less severe period must have occurred during the formation of the upper part of the 2nd Terrace at Folkestone, but such periods appear to have been rare.

The basis for the elucidation of the terrace system of the Great Stour was laid by Wooldridge and Kirkaldy (1936) ; these authors established the concept of each terrace level passing upstream into a knickpoint on the thalweg, the knickpoint having been initiated during a period of down-cutting which post-dated the formation of the terrace. Coleman (1952 ; 1954) made a geomorphological study of the development of the drainage pattern of the Blean district, the southern part of which extends into the present area. Wooldridge (in Dewey and others 1925) postulated the presence of a 200-ft platform south of the Chalk outcrop (p. 3) and Dewey and others (1925) described for the most part individual terrace occurrences at Chartham Hatch, Canterbury and Sturry. Terraces at the two last-mentioned localities and the implements found in them were also described by Dewey and Smith (1925). Smith (1933) was concerned with the implements from gravel near Fordwich. Topley (1875) described the terraces of the upper reaches of the Stour and Mackie (1851) that at Folkestone. A sketch section (in Geological Survey files) drawn by Whitaker shows gravel at four different levels near Chilham.

Wooldridge and Kirkaldy (1936, pp. 3–4) contended that tributary streams, with less erosive power than the main river, offered better opportunities for the determination of the terrace sequence. This precept is particularly applicable to the Great Stour drainage system, where the bewildering amount of gravel in the main valley, at so many different levels, presents a virtually insoluble problem of classification. The Lampen Stream (p. 2) and the dry continuation of its valley through Upper Hardres have three well-marked knickpoints on their thalweg, each corresponding with a terrace ; the second of these is common to the Lampen Stream and the Great Stour and so forms a basis for classifying the remaining terraces of the latter, particularly as it has a characteristic two or three-fold development. In addition the highest gravels of the Blean and of the Chalk outcrop are grouped together as the oldest terrace. The lowest terrace of the Great Stour has not been seen in the Canterbury district. The terraces on the present map are therefore numbered in descending order from 5 to 2. In the following account the level of the upper surface of gravel is used in each case.

The Blean gravels of the Canterbury (289) Sheet probably mark a very early course of the Great Stour, flowing in a northerly direction from Chilham. Successive terraces show that this part of the river course migrated in an easterly direction until it became entrenched in its present valley some time between the formation of the 3rd and 2nd terraces. Wooldridge and Kirkaldy (*in* Smith 1933, fig. 1, p. 166) portrayed the early courses of the Great and Little Stour rivers and Coleman (1952) described the easterly migration of the Blean rivers, which she attributed to secular tilting.

The 4th Terrace is an aggradation of coarse gravel and sand ranging in level from 210 to 310 ft above O.D. The extensive outcrops of this terrace northwards of Bagham, at Rough Common and Hothe Court, mark the course followed by the river. Its knickpoint is near Lenham, on the Maidstone (288) Sheet, according to Wooldridge and Kirkaldy (1936, p. 6) and Worssam (1963, p. 115).

The 3rd Terrace consists of up to three distinct levels of gravel, each corresponding to a phase of deposition by the rivers, and commonly separated by marked terrace bluff features formed during the intervening periods of downcutting.

In the Ashford area it comprises an upper group, up to 60 ft above the present flood plain, and a lower group adjoining the flood plain, with their surfaces in most cases not more than 20 ft above it. In the latter group are some deposits of sand near Wye which, though they have a terrace-like form, may have accumulated as sand bars in and alongside the present flood plain.

Near Godmersham the upper development lies at about 160 ft O.D. ; the lower, with its highest point at about 120 ft O.D., passes down below the alluvium at about 90 ft O.D. The full threefold development is present in

the vicinity of Canterbury. The uppermost level lies between 150 and 165 ft
O.D., capping high ground at Harbledown and east of Canterbury. Gravels
at the two lower levels occupy benches or cap spurs in or adjacent to the
river valleys down to a level of about 100 ft O.D. or, in two cases, slightly
lower. Usually the spreads at the lowest level are small. The patch of gravel
assigned to this terrace at Kenfield Hall lies at about 200 ft O.D. The
terrace begins to pass below the alluvium of the Great Stour near Godmersham, but the main knickpoint seems to lie well upstream, near Hothfield
(Maidstone Sheet 288). In the Little Stour the knickpoint appears to be
just south of Derringstone and in the Lampen Stream it lies north-east of
Nackington church.

The 2nd Terrace, like the 4th, comprises a vast aggradation of gravel. At
Littlebourne (Fig. 13), where the terrace reaches its highest development, it
ranges from below the alluvium at about 33 ft O.D. to about 110 ft O.D.
In the lower reaches of the Lampen Stream it forms a broad bench, similar
to that of the 3rd Terrace. The lack of gravel below is attributed to erosion
of the lower levels of the aggradation. In the Great Stour valley the aggradation rises mainly to about 60 ft O.D., with only a few patches reaching the
height attained at Littlebourne; fans of gravel were formed at the con-
fluences with tributary streams at Shalmsford Street and Fordwich. The
knickpoint occurs in the Great Stour valley at Godmersham, in the Lampen
Stream valley immediately north of the railway line near Bekesbourne Hill,
and in the Little Stour valley between Patrixbourne and Garrington Farms.

At about the time of formation of the 4th Terrace an escarpment of
Eocene rocks probably extended along Chartham Downs and Barham Downs
to, or near to, the limit of the sub-Eocene plain described by Wooldridge
and Linton (1955, p. 48). The Little Stour flowed in front of this escarpment,
following roughly its present course to Bridge, where part of its drainage
extended westwards to Lower Hardres and Swarling Farm. The gravel in
Whitehill Wood and Lower Hardres was deposited in this channel, that at
the latter locality partly as a fan at the mouth of the valley draining Upper
Hardres. The southern margin of this channel is marked by a sudden drop
of nearly 50 ft in the ridge top at Cook's Farm, but elsewhere the southern
margin is lost. There is no evidence to suggest the direction of water movement along this channel, for it is not until 3rd Terrace times that the confluence of the Great and Little Stours can be ascertained. The channel
could have been a continuation of the Petham valley which then joined
the Little Stour, to be captured subsequently by a tributary of the Great
Stour working eastwards from Shalmsford Street. On the other hand the
Little Stour itself may have flowed south of the escarpment to join the
Great Stour at Shalmsford Street; such an arrangement would account for
the enormous quantity of gravel at Chartham Hatch, just below the supposed
confluence.

By the time of formation of the uppermost level of the 3rd Terrace the
Little Stour flowed due north from Bridge and then swung north-westwards
to include in its course the gravels on Bekesbourne Hill, at Hoath Farm, and
immediately north-east of the Mental Hospital on the Canterbury–Wingham
road. It joined the Great Stour in the vicinity of the latter locality. The
Little Stour was then captured by a stream cutting southwards along the line

of the present valley between Wickhambreux and Bridge. After this capture the Lampen Stream and other minor streams near the Mental Hospital and Golf Course were initiated on the old gravel flat formed at the junction of the Great and Little Stours. The Lampen Stream cut back and intercepted the drainage of the Lower Hardres valley before the middle division of the terrace was formed. By this time all the streams were entrenched in their present valleys except the Great Stour, which in part lay somewhat to the south-east of its present course. This river became entrenched in its present valley by the time the late stage of the terrace was deposited.

Dry Valley and Nailbourne Deposits

The term Dry Valley and Nailbourne Deposits has been given to alluvial deposits that floor the Chalk valleys in which there are no permanent streams. The deposits consist of calcareous flinty loams, overlying gravel composed primarily of flints but with subordinate ironstones and 'sarsens'. The gravel can be made of angular, subangular or rounded fragments and usually rests upon disturbed or powdery chalk. The top of the deposit is commonly composed of 3 to 4 ft of brown calcareous flinty loam with chalk fragments ; this contrasts with the neighbouring Head deposits, which are rarely so calcareous except in their lowest levels.

The deposits grade on the one hand into Head and on the other into Alluvium. They are regarded as an older alluvium, formed when the water-table was higher than at present, but when flow in the streams was insufficient for erosion to take place. As the water-table fell, deposition ceased in most of the valleys and is now restricted to those with nailbournes. Most of the material was probably supplied by solifluxion down the valley sides and was re-sorted by the streams. Exceptional weather conditions probably accelerated deposition.

Peat

Peat occurs in the alluvial tracts of some streams and also at springs of the type which are common at the junction between the Folkestone Beds and the Sandgate Beds, where the water tends to ooze through the ground along an extensive belt. In some such places vegetation has impeded the concentration of the spring water into well-defined channels, and has led to the accumulation of extensive peat deposits.

Peat is an important element of the deposits of Romney Marsh, where it is called 'moor' or 'moor log'. It occurs at or close to the surface in The Dowels, in the valley south-west of Warehorne, to the west of Snargate and in a small patch north-east of Little Cheyne Court. It also extends eastwards of these outcrops, where it is concealed by later marsh deposits, and it is perhaps only lacking among and east of the New Romney shingle fulls, among the shingle near Belgar and Greatstone on Sea, and possibly at Hythe. Thin beds of peat at a lower level were proved in the Langdon Borehole (pp. 282, 284), but their extent is unknown.

R

The sands of the creek-ridge system (p. 254) have been found locally to rest in channels eroded through the peat, which is therefore not developed in a continuous bed. It is generally about 3 to 10 ft thick in the western parts of the marshland but thins in the east to 1½ ft or less, perhaps partly because of compaction by the weight of overburden.

The compaction of the peat at or near the surface, due mainly to artificial drainage, has played a major part in the formation of the characteristic micro-relief of the creek-ridge deposits and has also resulted in the formation of relatively low-lying districts such as The Dowels and the area west of Fairfield Court.

There is no evidence to show how much the marshland has subsided as a result of the compaction of the peat, or, for that matter, of the clays and fine sands, although the irregularities of the top of an old sea-wall near Fairfield Court (p. 282) reveal something of the relative amount of compaction within the creek-ridge marshland south-east of the wall.

It is usually accepted that the land stood well above sea-level during the period of peat formation. The later rise in sea-level of over 25 ft, suggested by Gilbert (1930, p. 94), is probably a combination of a rise in sea-level and of land subsidence due to compaction of the peat.

The peat was formed during a period when the marshland was wooded and swampy, for oak trunks and roots, the latter in position of growth, remains of alder and hazel nuts were recorded by Drew (1864, p. 15). However, Lewin (1862, p. 64) and Burrows (1885, p. 340) considered the deposit to have been carried in by the rivers draining the Weald. Dowker (1897, p. 212) stated that some of the stumps had been cut by an axe. The woody parts of the specimens collected for radiocarbon dating (below) were identified as birch by Dr. W. G. Chaloner.

The marshland peat has long been considered to have formed in the post-Neolithic period, for a flint flake of Neolithic to early Iron-Age date was found amongst the submerged forest off Pett Level (Milner and Bull 1925, p. 320), which is a geographical continuation of the marshland peat. This dating is confirmed as Bronze Age by C^{14} determinations. A specimen of peat (NPL 23)[1] from a ditch-side (977308), 1½ ft below the upper surface of the peat horizon and 6 ft below the ground surface, 720 yd S. 10° W. of Higham Farm, and a portion of a tree trunk (NPL 24)[1] trapped in the peat in a ditch-side (032243) 770 yd E. 42° S. of Court Lodge were submitted to the National Radio-Carbon Dating Service at the National Physical Laboratory, who reported that, relative to 1950, these samples were 3020 ± 94 years and 3340 ± 92 years old respectively.

Further radiocarbon datings of peat from two inland localities within this district were given by Godwin and Willis (1960, pp. 68–9). They recorded two datings from an indefinite patch within a tributary valley of the Little Stour at Wingham (p. 280): " Fine detritus organic mud with fine sand " was found to be 3105 ± 110 years old and " Coarse detritus mud of reed swamp from valley fen deposit " was 2340 ± 130 years old, the results being

---

[1] Laboratory reference of the National Radio-Carbon Dating Service.

determined in 1959. Godwin and Willis commented that "the two radio-carbon dates give a basis for the chronology of the whole deposit; extra-polation of this with pollen-analytic results suggests that from about 1600 B.C. to Roman time the Chalk here was open disforested country with arable land and pasture. Evidence of brackish water at the base of the deposit shows that the sea had reached its present height at or before 1600 B.C."

The second locality was in Asholt Wood (about 173383), where three datings were recorded from "coarse detritus organic mud". The basal part, from between depths of 90 to 100 cm, was found to be 2980 ± 130 years old, that between 20 and 22 cm 2640 ± 110 years old and that between 0 and 2 cm 2490 ± 130 years old. Godwin and Willis commented that pollen analyses from the basal sample indicated considerable agriculture and that cut wood, charcoal and introduced stones showed local prehistoric occupation. Analyses from the surface deposit proved widespread disfores-tation and extensive agriculture of the surrounding Chalk downs.

Marine Alluvium

The Marine Alluvium grades from fine sand to stiff tenacious clay. It was formed on tidal flats or saltings lying on the landward side of the Storm Gravel Beach Deposits and within this district it occurs on Romney Marsh proper, parts of Walland Marsh, parts of The Dowels, and on Romney Salts. These areas and much land lying outside the present district are loosely known as Romney Marsh. Only a very small proportion is true marsh; the remainder is highly productive farm-land. Most of the ground is flat and lies at a level between the high and low tide marks, for the Marine Alluvium cannot accumulate above the level of the highest tide. (Mean high water of spring tides is at 10·5 ft above O.D., according to Lewis and Balchin 1940, p. 270). A distinctive micro-relief has developed in the deposits as a result of compaction (p. 254): banks of shingle, sand, dunes of blown sand and sea-walls constitute a somewhat bolder relief. The sea is now excluded by walls and natural banks of shingle.

The scientific study of the marshland is largely connected with the forma-tion of Dungeness foreland which lies south of the present district. A vast and varied literature stems from J. Elliott (1847, and *in* Lewin 1862): pre-viously, Dugdale (1662) had deduced that the district was reclaimed by the Romans and quoted Patent Rolls referring to the Rhee Wall (p. 250), and Nennius had written a colourful description of the district in the 9th century. Notable contributors on the deposits were Drew (1864), Gulliver (1897) and Dowker (1897), the latter critical of Drew. Topley (1875) only repeated Drew's contribution with some additional comment. On the historical aspects of the district Holloway (1854), Furley (1874) and Burrows (1885) were early writers and Rice Holmes (1907) made an interesting assessment of the subject. More recently Lewis (1932) established the accepted theory of the formation of Dungeness, and Lewis and Balchin (1940), by means of a levelling survey, introduced as much precision as appears possible into the recognition of previously existing sea levels; their work contained much historical detail and the discussion to their paper is of great value. White

(1928b) and Gilbert (1933) were further authoritative writers of this period. Modern historical work comprised that of Homan (1938) and outstandingly that of Ward (1931–6). The whole of the previous information was admirably summarized by Steers (1946). Andreae (1958) made several interesting comparisons with Holland, and Green and Askew (1955–60) have published the first detailed accounts of the deposits and some of their deductions on origin.

Old Sea-Walls. Old sea-walls built to reclaim parts of the marshland are a conspicuous feature. They commonly delineate changes of ground level due to the curtailment of natural deposition within the innings. Nearly all lie within Walland Marsh and Romney Salts (Fig. 14). Remnants within Romney Marsh had not been recorded before Green and Askew (1958a, p. 30) mentioned those east of Rheewall Farm and further traces immediately north of the Rhee Wall from about Appledore Station westwards (1958b, p. 25). The same authors (1959, p. 25) also deduced possible walling along the roadway from Newchurch to West Hythe. Many of the roadways on Romney Marsh may have originated as walls, but direct evidence is lacking.

The old sea-walls are single earthen banks, apart from the prominent feature extending from Appledore Station to New Romney which is known as the Rhee Wall. This consists of a double wall enclosing a canal about 40 yd wide near New Romney, 50 to 60 yd wide at Snargate and somewhat wider at Appledore Station, where the walling features are confused on its southern side. Deposits of fine silty sand fill the canal, forming a flat surface lying about 4 to 6 ft above the surrounding marshland, except near Appledore Station where it lies at about the same level as the deposits south of the walls and 8 or 9 ft above those in The Dowels. Traces of a narrow channel meandering within the canal flat occur. On the evidence of their level, the deposits within the canal must be approximately contemporary with those lying south of the walling near Appledore Station, i.e. 14th century (J. Elliott *in* Lewin 1862). The date of construction of the Rhee Wall is controversial. In its present form it post-dates the compaction of the creek-ridge deposits (pp. 254–5), and therefore, contrary to popular belief, it is of post-Roman age, as was suggested by Ward (*in* Lewis and Balchin 1940, p. 282) and Green and Askew (1958a, p. 30). These authors[1] considered it to have been constructed in the 13th century, that is after the diversion of the River Rother from New Romney (p. 259). However, the older wall relics near Appledore Station reveal that the present Rhee Wall there is a reconstruction.

At the present time a sea-wall protects part of the Hythe shingle mass and forms the modern promenade. Another wall protects the east side of Romney Marsh and extends from the Redoubt (128320), north-east of Dymchurch, to between Littlestone on Sea and Greatstone on Sea. This latter wall is an extension of the original Dymchurch Wall, rebuilt under Elliott's supervision. Elliott (1847) considered that the threat of inundation on the east side of Romney Marsh had developed since the time of Edward III's reign by erosion of a natural shingle barrier, basing his opinion on the absence of historical references to expenditure on walling hereabouts before this period.

---

[1] Patent Rolls cited by Green and Askew as referring to the construction of the Rhee Wall are quoted in Homan (1938), Furley (1874) and others.

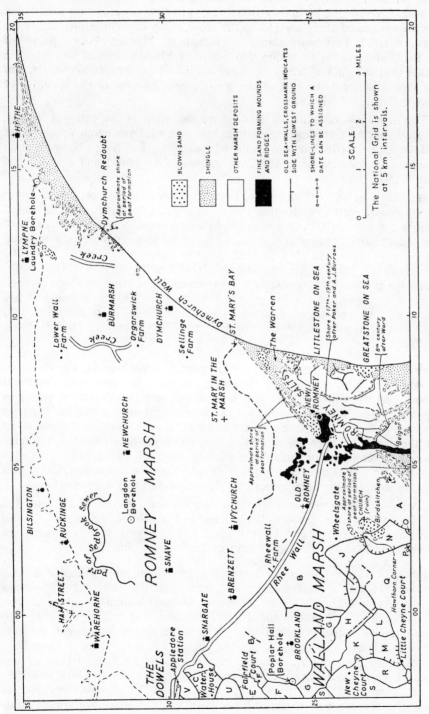

FIG. 14. *Sketch-map showing the principal features of the Marshland*

In an early attempt to check the shifting of the shingle, some groynes or 'knocks', consisting of masses of large loose rocks secured between timber piles, were placed at right angles to the coast, many of them bifurcating at the seaward end. They were shown on a map by Matthew Poker, published in 1617, and existed to within fifty years of Elliott's time of writing. He stated that the sites of many of them could still be seen at low water, but far separated from the line of the wall, giving an idea of the quantity of shingle that had been lost.

The first engineer called in to supervise the repair of the wall was Rennie, in 1803–4. At that time the wall to the east of Dymchurch required but little attention, as its foot was secured by shingle. In Elliott's reconstruction of the wall, commenced in 1839, its line was straightened and it was faced with Kentish Rag.

A breach in an old sea-wall is recognized by a sharp bow up to about 75 yd away from its regular line and up to about 150 yd across. The bowed wall is a repair wall, built upon either the landward or the seaward side of the breach, and it was made, presumably, to avoid dumping repair material into the deepest part of a flooded scouring, such as would occur in the breach itself. The repair material available locally might not have withstood the pressure of the next tide if so placed. The remains of the scoured channel are usually preserved on the landward side of the breach and are occupied by fresh-water alluvium or a fleet (lake). Particularly good examples of repairs occur at Flats Houses (974247) and at Midley House (017236).

Fleets. Most of the fleets within this district appear to be the result of scouring at breaches; others, however, particularly those near Fairfield Court, one of which extends into this district, seem to occupy depressions caused by peat shrinkage. The shrinkage occurred mainly because the peat was drained, but at Fairfield Court it seems also to have been compressed by the weight of later deposits lying north-west of the sea-wall. A similar process of compression may have affected parts of The Dowels area, but flooding here is controlled by pumping into the Royal Military Canal and this, by draining the peat, has caused the major part of the compaction.

Fleets occupying old salting creeks and channels appear to be restricted to the later marsh deposits lying west and south of the district herein described. As far as possible the fleets and low-lying marshy ground are distinguished on the geological map as fresh-water alluvium.

Drainage. The marshland is intersected by innumerable drainage ditches and sewers which present every variation from the ruler-straight modern excavation to the highly sinuous converted salt-marsh runnel. Minor streams which drain the Wealden and Lower Greensand upland region pass into this artificial system, which discharges into the sea by a complex arrangement of sluices and pumps. All the natural drainage courses have been modified by man, but an ordnance map reveals their sinuous traces. Three courses, however, deserve special mention: firstly, a channel which meanders from Brenzett (003273) to the north of Rheewall Farm, and thence passes near the old sea-wall remnants to a point on the Rhee Wall itself west-north-west of Old Romney church; secondly, the Sedbrook Sewer, which courses from Stockbridge House (010311) to the north-east and then meanders towards Peartree Farm (043330), beyond which its course becomes indefinite, though it may sweep to the north of Will's Farm (042315) ; thirdly, a channel which lies immediately west of New Romney at the eastern end of the Rhee Wall. The lower sharp meander of this third channel carried the waters from the Rhee Wall and it dies out among the irregular sand deposits which lie south-west of New Romney (p. 255). Northwards from the end of the Rhee Wall it crosses the sharp bend in the roadway and is lost among the sandy mounds and clay deposits thereabouts. Two further natural channels (p. 284), which are not discernible from maps alone, empty northwards into a broad depression west of the Hythe shingle mass. Present day ditches lie in the bottoms of these two channels and the drainage in the eastern one appears to have been reversed because it now passes southwards to the sluice at Dymchurch.

Numerous roadways, particularly some in the Newchurch area, have a sinuosity which suggests that they follow natural water-courses. That mentioned by Green and Askew (1959, p. 25) between Newchurch and West Hythe (125342) is perhaps a continuation of the second drainage course mentioned above. Other examples are the roads from Burmarsh (101320) to Botolph's Bridge (121335) and from Hurst Farm (071330) to Dymchurch. The channels described above, and a great many others, formed parts of the water-courses from which the majority of the marsh deposits were laid down.

The Marsh Deposits. The deposits lying at depth are but sketchily mentioned in the literature on the marshland. Drew (1864) recorded such information as was then available and recognized that Recent sands underlay the surface deposits. Since Drew's day four boreholes are known to have penetrated the full thickness of the deposits. A water-bore at Poplar Hall (976258) penetrated, according to the record, which was based upon verbal information from the drillers, " vegetable debris " 24 ft thick, resting upon " soft blue clay with dark green sand beds " 68 ft thick, with Hastings Beds below. The level of the Hastings Beds surface was 84 ft below O.D. It is believed that the recorded " vegetable debris " was composed at least partly of very dark silty clay similar to that found in more recent boreholes at Dungeness. Furthermore the borehole was situated on a broad sand outcrop, so that a considerable thickness of sand should have been proved at the surface. A borehole at Langdon (p. 284) proved 66 ft of marsh deposits with the Hastings Beds surface at about 58 ft below O.D. A borehole (149344) made in 1935 at a laundry on the Dymchurch road, just west of Hythe, entered what may have been Weald Clay at a depth of 48 ft (32 ft below O.D.). The uppermost 6 ft were reported to be made ground, which left 42 ft for the total thickness of marsh deposits (pp. 287, 289). The record of another borehole at Brookland, which is not precisely located, is difficult to interpret. The recorded succession is " Bog with tree roots ", 18 ft ; " light blue sand ", 30 ft ; " dark blue clay ", 60 ft ; " light grey sand and gravel ", 52 ft ; " blue clay ", 68 ft ; " Hastings Beds or similar ", 12 ft. In view of the thickness recorded in the borehole at Poplar Hall the base of the " dark blue clay " appears to be the most likely bottom to the marsh deposits.

Apart from the Storm Gravel Beach Deposits (shingle), blown sand and fresh-water alluvium, which are described elsewhere, the surface deposits comprise fine-grained sands and clays which in view of their association with creeks and channels can only be regarded as having accumulated on tidal flats and salt-marshes. Their fauna, which consists mainly of *Cardium*, the creek-dwelling *Hydrobia* and *Scrobicularia*, confirms the depositional environment.

Only the end members of the graded series clay to fine sand are shown upon the geological map. Generally, however, every member of this series is present, in variable quantities according to the local depositional circumstances, in the vicinity of the deposit boundaries shown.

A distinctive feature amongst the clays of the oldest reclaimed land is a dendritic system of low sand ridges varying in breadth from a few feet to over a mile. The broader ridges taper and eventually break up into the smaller, which also ramify from their margins. Laterally the fine sand in the ridges passes into the intervening stiff clays. The sands, therefore, appear to have been laid down in a system of salt-marsh creeks and runnels. Because none of the broad sand masses approaches the inland edge of the marshland and the finest dendritic development (south of Warehorne) lies close to the inland edge of the marshland and to the known Rother channel at Appledore (Tenterden Sheet 304), and also because some of the sand ridges in the Dowels pass into ridges of silty clay in their upper reaches, it has been concluded that the major proportion of the sand has been driven up the creeks from the sea rather than carried down by rivers draining the Wealden uplands. However, an indefinite proportion at Ham Street and near the Royal Military Canal may have been carried from the uplands by streams (Green and Askew 1956, p. 16).

Exposures have shown that the channels occupied by the fine sand have been cut into and in some cases through the underlying peat (Green and Askew 1959, p. 23). The sand is thought to have been carried back and forth by the ebb and flow of the tide until it was built up into levees on the channel sides ; eventually it must have nearly choked the creeks. The finer fraction was laid down as clays upon the surrounding areas of un-eroded peat, which were probably slightly compressed by its weight. After reclamation, inversion of relief from creek to ridge-feature was caused by the different amounts of subsidence resulting from the compaction of peat when drained. The process was probably assisted by the compaction of the sediments themselves, for under natural salting conditions the coarser levee material, situated on the creek margin, drained markedly better than the clays, which were farther from the drainage ' fall ' into the creek and also tended to retain water. Probably some compaction of the levees took place during each tidal low, but that of the clays is unlikely to have commenced until they were drained. Similar sand ridges are known in Holland where they were described by Edelman (1950, pp. 95–7), who introduced the terms " creek-ridge soils " and " pool soils " respectively for the deposits in and away from the ridges. With reference to the Romney Marsh deposits these terms were modified by Green and Askew (1959, p. 23) into the geologically more appropriate ' creek-ridges ' and ' pool areas '. The formation of such creek-ridges is figured by Guilcher (1958, p. 108). The broad sand masses about Brookland and thence to Ivychurch and that from Wheelsgate to St. Mary in the Marsh are considered to be the infillings of the major channels of the original salting, because of their intimate connection with the creek-ridges.

A C[14] age determination (NPL 25)[1] on *Scrobicularia plana* from 3 to 6 ft below the surface of a creek-ridge (024237) 500 yd S. 10° E. of Wheelsgate was carried out by the National Physical Laboratory. This proved an age for the shells, and thus for the infilling of the creek course, of 1550 ± 120 years before 1950, that is Late Roman time.

Not all the sands shown on the geological map are part of the creek-ridge deposits. Some occur in the land which was reclaimed in the 13th century or later south of Appledore Station, in the triangle south of Water House and west of Little and New Cheyne Courts (p. 283), and these form levees to clearly-preserved salting creeks and runnels and are associated with clays in the pool areas and also levees of silty clay. The latter are especially well developed on a small scale in area T (Fig. 14). The large spread of sand from New Cheyne Court to Little Cheyne Court appears to have formed as a tidal flat; the southern part of this expanse (R) is bounded by a sea-wall and appears to have been reclaimed before salt-marsh conditions were established, although an approach to such conditions is indicated by the clays alongside the eastern wall. The sand flat in the innings S passes westwards into one low creek-ridge system in which inversion of relief is incomplete; the original channel feature is preserved and the levees of sand pass up-channel into levees of sandy clay. The larger expanse about Upper Agney (011226) and that near Midley House also appear to have originated as tidal flats. Early innings of the 3rd group on p. 260, J, N, O and P, do not contain mappable clays and must also have been reclaimed before salt-marsh conditions were established therein. At the time the main mass of the sands was reclaimed a salting was probably present near Little Cheyne Court and extended well to the south of the district described. Elsewhere conditions must have approximated to those prevailing off Graveney Marshes in north Kent today, where clays occur adjacent to the sea-wall and vegetation, forerunner of a salting, is just beginning to form on the coarse silt of the ' mud-flats ' at a level of 7 to 8 ft below high spring tides.

Some of the sand at Brookland appears to have been deposited during flooding from breaches in the walls since reclamation (see also Green and Askew 1958b, p. 23).

The mounds and ridges occurring among the sand-flat of Romney Salts and extending westwards among the creek-ridge system contain sands which are finer in grain than the present Blown Sand and so it appears that they are unlikely to be of aeolian origin. The ridge west of Belgar is banked against the shingle fulls; this sand and also that at New Romney itself are considered to be beach deposits formed at the estuary mouth (p. 259). Both appear to have been partially eroded by waves. The low mounds (Fig. 14) are thought to be estuary sand-banks and to have been formed at about the same time as the creek-ridges, although some at New Romney may have been deposited by water flowing from the Rhee Wall. The form

---

[1] Laboratory reference of the National Radio-Carbon Dating Service.

of the mounds in at least the inland occurrences is thought to have been accentuated by the compaction of the surrounding deposits and by later channelling (p. 285). Elliott (1852, p. 44) and Drew (1864, p. 21) thought the mounds on Romney Salts were forts built to protect the harbour : several sea-walls which abut against them may add to the suggestion that they are artificial, but they are, however, here considered to be the natural extension of the inland sand-banks, which could not be artificial. Elliott (1847, p. 469) and Lewin (1862) stated that the New Romney mounds consisted of Hastings Beds, an error which, although corrected by Topley (1875, p. 304), was repeated by Burrows (1885, p. 335).

Fine sands which are older than the peat have been detected at the surface by Green and Askew (1958b, pp. 21–2) west of Westbroke Farm. The present writer cannot agree that the bank shown as Blown Sand on the geological map consists of this old sand (p. 290), but has found that such sand has a limited and apparently indefinite extent amongst the creek-ridge sands thereabouts. A similar occurrence of sand in Romney Marsh, near Moat House (p. 283), is perhaps older than the peat. In a letter, Mr. R. D. Green has supplied us with details of other areas where pre-peat deposits are known, as follows : sand and shingle to the south-west of Dymchurch Redoubt, extending for about ½ mile inland from the sea-wall (see also pp. 288–9) ; a narrow strip of sand fringing the north side of the sewer which runs parallel to the sea-wall south-west of Dymchurch ; and clay deposits at or close to the surface (099302) about 570 yd N. 33° W. of Dymchurch church. Significantly, in view of the date of deposition of the creek-ridge sands, these lie in the vicinity of Dymchurch where Belgic and Roman pottery have been found (p. 286).

The northern part of Romney Marsh is occupied by tenacious clay, reddish brown at the surface and greyish brown to blue below, overlapping silt and fine sand. Over wide areas, particularly near Burmarsh, the silt and sand lie below 3 to 5 ft of the clay, but elsewhere up to 10 ft of clay have been encountered. The present-day drainage system contains numerous creeks and runnels from which the clays were originally deposited, and sinuous lines of darker soil also mark former runnels. The original drainage is thought to have been northwards and eastwards into a broad depression lying to the north of Burmarsh and extending towards Hythe. This itself is considered to be a relic of a broad salting estuary which was progressively closed by the Hythe shingle fulls and which is confused with the River Rother in the literature of the marshland. Levees to the drainage pattern of the broad clay outcrop were recorded by Green and Askew (1956, p. 16), but are rarely of coarse enough material to be shown separately upon the geological map. It has not been ascertained how far the creek-ridge deposits extend below the clay outcrop or whether such sand outcrops as occur are part of this system or whether they are part of a simple tidal flat. However, the peat occurrences at 7 ft depth at Orgarswick Farm (p. 282) and near the Royal Military Canal (p. 282) suggest that creek-ridge deposits may be widely distributed below the clay. The appearance of such deposits southwards and westwards limits the area of clay deposition: the boundary on the west appears to be natural, but deposition was possibly controlled by a sea-wall which is now obliterated. The southern limit is thought to have been controlled by a wall, although much of its course is indefinite (pp. 284–5).

Natural Formation of the Marshland. The formation of the marshland commenced with the deposition of marine sands in a broad bay lying between Fairlight Head and Sandgate. Evidence from boreholes (p. 253) is insufficient to prove whether the surface upon which the marine sands were laid is a buried land surface or a broad wave-cut platform. The record of the borehole at Langdon (p. 284) reveals that, at least over parts of the bay, marine conditions were interrupted by emergence and formation of peat.

As Elliott (1847, pp. 466–7) realized, shingle spits grew north-eastwards from Fairlight Head into the bay. Little is known about the relationship of the shingle and the marine sand which forms the basis of the marsh. It appears probable that some shingle lying south of this district accumulated during the deposition of at least the upper part of the sand. How far, if at all, such spits extend into this district is not known because they plunge below later deposits. On the evidence of their alignment with those at Scotney Court (p. 287), the shingle spits lay eastwards of the present farm known as Birdskitchen during the Bronze Age. Emergence, after the formation of these spits, allowed the forest to grow which gave rise to the widespread peat horizon. The shingle immediately east of the Isolation Hospital rises to a higher summit level than the Birdskitchen spits and truncates one low spit (p. 287) and a large creek-ridge (see Green and Askew 1958b, p. 24). The higher shingle appears to have been driven inland: it was formed with presumably a higher sea level than the lower shingle and at a sufficiently later date to allow the intervening accumulation of a tract of shingle over 2 miles wide, lying south of this district on Lydd and Holmston Ranges. The higher shingle, therefore, probably post-dates the peat, since none has been proved amongst these fulls or to the east of them.

The configuration of the Bronze Age coastline northwards can only be ascertained by analogy with the Birdskitchen–Isolation Hospital area, for control by the peat fails. It is considered that a broad estuary, which probably was the Rother mouth, stretched from about Birdskitchen to New Romney. The recognition of this estuary at this early period in the natural formation of the marshland is based upon the presence of a low ridge of shingle on the immediate north-western side of New Romney, which is not in alignment with that at Birdskitchen. As at the Isolation Hospital, higher ridges of shingle, which occur in New Romney and along the line of the roadway to St. Mary's Bay (Fig. 14), truncate the lower spit. The higher ridge again is not in direct alignment with the Isolation Hospital fulls. Both higher and lower shingle occurrences in these localities are regarded as forming part of recurves into the estuary. As at Birdskitchen, the lower shingle at New Romney is regarded as older than the peat and the higher as post-dating the lower by a considerable period.

During the later stages of deposition of the marine sands the low shingle at New Romney probably continued as a bay bar into the present bay off Dymchurch. Shingle recurves with a low summit elevation occur again along the coast north-east of Dymchurch, where they were noted by Elliott (1847, p. 467). Some ten low recurves have been mapped on this part of the coast as far north-east as The Redoubt, i.e. to the junction of the roadway and coast (128320). The summits of the shingle fulls immediately north-east of this point again lie at a markedly higher level than those to the south-west.

The most north-eastern lower shingle recurve is therefore identified as approximating to the limit of shingle before the forest conditions were established.   One of the low recurves is a hooked spit, revealing that a considerable fetch to the north-west must have existed at the time of its formation. The angles between the low recurves and the present coast suggest, however, that the spits or bay bars formed initially a broad bow to the seaward and that this was somewhat flattened during the formation of the later low recurves.   This process was repeated during the formation and ultimate destruction of the later and higher fulls in Dymchurch Bay.

The amount of shingle with a low summit level which has been destroyed is not known, so that the above three localities can only approximate to the line of the coast preceding the forest period.   It appears, however, that the greatest loss has occurred at New Romney.

By late Roman time (p. 255) the land had subsided and allowed the sea access ; a dendritic system of channels had been scoured out of the peat and underlying sand, and salt-marsh conditions had been established with deposition of sands in the creeks and clays in the ' pool areas ', as described on p. 254.   The higher shingle fulls were now deposited : they cut off one large sand-filled creek near the Isolation Hospital (Green and Askew 1958b, p. 24) and recurved as already described into the estuary south of New Romney and the large salting estuary near Hythe.   Later, by Saxon time (p. 260), man began to control the natural accretion of the land, with the result that, apart from the coast, the history becomes confused.   It is convenient, therefore, to complete the account of the formation of the coast up to the present and to describe the probable sequence of reclamation later.

The accumulation of shingle in Dymchurch Bay was probably extensive. It withstood erosion until the reign of Richard III, when, according to Elliott (1847, pp. 469–70), walling became necessary.   This erosion was caused by the growth of Dungeness towards its present position, tapping the supply of shingle.   The Dymchurch Bay shingle was probably mainly reworked into the present fulls of the Hythe area, while a local reversal of long-shore drift during north-easterly winds carried some shingle southwards and formed the northerly recurves into the New Romney estuary, near The Warren (Fig. 14).   The long shingle spits near Belgar and Greatstone on Sea formed on the south side of the estuary as part of the Dungeness accretion.

Holloway (1854) recorded that in the reign of Henry VIII shingle growth in the Hythe area had reduced Hythe Haven, itself a relic of the Hythe salting estuary, to a narrow gut or channel running for nearly a mile eastward towards Folkestone, and that even this was lost by Elizabeth's reign. Probably the ultimate blocking of the Hythe salting estuary resulted from the reclamation of Romney Marsh, which would reduce the scouring effect of the tide ebbing from the saltings.   The time taken to close the estuary cannot be ascertained.   Traces of the early harbour, perhaps that of Elizabeth's reign, appear to remain among the Hythe shingle mass (p. 289).   A wall which formed the protection to the shingle in 1820 has been replaced by the modern promenade built about 100 yd on its seaward side.   Relics of this old wall remain.

South of New Romney part of the 8th-century shore was located by Ward (1931a) and is shown in Fig. 14. There is no further evidence for dating the position of the coast until Poker's map of 1617 (reproduced in Lewis and Balchin 1940, facing p. 264). The shape of the coast on this map has been adjusted to fit known shingle and walling features and the line on Fig. 14 must therefore be regarded as approximate. The coast seems to have remained in much the same position for some considerable time for the same line appears as the 1885 shore in Burrows (1885, map 4, facing p. 376). The final natural processes were observed by Redman (1854) and Gilbert (1933, p. 270), who both noted the growth of shingle on the north side of the estuary at the expense of shingle lying near The Warren. The gap between the northern and southern spits was closed artificially late in the 19th century.

The River Rother. It is believed that the broad estuary at New Romney carried the River Rother up to the date of the reclamation of the creek-ridge deposits, but only the lower reach of its course can be identified. This is thought to be shown by the mounds of sand leading from New Romney to the broad creek-ridge which stretches from Wheelsgate towards St. Mary in the Marsh, where direct evidence for its course ceases. It is, however, considered to have turned southwards, since during the growth of the shingle spits northwards the Rother and its tributaries would probably have been trapped on their landward side and would have coursed north-wards as the spits grew. It is therefore possible that the sand ridge through-out its length formed part of the Rother course at a very early date, leading towards the Hythe estuary, where a Rother mouth is often mentioned in the literature. In this case it must have been contained behind shingle of which there is no trace today. It appears to be likely that the northern part of the St. Mary in the Marsh sand ridge is the infilling of a large creek similar to that through Brookland and Snargate. The Rother channel therefore is thought to have swung southwards to Wheelsgate, where its course is lost below the later sand-flat deposits. Most of the sediments adjacent to this course consist of creek-ridge sands and clays and must have been reclaimed with those deposits by the 8th century (see below). But, on historical evidence the Rother is usually thought to have forsaken this estuary in the 13th century, following a series of tremendous storms. Therefore any Rother water which reached New Romney between the 8th and 13th centuries must have followed either the Rhee Wall in its present form or an unknown previous form, or it must have followed the channel features (p. 252) from Brenzett to Old Romney and thence through the sand mounds to New Romney, retracing its original 'creek-ridge' course in its lowest reaches. These channel features do not seem bold enough to have carried the full Rother and perhaps were used by only part of its waters, either artificially controlled or as a distributary. The erosional channelling noted in the sand mounds probably dates from this period in the river's history and the relics of old walling near Rheewall Farm and Appledore Station (p. 250) are thought to have bordered the north side of its channel (see also Green and Askew 1958b, p. 25).

Green and Askew (1959, p. 25), like Ward (1933) beforehand, mentioned the possibility that the Sedbrook Sewer near Stockbridge House (pp. 252–3) was a remnant of an early course of the Rother. Green and Askew traced further channel features from Stockbridge House westwards nearly to Appledore (Tenterden Sheet 304) and following the broad ridge of clays through the peat of The Dowels. These channel features also do not appear to be bold enough to have accommodated a full Rother and they are probably part of the salting creek system draining towards Hythe.

Reclamation. There is little archaeological evidence which is of value in dating the reclamation. According to Lewin (1866), coins of the 3rd century A.D. are known from Dymchurch and Eastbridge. Pottery remains are known from Dymchurch (p. 286), Ivychurch, Newchurch and Eastbridge. If all this pottery is correctly identified as Roman, then the marshland would have been habitable over a wide area at that period, and it is probable that sea-level, which must have been low during the formation of the peat in the Bronze Age, did not rise sufficiently to flood the marshland until some time during the Roman occupation. Belgic pottery has also been found near the Holiday Camp at St. Mary in the Marsh (081276) (R. D. Green, in a letter).

The hitherto widely accepted dates of the various innings stem from Elliott (*in* Lewin 1862, pp. lxvi–ii). The maps from this work are reproduced in Steers (1946, pp. 328–9) and elsewhere. Some of these dates are contradicted by evidence presented by the deposits, as they have been previously by the work of Green and Askew and also by the historical researches of Ward. The reclamational sequence described herein is based upon the type of deposit enclosed and the relative levels of adjacent innings.

The deposits are sufficiently distinctive for three broad groups to be made and as there is evidence that their boundaries are mostly artificial it follows that the deposits exhibit the broad sequence of reclamation. The three groups are presented in their probable order of deposition, although the relationship between groups 2 and 3 cannot be proved directly. The groups are: (1) the ' creek-ridge ' and ' pool area ' deposits ; (2) the clay deposits of salt-marsh origin found in the vicinities of Newchurch, Burmarsh, Ivychurch and St. Mary in the Marsh, hereafter called the Romney Marsh clays ; (3) the salt-marsh — tidal sand-flat deposits lying on the western and southern margins of this district. Differences of level between areas bounded by sea-walls reveal something of the reclamational sequence within groups 1 and 3. It is suggested, however, that some sea-walls, particularly near Brookland, are not the limits of reclamation but the limits to which later inundation was controlled. The age of an innings can only be determined historically and the researches of Ward upon Saxon Charters showed that land existed in the 8th or 9th centuries as follows : (i) Ward (1931b) located but did not delineate an area west of Lydd (outside this district) which was mentioned in a charter of A.D. 770. (ii) Ward (1931a) delineated an area east of Lydd which was alluded to in a charter of A.D. 774 ; from this he drew the coast-line of the period (Fig. 14), which was extended along the shingle fulls by Lewis and Balchin. (iii) Ward (1933a) considered that the land on either side of the Sedbrook Sewer west and north-west of Oxpound (030318) comprised a gift of land to the Church in A.D. 724. (iv) Ward (1933b) recognized that the roadway (p. 250) through Lower Wall Farm (084335) formed the northern boundary of land mentioned in a charter of

A.D. 850. (v) Ward (1936) considered that a charter of A.D. 700 referred to land adjacent to Sellinge Farm (086293).

The deposits within each of these areas are: (i) sand-flat and shingle-full and part of the Belgar ridge of sand (p. 285); (ii) shingle-fulls, clays and sand; (iii) creek-ridge deposits; (iv) and (v) Romney Marsh clays. Hence it becomes probable that the whole of the creek-ridge deposits and Romney Marsh clays were land by the 8th and 9th centuries, in fact by the 8th century in view of charter (v). This date is contrary to Elliott's widely accepted reclamation dates for ground south of the Rhee Wall and east of Snargate. It is also contrary to the 14th-century date which Green and Askew (1959, pp. 24–5) gave for the deposition of sediments near Newchurch, basing their conclusion on the calcareous content of the soil.

On the evidence of relative levels the following sequence of reclamation in the creek-ridge deposits can be made out. The deposits of the large area A and B (Fig 14) exhibit little or no variation in level and although some sea-walls have been recognized, and others may follow the roadways, no sequence of reclamation can be distinguished. Furthermore A and B are not separable either from the ground lying north of the Rhee Wall up to the borders of the Romney Marsh clays, or from the Dowels region. A and B include innings which Elliott (*in* Lewin 1862, pp. lxvi–ii) regarded as reclaimed by the Church in the 12th and 13th centuries, and which are named after Archbishops of Canterbury. In four cases the dates noted (Furley 1874, pp. 250–1) are those when the respective Archbishops held office. Even Elliott (op. cit., p. cxi) was uncertain whether Baldwin's Innings was named after Archbishop Baldwin or one Baldwin Scadeway, although the rough dates quoted would be applicable in either case. This doubt of Elliott's seems to justify a charge that he based his reclamation dates solely upon the names of the innings.

The area A contains some sand which pre-dates the peat. It must therefore be an early innings. Elliott stated that it was bounded by the line of the roadway past Westbroke House (040218). No wall has been noticed here. The remaining walled creek-ridge deposits reveal something of a succession of innings; near Brookland, however, the sequence is complicated and several interpretations appear feasible. Area C is generally very slightly lower than D and they are both slightly lower than B. Innings E is again lower than F. The areas C, D and E are regarded as early innings, separated by only relatively short periods of time, bordering the large creek F and area B. Perhaps originally they extended farther west under the areas U and V. The large area B contains the innings mentioned above, which were supposed to have been reclaimed by the Church in the 12th and 13th centuries but which appear to have been won much earlier at about the same period as C, D and E. Innings G and H were probably first won at about the same period, but were subsequently lost and are early innings containing the group 3 deposits (p. 260) of sand-flat origin. The western part of G consists of sand deposits at the surface with fresh-water alluvium occupying breach scourings. The ground surface is uneven and the surface sands, which extend for an indefinite distance beyond the north-east boundary of the innings, apparently rest upon creek-ridge deposits and were probably laid down during the breaching. Green and Askew (1959, p. 23), in noting sands of two depositional periods near Brookland, were probably referring to these two deposits. This inundated ground appears to have been soon regained. The eastern half of G, however, consists of a sand-flat (i.e. group 3 deposits) complete with some of its drainage channels ; a breach occurs at the north-eastern corner of the innings. This breach, therefore, is thought to have occurred later than those mentioned above. The sand-flat deposits are found also in areas H and I. At their southern edge, in the eastern part of H, the sands extend to a sharply defined margin which may have been an old sea-wall, now levelled. A later wall about 50 yd to the south looks as if it was built to provide additional protection for the northern boundary of K and L. Breaches in this wall are the continuation of the fleets of K into H. The sand of innings I lies at a lower level than the adjacent ground to the south and east. The easternmost part of G and the whole of H and perhaps I appear to be early inned ground which was lost, and where tidal sand-flat conditions were firmly established before they were regained. Their recovery appears to have been at approximately the same time as the innings of N and J ; they must therefore be regarded as innings containing group 3 deposits. Another interpretation is to regard I, H, G and F as a possible creek course, eventually choked by sand and bordered by innings K, L and part of B. Such a creek could not have carried the Rother because a connection to Snargate, where it could continue below the later deposits, is lacking.

The area J consists mainly of sand deposits and shows the effects of breaching : it is considered to be an innings containing deposits of group 3. The innings K and L were very early reclamations which comprise a 'window' of early creek-ridge deposits, almost surrounded by later sand-flat and salting deposits. Large fleets, formed by breaching, traverse the creek-ridge deposits and another wall to the north. They are bordered by levees composed of sandy clays and sand. There is, therefore, a complicated system of sands of two periods of deposition. Thin sand also occurs over much of the ground mapped as clay. These two innings are regarded as early reclamations of about the same age but which were lost at probably the same time as the flooding of G and H. In K and L, however, it seems that the ground was regained before the tidal flats were fully established. Similar levee and creek-ridge deposits occur in M, which is considered to have had a similar history of loss and gain, but as the general level of the deposits is higher than in K and L, the initial inning was probably later.

The creek-ridge deposits bounded by the Romney Marsh clays in the north and by the walled deposits of group 3 and Storm Gravel Beach Deposits in the south form the early reclaimed ground of the marshland. From the evidence near Brookland it can be considered to be the relic of a wider reclamation extending beyond its southern margin, and it may also have had a wider extent northwards. The creeks in the Romney Marsh clays are neither choked with sand nor inverted. The marginal relationships of these clays are indefinite and they may not have been walled (p. 284) throughout their period of deposition. The Romney Marsh clays may, therefore, have been deposited during inundation for a considerable period, in a similar manner to the third group of deposits in the south, and both must have occurred as a result of the rise in sea-level since the Roman period. It must be admitted, however, that no direct evidence is available on whether the creek-ridges and Romney Marsh clays were reclaimed at widely different periods or approximately together, but only that both were land in the 8th century.

The method and sequence of the reclamation of Romney Marsh is as doubtful as the date. It appears to have been achieved almost *en bloc* by the erection of walling across the salting estuary at Hythe, perhaps in the vicinity of the drainage gut from the Royal Military Canal to the sluice near Dymchurch Redoubt, whereabouts Elliott (*in* Lewin 1862, p. lxiii) mentioned a wall. The line of walling from St. Mary's Bay to Ivychurch (pp. 284–5) may have protected the creek-ridge reclamations from the inundation during which the Romney Marsh clays were deposited.

The third and later group of innings lying on the western and southern margins of the district are reclamations which Elliott recognized as 13th century or later in date. This group comprises areas labelled N to V inclusive (Fig. 14) and, in this interpretation, G (part only), H, I and J. Also included with this group are the innings on Romney Salts, which reveal a progressive eastern extension of the reclaimed ground, commencing with that recognized by Ward (1931a). The only control on the dating is afforded by Poker's map of 1617 (p. 259). A surprising feature of Romney Salts is the very slight difference in level on the tidal flat which spans innings from before the 8th century to the late 19th century. Deposition here must have been extremely slow.   J.G.O.S.

Storm Gravel Beach Deposits

Storm Gravel Beach Deposits consist of gravel, with some sand in places, which has been thrown into a ridge by waves. The deposits are colloquially known as ' beach '.

The gravel is driven up the beach by waves during flood-tide and accumulates as a ridge during the high-tide standstill period at the highest point to which it can be driven. Not all such ridges are preserved ; lower and weaker ridges are destroyed during higher tides or by storms. Every so often, however, by an accidental combination of weather and tide, a ridge is preserved from destruction by the accretion of shingle upon its seaward side. This protecting shingle is itself thrown into a ridge and in the process so dissipates the power of the waves that the earlier ridge is untouched. The balance of formation and destruction is delicate ; if the amount of protecting shingle is insufficient to check the waves, then the material is added to the earlier ridge, which is re-sorted partially or completely. When the supply of shingle is small a storm beach is formed which is re-sorted and replenished during storms, but if there is a ready supply of shingle the beach accretes seaward by the addition of sub-parallel ridges formed in the manner described. Such ridges of shingle are termed ' fulls ' and in certain localities they may extend for some miles. The fulls curve inland as they approach an estuary and such recurves are common upon the marshland. Their upper surface falls in level until the shingle is buried below the sands and clays of the marshland. The shingle is poorly bedded except upon the seaward slopes of the

S

fulls and where sand is present. Cobbles larger than about 5 inches in diameter have not been noticed and the usual size is about 3 inches. Brown clay, probably derived from dust and washed down by rain-water, commonly coats the pebbles below about 3 ft depth. The constituent material is almost entirely well-worn flint cobbles. White (1928b, p. 88) also recorded cherty sandstone, vein quartz, Wealden sandstone, reddish, grey and liver-coloured quartzites and dark quartz-tourmaline grit pebbles. Some of these may have originated as ships' ballast.

MARINE BEACH DEPOSITS

Sands, silts and muds found upon the present day beaches and tidal flats below the level of the Storm Gravel Beach Deposits are classified together as Marine Beach Deposits. They commonly have a scattering of itinerant pebbles lying upon their surfaces.

BLOWN SAND

Patches of Blown Sand in this district are of small extent and little importance. Topley (1875, p. 303) observed that the Blown Sand occurred to leeward of the marshland harbours, referring to the prevailing south-westerly winds; there are, however, exceptions, in particular the dunes at Greatstone on Sea. In fact the blown sand tends to accumulate on the north-eastern side of estuaries or tidal flats, from which sand can be removed by wind when uncovered by the tide.

The sands are bedded and are generally coarser in grain than the sands of the marshland. *Helix* shells are frequently found and *Cardium,* presumably dropped by birds, is also present. The bones of rabbit occur.   J.G.O.S.

LANDSLIP

Landslips are associated with all the ' solid ' clay formations in the present district. They take the form of earthflows, slumps or rockfalls, or combinations of these, in the classification used by Sharpe (1938). Toms (1946), Wood (1955) and Viner-Brady (1955) have described fundamental researches into the nature, origin and means of prevention of the landslips in Folkestone Warren (pp. 293–6), and these and other coastal landslides in Kent have been studied by J. N. Hutchinson (in preparation).   G.B.

DETAILS

RIVER GRAVELS

5th Terrace. Gravel has been worked on Godmersham Common (084508), 700 yd E. 30° N. of Eggarton Manor. Its constituents are water-worn and subangular flints, angular chips of flint, medium and coarse-grained sandy ironstones, Tertiary-rolled flint pebbles and micaceous loams. Capping the London Clay at Rhodecommon, Dunkirk, and Fishpond Wood are numerous pockets and patches of gravel which are referred to this terrace and which testify to its original wide extent. Disturbed pockets and masses of gravel are present, often without distinguishing surface features, on the valley slopes and escarpment face. A temporary section in one such patch (067573) 500 yd S. 25° E. of Winterbourne exposed 5 ft of crudely-bedded gravel with flints up to 8 in long, many Tertiary-rolled flint pebbles, and ironstones up to 18 in long, set in a brown gritty matrix. Pockets of gravel are plentiful in the fields surrounding Forester's Lodge Farm.   J.G.O.S.

4th Terrace. North-west of Ashford several small gravel spreads occur between 200 and 250 ft above O.D., on Folkestone and Sandgate Beds. A temporary excavation (977449) on the hill-top 200 yd E. of Home Farm showed gravel consisting mainly of angular flints, but including also tiny quartz pebbles derived from the Folkestone Beds and fragments of chert and limestone from the Hythe Beds. About ¼ mile south of Potters Corner a spread of gravel, probably not more than a foot or two thick, occurs at 250 ft above O.D. A temporary exposure showed it to consist mainly of flints, angular and sub-angular, with a large proportion of pebbles of ½ to 1-in diameter, and to include also sub-rounded ironstone and chert fragments, tiny quartz pebbles and Wealden siltstone pebbles. The Wealden pebbles were noticed by Topley (1875, p. 169). Their presence suggests a tributary from the south.

The railway cutting at The Warren (001443) shows up to 4 ft of gravel, mainly of flints, on Folkestone Beds. In Ashford gravel of the 4th Terrace is at about 220 ft O.D. A temporary excavation (005431) showed, on Sandgate Beds, 4 ft of gravel, of which the lowest 1 ft consisted of closely packed small Wealden pebbles and the upper 3 ft of dark brown coarse sand with scattered angular flints and flint pebbles. At Kennington is an extensive spread, about 220 ft above O.D. at its northern end but less than 200 ft in the south, dug to about 5 ft depth in old pits just north of the church. Topley (1875, p. 168) noted this gravel to contain a very large quantity of Wealden pebbles. Near the park gates at Goat Lees a temporary section (015459) showed 5 ft of gravel of angular and subangular flints, flint pebbles and chalk pebbles in a clayey matrix. The gravel hereabouts appears to be sunk into the Gault clay in pockets. Gravel evidently deposited by the East Stour River and containing, as Topley (1875, p. 170) emphasized, no Wealden pebbles, caps a hill-top north of Willesborough Street. Topley recorded that some pits in this spread gave a thickness of 8 ft of roughly stratified gravel. A sand-pit (036420) 100 yd N.N.W. of Lacton Hall showed up to 5 ft of gravel, of brown-stained angular flints, flint pebbles and tiny quartz pebbles, in a coarse sandy matrix, with a sharp though irregular junction on Folkestone Beds. Local disturbance of the gravel is indicated by a pit (035421) 150 yd to the north-west, where 2 ft of river gravel as above were overlain by 3 ft of reddish clayey coarse sand with streaks and festoons of grey clay containing white-patinated small angular flints, and at its base a lens of gravel of the white angular flints, up to 6 in thick. Wooldridge (*in* Dewey and others 1925, p. 276) noted these white flint fragments at Willesborough and Brabourne Lees, and regarded them as characteristic of the gravelly material on the ' 200-ft platform '. He supposed this material to be " essentially a decalcified land-wash dating from Glacial times ".   B.C.W.

The small spreads of this terrace at Ouseley and 500 yd N. of Quarrington are composed of coarse brown sand, flint chips and Tertiary-rolled flint pebbles. Similar river drift which caps the ridge from west of Mersham Le Hatch to Brabourne Lees, at a level rising from 230 ft O.D. in the west to over 300 ft in the east, contains beds of flinty sandy clays and loams. This deposit seems to be more gravelly at the eastern end. It was worked 500 yd N. 35° E. of Smeeth church on ground now levelled and in use as playing fields.

Topley (1875, p. 171) noted that the gravel north of Lilyhole was dug to a depth of 10 ft in one place ; it contained no pebbles of Wealden sandstone and the flints were white and rather angular. That now exposed at a locality (090398) 500 yd E. 30° N. of Lodge House shows the flints to be less than 2 in long, with some larger near the base. The deposit contains small quartz pebbles probably derived from the Folkestone Beds and Tertiary-rolled flint pebbles, some of which are split. The matrix is a sandy loam and the bedding is crude. This terrace is also exposed in the sand-pit 450 yd S. of Pound House, where up to $7\frac{1}{2}$ ft of brown and grey silty clay rest upon $1\frac{1}{2}$ to 3 ft of mainly subangular flint gravel. Streaks and seams of similar gravel occur within the clay. The clay is partly derived from the Gault, but its relationship to that formation is obscure. Similar gravel to that at Lilyhole is present at Stone Hill ; Topley (1875, p. 171) noted, however, that it was more distinctly bedded. J.G.O.S.

At Pedlinge are three isolated patches of gravel between 280 and 300 ft above O.D. Some old pits on the northernmost patch (141361) showed 2 to 4 ft of gravel of closely packed angular flints, on Folkestone Beds sands. These gravels were most likely laid down by a former headstream of the East Stour River, since captured by the coastal drainage. B.C.W.

Whitaker (in MS.) recorded a section (089567) 7 ft deep, in Denstead Wood, 630 yd No. 29° W. of Nickle Farm. It consisted mainly of coarse gravel with a little coarse and fine sand and a partial ferruginous cement. Some of the land was false bedded.

The gravels on the ridge from Chartham Hatch to Bigbury Camp rests on an irregular and channelled surface cut in Woolwich Beds and Thanet Beds. The following description is from Dewey and others (1925, p. 277) : " Structurally they comprise a tumultuous assemblage of torrent-bedded sand and gravel lenticles, extremely variable in form and size and without discernible order or sequence. The sandy lenses are of coarse grain and contain a number of small angular stones and chips scattered throughout them. The constituents of the gravelly seams include large, little-worn flint nodules, ' mahogany ' flints, blocks of sarsen, and Lower Greensand ironstone, Tertiary flint pebbles and small quartz and lydite pebbles from the Folkestone Sands ". These gravels are typical of the deposits of the Great Stour north of the Gault outcrop ; their tumultuous appearance is now thought to be largely the result of frost-churning.

In a pit (105570) 300 yd N. 20° E. of the cross-roads at Chartham Hatch 0 to 6 ft of gravel with current-bedded gritty sand lenses were noted resting mainly on Woolwich Beds (p. 192) ; the terrace base was channelled down to Thanet Beds in one place. The gravels seem to have crept slightly northwards since deposition, for there is some orientation of the individual flint cobbles. At the western end of this pit the terrace comprises $7\frac{1}{2}$ ft of well-bedded brown sands with gritty and fine gravel and clay seams, with gravel below. Seams of contorted clay which underlie undisturbed sand lenses suggest deposition during rigorous climatic conditions. In another pit (110569) 750 yd E. 10° N. of the cross-roads, the terrace deposits are 6 ft thick, but northwards towards the roadway they are channelled down to a total thickness of 15 ft ; they rest throughout on Thanet Beds. Lenses and seams of coarse light to dark brown current-bedded sand up to 4 ft in thickness, impersistent and irregular light grey clayey seams and dark manganiferous bands are present. A few chalk pebbles were seen in the gravel and a ' sarsen-stone ' measuring 2 by 2 by $\frac{1}{2}$ ft was reported to have come from it. The uppermost 4 to 5 ft of the terrace have been disturbed and contorted by frost action, with the result that small but sharp flexures are developed in the sand and clay seams.

On the north side of the roadway (111573) 1000 yd. E. 28° N. of the cross-roads at Chartham Hatch the gravel is 16 ft in thickness, with contortions in the upper layers. Whitaker (in MS.) recorded that the 12 ft of gravel south of the road near Bigbury Camp (119575), 850 yd W. 40° N. of Tonford Manor, contained a boss of Woolwich Beds sand protruding into the gravel from below the floor of the pit. The occurrence of manganese in the gravel here was noted by McDakin (1888, p. 134).

As emphasized by Wooldridge (in Dewey and others 1925), the deposits of this terrace occur northwards of Chartham Hatch. The spreads at Homestall Wood and Rough Common rest on London Clay and have therefore been extensively disturbed by freezing and thawing. The bases of these gravels are so irregular that observed thicknesses are of very little value in determining the amount of gravel present, and locally stiff disturbed London Clay or gravelly and gritty clays reach the surface.

In the roadside (124584) 420 yd W. 27° N. of Hall Place 3 to 5 ft of gravel are exposed ; in a trench nearby the gravel consisted of unbedded subangular flints (up to 6 inches in length) and numerous Tertiary-rolled flint pebbles in a matrix of gritty brown sand ; brown sand with flints was also seen, and some of these flints were associated with small grey sandy clay patches. Debris from trenches (127594) 340 yd W. 29° S. of Hothe Court consisted of gravels, sands and gritty clays ; the trenches were reported to have been 8 ft deep and not to have bottomed the terrace. Up to 4 ft of coarse brown gritty sands with angular flints and Tertiary-rolled flint pebbles were noted in another trench (133595) 300 yd E. 40° S. of this place. A borehole (134592) at a school (St. Edmund's) proved only 3 ft of gravel resting on London Clay (p. 198). A patch of gravel seen by F. H. Edmunds at Newhouse Farm (130549) indicates the broad extent of the original spread.

The dry valley at Lower Hardres is incised through the terrace gravel. On his map F. H. Edmunds noted 4 ft of dirty gravel below 1 ft of loam in the road cutting (151531) 300 yd W. 20° S. of the church.

The high patches of gravel along the course of the Little Stour River range too high to be correlated with the 4th Terrace with certainty and too low to be included with the 5th. Accordingly they are shown on the map as 4th–5th Terrace. The surface of the gravel ⅓ mile north-east of Park Farm (182515) lies between 300 and 320 ft O.D. and is strewn with water-worn flints.   J.G.O.S.

On the spur between Heart's Delight and Barham (200500) the ground surface is relatively flat and declines from about 360 ft O.D. in the south to about 290 ft in the north, with a suggestion of a bluff limiting it on the south-west. The soil is sandy loam scattered with flints, mostly rounded and in concentrations which, in places at least, justify classification of the deposit as gravel. It is underlain by Clay-with-flints.   G.B.

3rd Terrace. Gravel adjoining the Great Stour flood plain near Bucksford (988425), with surface level about 140 ft O.D., has been dug in extensive pits 10 to 12 ft deep, reaching the underlying Weald Clay. It contains small angular flints, some chert fragments and siltstone pebbles, in a coarse sandy matrix. In the East Stour valley, gravels at about 160 ft O.D. include the patch just west of Kingsnorth church (005394), three others respectively 500 yd S.S.W., 900 yd W. 10° S. and 1100 yd W. of it, and two respectively at and 600 yd north of Stanhope School. A lower group, with surface at 130 to about 140 ft O.D., appears to grade into a terrace with surface at 125 to 130 ft O.D., bordering the flood plain south of Ashford (010417) ; to this group may be related the patch of brickearth west of Bartlett Farm (p. 232). Although these gravels contain mostly Wealden siltstone pebbles, spreads at both levels contain also angular flints, flint pebbles, and fragments of brown ferruginous sandstone and of Kentish Rag.   B.C.W.

A tiny patch of the lower series of this terrace caps the outlier of Atherfield Clay (046389) 150 yd S.W. of The Den. In the tributary valley north of Hatch Park the terrace is represented by two small spreads of gritty sands and flints at about 180 to 200 ft O.D., situated 250 yd E. 30° S. and 500 yd S.E. of Quarrington respectively. Downstream there are several drift-free bench features cut in Folkestone Beds at about the same elevation.   J.G.O.S.

Temporary excavations in Park Street and Tufton Street, respectively 150 yd N. and 150 yd E. of the church in the centre of Ashford, showed 8 ft of sandy gravel with flints and Wealden pebbles. The surface of the spread is about 185 ft above O.D. Topley (1875, p. 169) mentioned that the gravel at Ashford had been largely dug on the north side of the town. The old pits are now presumably built over. Gravel south of Kennington attains 165 ft above O.D. in the cemetery 400 yd N.W. of Bybrook, where 6 to 8 ft were noted, on Folkestone Beds sand. At nearly the same level is the gravel in the floor of a large brickpit 300 yd west of Spearpoint Corner (p. 233). Other spreads of gravel at a relatively high level border the brickearth 750 yd west of Wilmington Farm and emerge from beneath it ¼ mile south of Kempe's Corner. Siltstone gravel shows at the southern fringe of the brickearth spread south of Perry Court.

The surface of the gravel spread at Conningbrook declines from about 140 ft O.D. in the north to 116 ft in the south. In pits 6 to 10 ft deep beside the railway it has been dug down to the underlying Folkestone Beds. This gravel may continue north beneath the brickearth to connect with that east of Kennington Hall ; the latter has been worked in extensive old pits to about 6 ft depth, and consists mainly of flints, with many chalk pebbles. Gravel at Spring Grove, presumably in the old pit south of the road (042465), was noted by Topley (1875, p. 168) to be rather chalky, with great quantities of flints ; Wealden pebbles were scarce.

In the two patches of the 3rd Terrace bordering the alluvium respectively ⅜ mile west and ½ mile north-west of Blackwall Farm, and in that which, flanked by brickearth, forms an 'island' in the alluvial tract ½ mile W.N.W. of the farm, no gravel has been noted, the soil being sandy and the material exposed in ditches and hedge-banks being a coarse sand. This sand possibly came from east of Willesborough Lees (about 040425), where Folkestone Beds form an escarpment some 50 ft high, capped by river gravel of the 4th Terrace. In a time of heavier rainfall than the present, the springs at its foot might deliver much sand to the River Stour, and the re-deposition of this sand as bars on the edge of the flood plain would account for the deposits near Blackwall Farm. Each of the sand bars partially blocks the mouth of a tributary valley and by impeding the drainage of spring water from these valleys to the River Stour may well have assisted the accumulation of peat (p. 281).   B.C.W.

Near Wye the terrace comprises both sand and gravel, the former predominating on the east side of the river. Coarse gravel 4½ ft thick, consisting of battered subangular flints up to 6 in long, numerous water-worn pieces of chalk, and occasional ironstones and Tertiary-rolled flints, was noted below 1½ ft of flinty loam at a locality (044466) 600 yd W. 30° S. of the level-crossing. Bedding in this gravel was shown only by dark manganiferous streaks and bands of the chalk pebbles. Nearby trench debris contained 'sarsen-stones' up to 1½ ft long. The sands of this terrace were exposed for 3 ft below Head (p. 227) in a trench (052466) 350 yd S. 35° W. of Wye church, and comprised fine laminated green sand with clayey patches and irregular brown ferruginous staining. They contained many rootlet tubes, some of which were lined with calcareous matter. These sands were proved by augering for a further 14 ft 9 in, before a hard seam, probably of gravel, was encountered. There was a thin light grey clay seam near their top. In the lowest 7 ft the sand was coarser and browner in colour. Sands present on the mound rising through the alluvium 1000 yd S. 20° W. of the church exceed 7 ft in thickness.

At Godmersham there are two spreads of this terrace gravel; the larger slopes gently westwards from about 110 ft O.D. to pass below the alluvium, while the smaller mantles a spur some 40 ft higher. The lower gravel has been dug to a depth of 8 to 10 ft in an old pit (068505) 650 yd E. 7° N. of Godmersham church, but only 3 ft of subangular flints, up to about 5 in long, and occasional ironstones, in a gritty brown matrix, are now exposed. Much of this gravel seems to have been carried down the now dry valleys to the east and deposited where they met the main stream (p. 277).

In a small disused pit (128526) 290 yd S. 30° W. of Swarling Farm F. H. Edmunds noted 3 ft of well-bedded water-worn flint gravel with ironstones, underlying 2 ft of flinty loam. A temporary exposure (126561) 550 yd W. of Cockering Farm exposed 3 ft of bedded angular and subangular flint gravel, with Tertiary-rolled flint pebbles and ironstones.

At Harbledown the terrace is again represented by deposits at two levels. The western part of the spread centred 640 yd S. 11° E. of the church with tower (St. Nicholas's) at Harbledown caps the ridge: the eastern portion is benched into the hillside and separated from the cap by a marked bluff, which itself carries a deposit of over 4 ft of gravel and sand. Many sections seen during construction of the housing estate here showed over 4 ft of subangular flints, Tertiary-rolled flint pebbles, Wealden siltstones, and ironstones, tightly packed in brown ferruginous and manganiferous sands. Local passages laterally from gravel to gritty clay and sand were noted. A section (131577) 475 yd S. 4° E. of this church with tower showed 3 ft of similar gravel, but containing large ironstone fragments, resting on Thanet sand. Temporary sections (134578) recorded by Whitaker (in MS.) show that the gravel 530 yd S. 36° E. of this church is 6 ft or more in thickness. Deposits capping the ridge at the church without spire or tower contain beds of clean whitish sand, which Whitaker (in MS.) was told reached a thickness of 8 ft. Flint gravel 6 ft thick and containing 'sarsen-stones' and ironstones was noted in the road cutting (133581) 175 yd E. 20° S. of the church with tower.

On the London Clay overburden of the Shelford Sand-pit (p. 193) there are disturbed relics of the 3rd Terrace. The face, which is worked in a west-south-west direction, exposed, in 1953, pockets up to 10 ft deep of compact gravel in a gritty sand matrix with streaks of manganiferous staining. Masses of contorted London Clay, up to 10 yd wide and containing numerous polished surfaces, separated the pockets of gravel. In one place the clay was overturned to rest upon the gravel to a thickness of 3 ft and for a distance of 4 to 5 yd. The bedding in the gravel was only markedly disturbed near the margins of the pockets and it was probably frozen hard at the time the clay was squeezed upwards. Flinty loam soil up to 3 ft thick rested on both the gravel and the disturbed clay. Quarrying operations revealed that the gravel was not continuous on this spur, but that it occurred in pockets similar to but larger than those of the 2nd Terrace at Harbledown (p. 274).

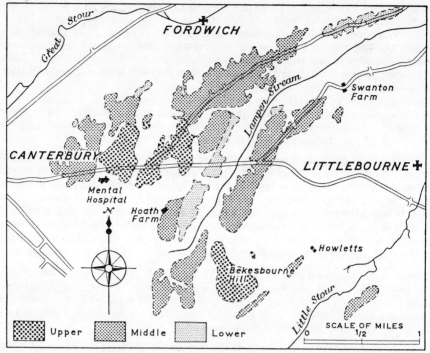

FIG. 15. *Sketch-map showing the distribution of the three levels
of the 3rd Terrace between Canterbury and Littlebourne*

East of Canterbury the distribution of the three levels of this terrace is shown
in Fig. 15. The lowest division is channelled into Thanet Beds in a pit (164585)
1000 yd N. 27° W. of the Mental Hospital near the Wingham road, where 3 to
6½ ft of tight unbedded gravel and current-bedded gravelly sands were noted.
The gravel contains a lens of laminated brickearth 1½ ft thick and a thin light
grey clay seam at the base. The unbedded gravel is itself channelled into the
gravelly sand in one place. Gravel of the middle division of the terrace, forming
a broad bench, was seen to exceed 5 ft in thickness in a temporary section
(165577) 240 yd W. 34° N. of the Mental Hospital. Whitaker (1872, p. 172) noted
" Brown false-bedded gravel, partly coarse, partly fine, with lenticular masses of
sand ; 8 or 10 feet ", resting on Woolwich Beds (p. 193), in a pit now much
obscured (165585), 870 yd N. 15° W. of this Mental Hospital. A marked bluff
capped by gravel of the highest division occurs at the eastern edge of this broad
bench. The upper gravel is continuous with further developments of the middle
and lowest divisions which occur successively northwards. The gravel surface
on the ridge falls northwards about 60 ft in ¾ mile and the three divisions are
ill defined. The upper two divisions are also represented in the patch 1100 yd
N. 37° E. of the Mental Hospital. There is a further development of the uppermost division ½ mile east-north-east of this hospital, which extends in a north-
easterly direction adjacent to the roadway leading to Trenleypark Wood. The
fall between this and the middle division which occupies much of Trenleypark
Wood itself, is for the most part concealed by a patch of brickearth. Gravel
workings south of Fordwich have obscured the relationship between the two
lowest divisions. An old working face in the middle division of the terrace
(179588), 400 yd W. of the western T road-junction ¾ mile south of Fordwich,
shows 2 ft of gravel overlain by 4 ft of coarse brown gritty sands with small

lenses and beds of angular flint chips, below 2 ft of gravelly soil. The total depth of gravel here varied from about 7 ft to over 20 ft according to Smith (1933, pp. 165-6), who, on the basis of his identification of Acheulian and Clactonian implements collected from this pit, regarded the gravel as the equivalent of the 100-ft terrace at Swanscombe.

A section (181593) between the two roadways 500 yd S. 6° W. of Fordwich church, shows what is believed to be the lowest division of the terrace, channelled into Oldhaven Beds (p. 197) and Woolwich Beds (p. 193). The channel is seen in two places: firstly, 3 ft of typical subangular flint gravel with sandy lenses cut vertically into Oldhaven Beds; a 'stringer' of gravel extends obliquely into the Oldhaven sand at the bottom of the vertical cut, revealing that part of the Oldhaven Beds sand is not *in situ*. Secondly, the undulating surface of the Woolwich Beds is overlain by 0 to 3 ft of tight crudely-bedded gravel, composed mainly of pebbles from the Oldhaven Beds, but with other flint cobbles up to 6 in long, in a matrix of gritty clay and sand, below 2½ ft of gravelly clay soil. At the western end of the section the crude bedding is lost in places and two large 'rafts' of clay and 'blocks' of gritty sand with flint chips are incorporated in the gravel. One of the clay rafts, 6 ft long and 9 in thick, transgresses the bedding of the gravel; the other, 12 ft long and 1½ ft thick, is concordant. The rafts are weathered to a brown loam and, like the blocks of gritty sand, they are thought to have been derived when in a frozen condition. Dewey and others (1925, p. 279) noted rafts and masses of friable sands in gravels at approximately the same altitude on the north side of the river at Sturry on the Faversham (273) Sheet. Rafts of clay at Sturry, however, are now known to have foundered from a buried river cliff of London Clay and Oldhaven Beds with which the gravels were in contact.

The middle division of the terrace is again seen, in faces ranging from 7 to 19 ft high, in a pit (185591) 900 yd S. 28° E. of Fordwich church. The base of the gravel generally lies at about the junction of Oldhaven Beds and Woolwich Beds, but is channelled lower. Many Tertiary-rolled flint pebbles and blocks of purple sandstone from the Oldhaven Beds are incorporated in the gravel. The gravel is sandy, reddish brown and crudely bedded, but contains sandy lenses in which beds of smaller flints are common; some of these lenses are current-bedded. Many of the Tertiary-rolled flint pebbles are split. Grey clay seams (not rafts) are also present.

The same terrace is dug in Trenleypark Wood (189593), 1000 yd E. 40° S. of Fordwich church, where 6 ft of gravel similar to the above were noted. Near the roadway a 3-ft bed of grey and red sandy clay and stiff clay, which contains small flint chips and thin seams of flints, overlies the gravel. This appears to be an old alluvium which was derived from the outcrop of London Clay.

In the valley of the Lampen Stream (p. 2) the two lowest divisions of this terrace are well represented. In the patches of gravel at and ¾ mile north-north-east of Hoath Farm, both divisions seem to be present. The middle division at the latter locality is probably continuous below the patch of brickearth with the large spread in Trenleypark Wood.

On the east side of the valley the middle division is represented discontinuously for over two miles by a well marked bench feature covered by gravel which has been extensively worked. In Oldridge Wood (189582), 340 yd N. 8° E. of the cross-roads, 6 to 8 ft of gravel in a sparse brickearth matrix, are exposed. Eastwards this gravel is overlain by brickearth which oversteps on to the Thanet Beds.

In Trenleypark Wood the middle division is seen by the entrance roadway to the sand-pits (194593), 1000 yd W. 27° N. of Swanton Farm; the section shows 2 to 3 ft of rough unbedded gravel in a gritty clay matrix, with long lenses of gritty ironsand and seams of grey laminated gravelly clay, below 2 ft of gravelly soil. The disused gravel-pit (196597) north of the road 900 yd W. 4° S. of Elbridge House shows gravel of the middle division channelled into Oldhaven Beds and Woolwich Beds: 2 ft of pebbly gravel are overlain by 0 to 1 ft of fine laminated sand, below 1 to 3 ft of coarse brown sands containing small white ½-inch chips of flint, particularly near the base ; these are overlain by a seam, 0 to 1 ft in thickness, composed mainly of pebbles derived from the Oldhaven Beds, beneath 1½ ft of flinty soil.

A water borehole (179563) 330 yd. N. 13° W. of Hode Farm proved 5 ft of gravel of the middle division on Thanet Beds (p. 186). Separated from this spread by a distinct bluff, the highest division of this terrace capping Bekesbourne Hill was reported to be 7 ft in thickness. On the slope of the Little Stour valley, the middle division is seen in the cutting at Bekesbourne Station (188561), where 6 to 10 ft of roughly-bedded gravel are channelled into Thanet Beds (p. 187).

At Bishopsbourne this terrace is closely associated with Head and Head Brickearth deposits and also passes below the alluvium of the Little Stour. A temporary section (187526) 110 yd W. 14° S. of the church, showing 2 ft of coarse gravel overlain by 2 ft of brickearth below 4 ft of dirty gravel, was recorded on his map by F. H. Edmunds.   J.G.O.S.

2nd Terrace. A patch of gravel forming the extreme tip of the spur at the union of two valleys near Leather Bottle (980572), 1100 yd N. 38° W. of Belmont, has been worked to a depth of 8 ft. A water bore sited in the base of this pit proved 4½ ft of drift, on 8 ft of rubbly chalk, resting on Upper Chalk (p. 143).

The fan of gravel at Shalmsford Street has been deposited at the confluence of the Petham nailbourne and the Great Stour. Much of the material seems to have been carried by the nailbourne, but some was probably transported down the valley draining south-east from Lower Ensden. The gravel, which is unbedded, is exposed 100 yd S. 10° E. of the level-crossing (094545), where F. H. Edmunds noted that it was up to 12 ft thick, resting very irregularly on chalk, and overlain by Dry Valley and Nailbourne Deposits (p. 277). The gravel and the underlying chalk have been churned by frost action, with the result that the chalk is squeezed upwards and nearly penetrates the gravel in one place. The gravel is composed of flints, which average 4 to 6 inches in length, in a light to medium brown clay–sand matrix except adjacent to chalk, where it is dark chocolate for up to about 1 ft. A raft of shattered chalk was noted within the deposit.

The gravels below the alluvium (093552) 500 yd E. 40° N. of the road-bridge crossing the river at Shalmsford Street are actively worked. They are from 6 to 22 ft thick and are composed mostly of flints up to 4 in long, with some cobbles very much larger, and also ' sarsen-stones ' and ironstones.

Near Howfield Farm the terrace has been fed with material carried down minor valleys draining the Chartham Hatch ridge. Over part of the outcrop the gravel is indistinguishable from that of the 3rd Terrace, with which it is apparently continuous. In an exposure (116562) 250 yd W. 11° N. of Howfield Farm 6 ft of very tightly packed gravel with a matrix of coarse sandy or gritty clay were seen below 5½ ft of brickearth (p. 237).

Two water boreholes (120562) 250 yd E. 28° N. of Howfield Farm encountered as much as 22 ft of drift overlying chalk. The major part of this comprised gravel, underlying alluvium.

Whitaker (in MS.) noted 5½ ft of gravel resting on Thanet Beds in a temporary section (135575) 840 yd S. 36° E. of the church with tower at

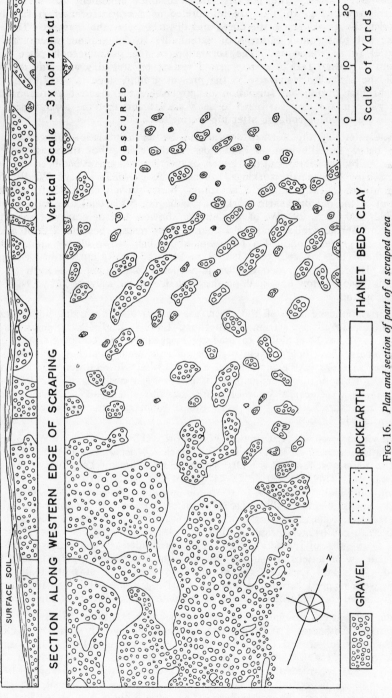

FIG. 16. *Plan and section of part of a scraped area 820 yd E. 43° S. of St. Nicholas's Church, Harbledown*

Harbledown. Levelling operations for playing fields (137576) 820 yd E. 43° S. of this church showed the 2nd Terrace gravel extensively pocketed into the Thanet Beds clays. The pockets had a distinct alignment across the slope of the ground (Fig. 16) and are interpreted as a crude form of frost-polygon in which the gravel forms the cores and disturbed clay the margins. These are unlike polygons found in areas of perennially frozen ground, where the finer material forms the cores. The formation of the pockets took place at the bottom of several feet of coarse gravel resting on stiff clay and why the result should be a pattern transverse to the present ground slope is not known. It is thought that during the formation of ground-ice the greatest accretion of ice took place in the clay surround to each pocket and that the arrangement now seen is the result of collapse after the ice melted.

At Canterbury this terrace appears to be an aggradation of gravel rising from below the alluvium on both sides of the river. For the most part it is concealed by brickearth, but north-west of the river the gravel has been uncovered by extensive brickearth workings. A borehole (144583) 350 yd W. 10° S. of the northern railway station (Canterbury West) proved chalk 13 ft below the bottom of the old brickearth working. A digging for a tank at garage premises (145582) 200 yd S. 29° W. of this station showed 4 ft of gravel of subangular flints and Tertiary-rolled flints in a coarse gritty matrix, beneath 2 ft of gravelly loam ; the gravel contained a 1-ft seam of reddish brown clayey sand. Dewey and others (1925, p. 282) recorded 12 to 14 ft of locally current-bedded gravel, which yielded two abraded Chelles implements, underlying brickearth (p. 238) in a pit (144589), now obscured, on the west side of the railway 650 yd N. 15° W. of this station.

On the opposite side of the river, gravel has also been dug south-west of the principal railway station (Canterbury East). Reid (1891) described this gravel as 12 to 23 ft thick, covered by brickearth, and composed of boulder-flints, coarse and fine gravels, with sand and loamy patches, generally with the finer material in the middle and upper levels. He recorded that a vertebra and fragments of teeth of mammoth, pieces of a molar tooth of *Rhinoceros tichorhinus,* a metacarpal bone of *Equus caballus* and oyster shells and silicified wood from the Tertiary beds were found when the pit was extended at its eastern end in 1890.

One of the wells at Thanington Waterworks (138567), 1070 yd W. 33° S. of this station, proved the drift to be 17 ft thick. Most of this is probably gravel but the uppermost few feet are Dry Valley and Nailbourne Deposits. A borehole in a car park (149575) 350 yd E. 44° N. of this station proved 16 ft of gravel resting on chalk, below 14 ft of made ground. A temporary section at a garage in St. George's Place (153575) showed only the uppermost 2 to 3 ft of terrace deposits underlying brickearth (p. 239). The terrace comprised loamy sands with an irregular bed of flint gravel containing chalk pellets. In one part a column of powdery grey chalk protruded through the gravel and connected with a lens of similar chalk, with a concave upper surface, $3\frac{1}{2}$ ft broad. This lens separated the gravel from 0 to 8 in of flinty loamy sand, which were trapped between it and the overlying brickearth. The frost-heaving which caused this disturbance seems to pre-date the deposition of the brickearth.

Near the eastern margin of the brickearth cover (156579) 510 yd E. 7° S. of Canterbury Cathedral 3 ft of gravel were noted below 3 ft of flinty brickearth and 6 ft of made ground. No gravel section was exposed in 1952 in the Vauxhall Brickworks (161589), 1500 yd N.E. of the Cathedral, but Dewey and others (1925, p 282) recorded a section 10 to 14 ft deep, consisting of current-bedded gravel and sand, resting on Thanet Beds ; two late St. Acheul implements were recovered from the gravel base. The same authors also stated that 10 to

20 ft of gravel resting on Thanet Beds were present below brickearth (pp. 238–9) 1200 yd N. 10° E. of the Cathedral (152590).

Exploitation has removed the gravel under the alluvium from much of the available ground north-east of Canterbury. According to Whitaker (1908, p. 205) there was a total of 8 ft of gravel below 7 ft of alluvium in a borehole (175600) 590 yd W. 17° N. of Fordwich church. The gravels centred 700 yd W. 30° S. of Fordwich church have the shape of a fan deposit formed at the mouth of a minor valley. The stream occupying this valley is deeply incised into the gravel.

A small patch of gravel capping a knoll in Trenleypark Wood is exposed for 3 ft in the sand-pit (192592) 850 yd W. 25° N. of Swanton Farm. Towards the margin of the brickearth (p. 240) the gravel is intensely confused and festooned into the underlying Woolwich Beds: it is continuous with the thin gravel bed underlying the brickearth.

On the south-east side of the Lampen Stream the 2nd Terrace forms a broad gravel-covered bench, with brickearth overlying it on the side away from the stream. A pit (204590) 400 yd E. 20° N. of Swanton Farm showed 3 ft of subangular battered flints, many Tertiary-rolled flint pebbles and small angular chips of flint in a sandy matrix, overlain by 2 ft of similar gravel in a loamy matrix. Another section (214598), 1100 yd E. 5° N. of Elbridge House, showed 5 ft of gravel with lenses of brickearth up to 1½ ft in thickness.

In the Little Stour valley gravels of the 2nd Terrace underlie the alluvium and parts of the brickearth spread between Patrixbourne and Wickhambreux. They penetrate the alluvium at Well Chapel and Littlebourne. Temporary sections (189556) 540 yd N. 5° W. of Patrixbourne church showed 6 to 8 ft of gravel underlying brickearth. The gravel has been worked in a pit (201569) 600 yd E. 16° N. of Howletts, where Smith (1918) noted that the 20 ft of gravel contained occasional seams of sand and were not marked by a conspicuous bench feature, so that they must have lapped against the Thanet Beds in a similar manner to that at Littlebourne. He described the gravel as being coarsely false-bedded, consisting of subangular flints from " about the size of a football to small pebbles ", with Tertiary-rolled flint pebbles and fragments of indurated Thanet Beds sand ; iron staining and partial induration were also noted. Smith identified unrolled Chelles and rolled St. Acheul implements and a single point of Le Moustier type, but in a later work (1926, p. 37) he stated that the unrolled implements were of St. Acheul type and contemporaneous with the formation of the gravel.

At Littlebourne the gravel aggradation and associated brickearth were exposed in trenches dug along the two roads running north-west from the village. The close association of these two deposits is shown in Fig. 13. The gravel in the trenches was unbedded and coarse, with flints ranging up to about 9 in long, usually very tightly packed in a slightly clayey or brickearthy matrix, but in places there was only a thin clayey skin to the flints. The absence of sand, apart from one seam of derived Thanet Beds, was probably due to the lack of sand sources upstream. For the most part the passage from coarse gravel to brickearth was sharp, but seams of flints and scattered flints occurred locally in the brickearth. Such occurrences were the only traces of bedding in the entire deposit. A faint bench feature occurs at about 50 ft O.D. where the gravel comes to the surface, and a similar bench follows the gravel outcrop mapped at Wickhambreux.

A temporary section (209577) 210 yd W. 30° S. of Littlebourne church was situated between the two trenches shown in Fig. 13. It exposed, below 1½ ft of flinty loam soil, 4 ft of poorly-bedded gravel, composed largely of subangular flints up to 10 in long, some showing signs of abrasion, a few Tertiary-rolled flint pebbles and abraded ironstones and siltstones, tightly packed in a brown clayey matrix. Such bedding as was seen consisted of tolerably even bands where the matrix was absent. In these bands the flints had a thin coating of red clay. The gravel here is similar to that seen in the two trenches and is believed to extend north-eastwards below the brickearth at least as far as Wickhambreux. Whitaker (in MS.) noted 8 ft of gravel of similar composition, but with a few small lenticular patches of sand, underlying brickearth (p. 240) in a pit (220576), now obscured, 650 yd S. 25° W. of Ickham church. This occurrence shows that the Littlebourne gravel mass extends under at least part of the brickearth lying east of the river.

Deposits of the 2nd Terrace have a restricted outcrop at Folkestone Battery (231359) 300 yd W. 30° N. of the station at Folkestone Harbour, where they occur in close association with brickearth (p. 234). The 7-ft section seen there in 1956 consisted of fine grey bedded silt with thin fossiliferous ·seams of chalk pellets up to about 5 inches in thickness, overlying a bedded gravel of flints, some of which are desilicified, ironstones, chalk and phosphatic nodules. The pebbles increased in size downwards to a maximum length of about 5 in. The section was first recorded by Mackie (1851), who noted that in the west the terrace was overlain by brickearth, while in excavations for foundations within the Battery he saw Folkestone Beds overlain by from 1 to 5 ft of gravel beneath calcareous sandy marl, the whole being covered by up to 6 ft of made ground ; like the local brickearth, the calcareous marl contained a rich microfauna derived from the Chalk. Falconer (1868) described some of the bones collected. The following section is based on De Rance (1868, p. 165):

	Ft	in
Made ground	6	0
" Small shingle "	–	4
" White loam, with *Helix concinna, Succinea oblonga.* Lower Chalk. *Terebratula,* and ' junction bed.' *Ammonites* and nodules "	6	4
" Angular flints, with bones of mammalia, and fragments of *Am. interruptus* "	1	6

Topley (1875, p. 164) listed the following mammalia as having been found in this deposit, on the authorities of Mackie (1865) and Falconer (1868): *Bison priscus, Bos primigenius, B. urus, Cervus tarandus, C. euryceros, C. elaphus, Elephas primigenius, Equus, Hippopotamus major, Hyaena spelaea, Rhinoceros haemitaechus, Sus,* and *Ursus?.* Mackie (1851) and De Rance (1868) recorded the occurrence of the mollusca *Helix nemoralis, H. concinna* and *Succinea oblonga.*

Topley (1875, p. 164) also noted patches of chalky gravel capping the East Cliff and near Folkestone Junction Station ; these are now obscured.   J.G.O.S.

DRY VALLEY AND NAILBOURNE DEPOSITS

Hanslett's Forstal—Godmersham—Petham valley—Thanington.

In the valley bottom 700 yd north-west of Hanslett's Forstal, gravel-pits 10 to 20 ft deep, now overgrown, extend for about ½ mile (978598).

Gravels have been worked at Hockley (980554) and near Kennaways (991584: a pit 150 yd W. 30° S. of the latter was dug to about 6 ft depth and exposed loam with sporadic angular flints up to 3 inches long, adjoining and possibly resting on 1½ ft of gravel composed of flints, 'sarsen-stones' and broken iron-stones in a matrix of brown sandy loam. The larger pebbles were mostly well rounded and presumably derived from an earlier drift deposit, while the smaller were angular and mostly flints. This valley is now dry above the intermittent spring ⅓ mile S.W. of Whitehill, but Whitaker (1908, p. 60) recorded that in 1890 there was water along the floor of the valley for some five miles above the spring; this flooding was caused by 'run-off' from melting snow with the ground frozen hard, and was not the result of springs breaking out higher up the valley.

The water bore (995559) 500 yd E. 30° N. of Throwley church proved the deposit to be 9 ft thick, of which perhaps only the lowest 3½ ft consisted of gravel. The record of a borehole drilled to observe water levels at Dane Court (041581) shows the deposit to comprise: soil, 1 ft, "clayey soil", 6 ft, and gravel, 2 ft, resting upon Upper Chalk.

South-east of Godmersham the relationship of the Dry Valley and Nailbourne Deposits to the 3rd Terrace indicates that the former are not always the same age. The terrace surface falls gently towards the Great Stour. The deposits of the minor valley at the southern end of the terrace grade into the upper surface of the gravel, while the large tributary valley in the north is incised into the upper levels of the terrace and its deposits grade into a lower level.

The Petham nailbourne rises at a farm near Petham church. Whitaker (1908, pp. 59–61) recorded that the bourne might rise after a very wet period at Duck Pit (now New Barn Farm), and exceptionally near Dean Farm some 3½ miles south of Petham; the stream rarely reached the Great Stour at Shalmsford Street, for it was diverted into a pond near Perry Court Farm and lost. The deposits of this valley pass into a gravel fan of the 2nd Terrace at Shalmsford Street; 100 yd S. 10° E. of the level-crossing there (094545), up to 8½ ft of flinty loam are exposed resting on gravel (p. 272). Gravel which underlies much of the deposit in the Petham valley was exposed in shallow trenches (132525) at the roadside south of Swarling Farm and has been worked on a small scale (123472) 600 yd S. 10° E. of Yockletts Farm.

Some of the valleys which fall east and south-east from the high ground near Chartham Hatch 'hang' above the level of the Great Stour alluvium. At Howfield, however, the deposits of three valleys grade into the 2nd Terrace gravel and its associated brickearth.

In a temporary section (129565) 300 yd S. 40° E. of the ford at Thanington gravel composed of large angular flints and small water-worn flints in a gritty loam matrix was exposed for 6 in below 1½ ft of brickearth with chips of flint, overlain by 4 ft of brown flinty loam with chalk fragments.

Pett Bottom—Elham valley—Alkham valley—Satmore—Farthingloe.

At Pett Bottom (162521) excavation for a soakaway exposed grey powdery chalk beneath 3 ft of light angular flint gravel underlying 3 ft of brown flinty loam with chalk pellets. F. H. Edmunds noted on his map the occurrence of 6 ft of loam in the valley floor (180544) 550 yd N. 40° W. of Bridge church.

J.G.O.S.

At Ileden (209523) brickearth 4 ft deep was seen in foundation excavations in the valley floor. At Marley (190502) angular flint gravel around deteriorated pits along the valley bottom suggested that the deposits there were dug for gravel, and similar gravel was also seen around a shallow depression (211494) at the valley-junction south-west of Gravel Castle, where the deposits grade into Head.

In the valley north of Elham a cesspit excavation (176444), 14 ft deep, was reported to have revealed 6 to 7 ft of " clay " resting on 7 to 8 ft of " gravel " with flints up to 6 inches in diameter. A similar hole (154412) ½ mile W. 28° N. of Lyminge church was 12 ft deep and reputedly proved loam passing down into sandy clay with occassional flints. In neither case was chalk touched.

The head-spring of the Elham nailbourne is at Etchinghill ; in July, 1955, the stream was seen to flow northwards, but it disappeared before reaching Lyminge ; other springs emerged at Lyminge but the flow ceased some way to the north. During the summer of 1951 Edmunds (1953, p. 24) noted a stream flowing strongly in the valley bottom in Bourne Park for about 1½ miles between two dry stretches of the bourne valley. The permanent source of the Little Stour River is between Bekesbourne and Littlebourne.

The main sources of the Alkham or Drellingore nailbourne are in a depression at Drellingore (242411) and at a place near South Alkham ; other springs occur at Lower Standen Pumping Station and at Lower Standen farmhouse. " In Drellingore Pit the water attains a depth of 9 feet and covers an acre before overflowing a natural bar . . . In very wet seasons the water joins the Dour, but sometimes it disappears into its bed between Alkham and Wolverton [on Dover Sheet 290] " (Reynolds 1948, p. 78).

On the Clay-with-flints between Capel Street and Satmore, valley-bottom deposits were traversed by a sewer trench (251390 to 254390) which showed up to 4½ ft of brickearth without touching clay. At one place (252390) a 6-in bed of reddish brown sandy and gritty clay with flint chips was seen in 5 ft of brickearth ; 400 yd E.N.E., beside the lane, a bed of very flinty clay 1¼ ft thick was similarly present, suggesting a break in the deposition of the brickearth.

The Farthingloe Borehole proved 10 ft of " clay " on Chalk. About 500 yd north-east of Shakespeare Cliff, at Aycliff (309400), flinty loam is poorly exposed at the cliff-top, resting on Melbourn Rock.   G.B.

ALLUVIUM

Brogdale—Boughton Street.

The log of a well (004601) 850 yd N. 43° W. of Brogdale shows the alluvium and underlying gravel to be 15 ft thick, comprising in descending sequence: loam and stones, 2½ ft ; gravel, 6½ ft ; chalky gravel, 6 ft.

The alluvium of the small stream west of Boughton Street extends only a short distance on to the Chalk outcrop, where it passes insensibly into Dry Valley and Nailbourne Deposits. Whitaker (in MS.) recorded a section at the spring-head (052599), 500 yd E. of Nash Court, showing light grey clayey sand below about 5 ft of coarse gravel composed of subangular flints, Tertiary-rolled flint pebbles and a few pieces of ironstone and flints from the Bullhead Bed, below 5 to 6 ft of brown loam ; the gravel was roughly bedded, with lenticular masses of sand chiefly in the lower part, where it was also partially iron-shot.

                                                                          J.G.O.S.

Shoddington—Kingsnorth—Sellindge—Horton Park.

A dark peaty soil was noted in the alluvial tract of the Westwell stream near Shoddington (996469). Peat was noted in the alluvium of the stream which cuts through a spread of river gravel north of Bybrook (014443) and of that below Eastwell Park lake (016463).

The alluvial belt of the tributary of the East Stour River west of Kingsnorth is much wider than the present misfit stream is likely to have produced, and the deposits are correspondingly thicker. A roadside section (999396) in the bank of the stream 500 yd S. 10° W. of Westhawk Farm showed 2 ft of well-bedded brown siltstone gravel with a few flints, overlain by 2 ft of alluvial clay. Siltstone gravel overlain by 3 to 4 ft of clayey wash was also seen in ditches near Court Lodge (993397). In the expanded part of this alluvial tract $\frac{1}{2}$ mile west of Court Lodge brown loamy clay wash more than $3\frac{1}{2}$ ft thick was proved by augering. In the wide alluvial flat north of Kingsnorth the soil is everywhere a heavy clay, except in a field $\frac{1}{4}$ mile N.N.E. of the church, where coarse-grained sand forms a low rise 250 yd long and 2 or 3 ft above the surrounding levels. Former courses of the rivers South Stour and East Stour, shown respectively by a meandering marshy hollow and by a sinuous ditch followed by the parish boundary, meet (at 012405) just north of this sand.   B.C.W.

The broad alluvial spread of the South Stour River was proved to be 6 ft in thickness 850 yd S. 10° E. of Buttesland Farm (017377) and to be composed of dark brown sand with 'shrave' (derived Wealden ironstone).   J.G.O.S.

In embayments of the East Stour alluvial tract east and west of Waterbrook (028402) the surface deposit is a heavy grey or brown clay which, at a depth of 2 to 4 ft, rests on coarse sand apparently derived with little sorting from the Hythe Beds. Nearer the centre of the flood plain in this vicinity the clay is underlain by flint gravel. A section in the river bank 500 yd S. 11° E. of Waterbrook showed, on Weald Clay, 6 in of flint gravel overlain by $3\frac{1}{2}$ ft of greyish brown alluvial clay, and ditches (043395) 200 to 400 yd south of Conscience Farm showed 1 to 2 ft of sandy wash overlain by 3 to 4 ft of clay.   B.C.W.

The fall of the East Stour from Sellindge to The Forstal is only about 10 ft per mile ; consequently the alluvial flat is broad considering the size of the stream. The deposit (063382) 100 yd W. of Evegate Mill consists of sandy loam with chips of flint, 3 to 4 ft in thickness, resting on grey sand. In the stream bank (089378) 340 yd N. 8° E. of Partridge Farm 1 ft of gravel was overlain by 5 ft of brown and grey mottled sandy clay. Near Horton Park (123399), where the alluvium hardly exceeds 4 ft in thickness, the deposit contains much chalky loam, derived chalk and flints.

Wye—valley of the Great Stour River.

The disproportionately broad alluvial flats of the tributary streams on the Gault outcrop south of Wye are covered with derived Gault clay containing chalky pellets, which contrasts with the silts and sands along the course of the Great Stour.

The alluvium of the Great Stour narrows where it crosses the gravels of the 3rd Terrace at Wye. In the bank (057485) 230 yd W. 11° N. of Olantigh Towers 2 ft of chalky angular flint gravel underlie $1\frac{1}{2}$ to 3 ft of derived brickearth with small chips of flint ; let into this gravel, probably as a result of 'piping' in the underlying solid chalk, is a pocket of pellety chalk. The alluvial flat narrows again as it crosses the Melbourn Rock at Trimworth Manor and again opposite a spur at Godmersham. In the Godmersham water-bore (073516) 350 yd W. 30° N. of Pope Street Farm 4 ft 9 in of flinty loam with derived chalk fragments rested on 14 ft 9 in of chalky gravel. A group of Geological Survey boreholes sited on the alluvium to the north-west of Chilham Mill (077534), which is 950 yd E. 20° S. of Chilham church, were drilled in connection with a hydrological study (Ineson 1962, p. 63; Ineson and Gray 1963, p. 213) ; they penetrated drift deposits ranging in thickness from 12 to 28 ft, and consisting of alluvial clay and silt overlying river gravel. North-east of Chilham and at Shalmsford Street

(p. 272) the 2nd Terrace gravel beneath the alluvium is excavated by dredging, but the workings immediately north-east of Canterbury are now abandoned.

At the abattoir (158595) 950 yd W. 6° S. of the Sewage Works west of Fordwich 1 ft of gravel was overlain by 4 ft of reddish brown clay passing up into reddish brown loam. An excavation (167594) near the road, 250 yd S. 23° E. of the Sewage Works, showed poorly-bedded gravel, seen for 3 ft, grading up into gravelly grey silt 2 ft thick. Whitaker (1908, p. 205) recorded that a water-bore (175600) 590 yd W. 17° N. of Fordwich church encountered 7 ft of alluvial soil resting upon 8 ft of gravel before entering Thanet Beds. The gravel below the alluvium is extensively worked north of Fordwich, where most of the dredgings lie on the Faversham (273) Sheet.   J.G.O.S.

Valley of the Little Stour River—Wingham.

The thickness of the alluvium in the Elham valley north of Lyminge is probably comparable to that of the Dry Valley and Nailbourne Deposits in the vicinity (p. 278); the surface alluvium consists mostly of sandy brickearth, but at Breach and Derringstone patches of angular flint gravel were observed. At Kingston (201513) and in Charlton Park (192520) surface irregularities suggested that gravel may formerly have been dug there and coarse angular flint gravel was seen in the banks of the dry stream bed.   G.B.

There is much gravel below the alluvium of the Little Stour River. At Bishopsbourne (189527) gravel is exposed for 4 ft 150 yd E. 20° N. of the church and 3 ft were noted below 3 ft of loam in an excavation (192551) 120 yd E. 12° S. of Patrixbourne church. At Littlebourne small patches of 2nd Terrace gravel are seen as slight mounds on the alluvial flat. A trench (209573) 100 yd S. of the cross-roads exposed 6 ft of tight gravel showing only a slight semblance of bedding and including a lens of angular flint chips 6 in long and 1 in thick, below 1½ ft of made ground; there were many dark manganiferous bands and patches in the gravel.

North of Littlebourne and in the valleys of the tributary streams near Wingham there is much peaty silt and peat in the alluvium. This peat is 9 ft in thickness north-west of Wingham church, but it is not shown separately on the geological map because of its indefinite boundary with the peaty silt of the alluvium.

Romney Marsh—Folkestone.

The alluvium of the streams bordering Romney Marsh grade insensibly into the marine deposits of the marshland. At Ham Street, in particular, the upland streams probably contributed much of the material found today in the sand-ridges of the nearby marshland (p. 254). Some of the masses of sand in the marshland near Bilsington and south of Aldington Knoll were considered by Green and Askew (1956, p. 16) to be the levees of streams which flowed prior to the construction of the Royal Military Canal.

Patches of alluvium have been mapped among the lower-lying parts of the marine alluvium of the marshland. Most of these deposits occupy scars formed by breaching the sea-walls (p. 252) or depressions formed by compaction of peat. The deposits consist of marine alluvium reworked during flooding by fresh water.   J.G.O.S.

The alluvium in the valley of the stream (Pent Stream) which crosses Folkestone Golf Course is exposed (219369) 600 yd W. 23° S. of Park Farm, where 2 ft of purplish brown sandy clay with occasional small flints, both angular and rounded, were seen to underlie 1¼ ft of soil and to rest on coarse green sand.   G.B.

PEAT

Westwell—Blackwall Farm—Horton Priory—Saltwood—Shorncliffe.

In a low-lying part of a meadow (991472) 300 yd S. 11° E. of Westwell church about 2 ft of peat rest on Lower Chalk. Peat occurs along a belt of springs thrown out by clayey seams in the Folkestone Beds between Willesborough Lees and Hinxhill. In the three patches shown on the map it is over 3 ft thick.

West of Blackwall Farm (037440), in the central part of a basin-like hollow separated from the Stour flood plain by 3rd Terrace deposits and floored with alluvium, there is a peaty topsoil 1 to 2 ft thick, and only in a patch near the exit, 400 yd west of the farm, does the peat exceed 3 ft in thickness. In all probability the peat here was thicker and more extensive before the hollow was artificially drained.

The peat in the valley north-east of Blackwall Farm appears to have wasted away at its western end as a result of drainage, since the surface of the main mass slopes down to the alluvial tracts of the stream courses along its margins and also towards the drainage ditches. Part of the peat remains (1954) as wild fen. Topley (1875, p. 171) noted that the peat here was at one place more than 20 ft thick. Only 5 ft of peat were encountered 350 yd N. 35° E. of Blackwall Farm, where they rested on 5 ft of alluvial silt and sand.

Two narrow deposits of peat occur along stream courses (110389) ⅜ mile S.E. of Horton Priory. Brown peat over 3 ft thick was noted at Gibbins' Brook (115385), a flat-floored marshy valley south-west of Hope Farm ; peat also occurs in the bottom of the narrow valley south-east of Hope Farm.

Peat occurs at springs from the Folkestone Beds near the heads of valleys in Sandling Park. A similar deposit ¼ mile N. 22° W. of Saltwood church was said to be more than 15 ft thick. The peat which occurs at springs (163365) ⅜ mile N.E. of Saltwood church, lies on a slope. The deposit in the wood ¼ mile farther north-east is on the valley side at its south end, but declines with the spring line to the valley bottom at its north end. On the steep west side of the valley (178369) ¼ mile E. 22° N. of Bargrove, peat over 3½ ft thick supports a dense fen vegetation of reeds and *Equisetum*. In an east–west valley ¼ mile E. 10° S. of Bargrove the banks of the stream showed 4 ft of peat.

In the wood 500 yd east of Scene Farm the ground in the valley bottom, water-logged by springs, has 2 to 3 ft of peat on clayey silt or sand wash. On the upper slopes of the valley north of Horn Street (190362) the ground below the Sandgate Beds–Folkestone Beds boundary is rough pasture with a foot or two of peat forming the top-soil. In two patches shown on the map the peat is over 3½ ft thick and forms low boggy mounds. A further patch is present in the landslipped valley east of Shorncliffe Camp.   B.C.W., J.G.O.S.

The Dowels—Walland Marsh—Romney Marsh.

On the marshland, Drew (1864, p. 15) described peat in sections, now obscured, near the railway north of Appledore Station. No thicknesses were recorded. Some 700 yd westward of this district the peat in The Dowels was found to be only 3 to 5½ ft thick (Shephard-Thorn and others 1966, p. 102). Thicker peat occurs west and south-west of Snargate, where Gilbert (1933, p. 253) recorded 9½ ft of peaty deposit, of which the lowest 8 ft comprised black leafy mould and tree stumps, to the west of Fairfield Court.

A ditch (980292) 750 yd S. 30° E. of Appledore Station exposed 3 ft of peat, partly channelled by blue clay ; a further 4 ft of peat below the section was proved by means of an auger. Ditching (982282) 1000 yd N. 19° E. of Cherrytree House exposed 4 ft of peat resting upon blue clay.

In the older marsh deposits north of Little Cheyne Court peat occurs at about 3 ft depth between the sand ridges ; at a locality (993224) about 1100 yd E. 20° N. of this Court, scouring has exposed the peat which exceeds 4 ft in thickness.

Drainage operations in Walland Marsh frequently encounter peat, except near the Rhee Wall, where it is buried deeper. In the sewer (039240) 1200 yd N. 17° E. of the church ruin near Hawthorn Corner, peat underlies 4 to 5 ft of clays (p. 285), and tree stumps and logs occur among the debris dug out. This peat was noted by Green and Askew (1958b, p. 21) to be 1½ to 3 ft thick. Peaty traces only were found in the same sewer (027217) 1400 yd W. of Westbroke House.

At Brenzett Green (011287) peat, reported to be 3 ft thick, underlies 5 ft of later deposits (p. 283). Peat bands of unknown thickness between 9 and 11 ft depth in the Langdon Borehole (p. 284) are considered to be the 'surface' peat horizon of the marshland, and those between 46½ and 55½ ft constitute the only evidence of an older peat bed.

Green and Askew (1956, p. 16) recorded peat at depths of 6 and 8 ft near the Royal Military Canal in the vicinity of Bilsington. They considered the peat to be absent below some of the old water-courses in the northern part of the marshland.

The most easterly record of peat is at Orgarswick Farm (090309), where it occurs at 7 ft depth (Hall and Russell 1911, p. 62). Over 70 trial boreholes, mostly of under 20 ft depth, sited near the coast between St. Mary's Bay and Dymchurch Redoubt, failed to prove any peat. However, a trial pit (120314) (p. 287) 2360 yd E. 20° S. of Burmash church encountered a " hard seam of decaying wood ", presumably the peat horizon, at 18 ft depth.   J.G.O.S.

MARINE ALLUVIUM

Warehorne—Snargate—Brookland—Little Cheyne Court.

To the north of Appledore Station and also in the valley south-west of Warehorne, brown silty clays occupy ridges within the peat, much as described by Edelman (1950, p. 74) in Holland. The ridges are similar to the sand creek-ridges into which they appear to pass imperceptibly. Up to about 3 ft of stiff clay overlie the peat in the pool areas. A ditch along the main clay ridge of The Dowels probably follows the original channel from which the sandy clays were deposited. A ridge of sand (983316) 1000 yd W. 20° S. of Bridge Farm, Warehorne, retains for about 300 yd the remains of the sinuous natural drainage channel.

South of the Rhee Wall at Appledore Station (v, Fig. 14) the deposits lie 10 to 12 ft higher than those in The Dowels. Much of the ground consists of tenacious reddish brown clay, but levees of fine silty sand and sandy clay flank the lines of natural drainage. Similar heavy clays occur bounded by sea-walls south of Water House (u). The upper surface of the sea-wall is no longer even, but rises 2 to 3 ft in line with the ridges of fine sand in the lower ground south-east of the wall and falls over the outcrop of inter-vening clays, revealing the amount of compaction in the lower ground since the wall was built.

There is an intricate pattern of sandy creek-ridges with clay and peat in the pool areas south of the Rhee Wall and west of Snargate. Near Water House a channel south of the sea-wall is thought to have been a drainage course connected by a sluice to the older gound north of the wall. Immediately north of this supposed sluice there is an indefinite small patch of sand which is not part of the creek-ridge deposits. A ditch (984282) 840 yd W. 28° S. of Snargate church exposed 5 ft of fine brown sand occupying one of the broader ridges. In the same ridge 200 yd N.N.E., peat was overlain by 4½ ft of grey silt. Some 430 yd W. 19° S. of the above church dark grey fine sand was overlain by 2 ft of peat, below 4 ft of fine grey and brown sand.

There is a particularly strong creek-ridge topography north and west of Fairfield Court (E); some of the ridges rise 4 to 5 ft above the intervening clays. At Brookland some sand enclosing irregularly shaped hollows infilled with sandy clays and stiff clays overlying sand was probably connected during deposition with that of the creek-ridges near Fairfield Court, although a relic of a sea-wall separates them now. The Brookland sand was reported to exceed 10 ft in thickness near Bowdell (992274) and debris from ditches (000264) near the level-crossing contained irregular weakly cemented ferruginous nodules. Peat underlies much of this sand area but its extent and thickness is unknown. Some rather loose sand at the surface was probably deposited from breaches in the sea-wall near Flats Houses.

To the south-west of the sea-wall from Dean Court to Little Cheyne Court (S and R), and thence expanding eastwards to the line of the sea-wall and roadway through Hawthorn Corner, as far as a further wall which follows the sewer (Jury's Gut) running south-south-west from the bridge 650 yd W. of Westbroke Farm (O, P, Q and N), there lies a series of sands and clays which are of tidal flat and salting origin. They are of relatively recent age (group 3 deposits, p. 260) and relics of channel courses and runnels commonly occur. West and north-west of Little Cheyne Court a channel system with levees of sand passes northwards into a weak creek-ridge formation. This tract contains broad expanses of sand of tidal flat origin which pass gradually into clays which tend to occur near the old sea-walls. The sea-walls were constructed during the reclamation of the Rother channel towards Rye and the Wainway Creek, both of which lie outside the district described in this memoir.

The broad sand deposit near Brookland passes into subdued creek-ridges near Snargate and Coldharbour, and it goes uninterrupted under the Rhee Wall towards Ivychurch, where in the west it passes into creek-ridges and in the east it is overlapped by the Romney Marsh clays.

Ivychurch—Snave—Newchurch.

A section (029277) 100 yd E. of Ivychurch church displayed fine khaki sand overlain by 4½ ft of soft grey silty clay, while in a temporary section at Brenzett Green (011287) bluish grey silty clay was overlain by 3 ft (reported) of peat, below brown and grey clayey sand, 3 ft, and brown and grey silty clay, 2 ft. Ditch clearance 400 yd E. 20° N. of Brenzett Green revealed dark reddish brown ironstone, 6 to 8 inches thick and containing *Cardium*, 5½ to 6 ft down in the sand. A new ditch (020293) 750 yd N. 17° W. of Moat House showed grey and brown sand, seen for 4 ft., passing up into brown sandy surface loam, 3 ft, and in an exposure (019299) 550 yd E. 10° S. of Snave church greyish brown fine sand was sharply overlain by grey and brown silty clay which was 3 ft thick with a few small calcareous nodules at the base. A ditch (023295) 1000 yd N. 10° E. of Moat House exposed silty and sandy clays 2 ft thick with traces of peat locally below, overlain by 2 ft of brown clay, again with calcareous nodules at the base; on one side of the ditch only, fine grey sand rose gradually westwards into the silty and sandy clays until 1 ft was exposed, and then plunged below water level. This sand is unconnected with the creek-ridge deposits and is probably older than the peat. In a section (997314) 700 yd W. 17° S. of Ham Mill Green, peat is overlain by 4 ft of bluish grey clay which passes up into brown clay 2 ft thick.

The creek-ridge system continues for a short distance east of the Ham Street—Snave roadway, whereupon the sand outcrops become indistinctly featured among the Romney Marsh clay. This may be the result of the clay overlapping the ridge system so that only the higher portions remain exposed, as at Ivychurch, or on the other hand deposition of the clays may have been controlled by a wall, now obliterated.

The borehole at Langdon (032312), 630 yd S. 9° E. of Oxpound, proved 66 ft of marsh deposits in descending sequence as follows: turf and mould, 6 in; clay with shells, dipping sharply east, 4 ft; loam sand, 1½ ft; fine grey sand (clayey)[1], 3 ft; sand and mud with bands of peat, 2 ft; soft light blue clay (grey sandy clay), 5½ ft; grey sand with a few shells, 30 ft; soft light blue clay with bands of peat, 9 ft; and brown sand mixed with pebbles and stones of various kinds, 10½ ft; resting upon ? Hastings Beds (after Whitaker 1908, pp. 235–6).

The heavy Romney Marsh clays extending from Ivychurch to Dymchurch and northwards to the Royal Military Canal are brown at the surface but bluish grey or greyish brown below; they generally exhibit a diced structure when weathered. The uniformity of the clay tract is broken by low mounds and ridges of sand near Newchurch, in a belt north of that place, at Forty Acre Farm and near St. Mary in the Marsh. Silt and sand are commonly found 4 to 5 ft below the surface of the clays (p. 256).

At Brooker's Farm (047311) 3½ ft of brown fine sand were overlain by 3 ft of sandy clay. In the bank of the Royal Military Canal (056343) 150 yd S. 30° W. of St. Rumwold's Church 4 ft of grey clay lay beneath 3 ft of brown and grey silty clay. Augering proved the Romney Marsh clays to be more than 9 ft thick at Hurst Farm (071330) and to exceed 8 ft 900 yd W. 13° N. of there.

Lower Wall Farm—St. Mary in the Marsh.

To the east of Sherlock's Bridge many ditches are particularly sinuous and meandering lines of darker soil can be seen in ploughed fields. They are both interpreted as the original creeks and runnels of a salting environment. An auger hole (089335) 450 yd E. of Lower Wall Farm proved 7 ft of stiff light grey clay and brown mottled silty clay overlying 4½ ft of dark grey silty clay and another (092330) 950 yd E. 33° S. of this farm found 5 ft of brown and grey mottled clay overlying 6 ft of dark grey clayey silt.

An old creek bifurcates at Forty Acre Farm (090318): one arm extends south-westwards for 600 yd before a further division occurs, whereupon the channels become indistinct; the second arm runs southwards from Forty Acre Farm towards Orgarswick Farm, then turns eastwards and is lost among a depression. Northwards from Forty Acre Farm the creek follows the line of a sewer until east of Lower Wall Farm, where the banks become vague. The reddish brown and grey clays of the channel floor merge with clays occupying a wide depression which extends eastwards to the shingle fulls. A similar channel which joins this broad depression near Botolph's Bridge, passes through one of the low mounds of sand east of Burmarsh (p. 286) and dies out at the line of possible walling (p. 263) which follows the sewer to the Dymchurch Sluice.

The clay 500 yd E. 15° S. of Sellinge Farm (090291) was reported to be 12 to 20 ft thick and rested upon running sand. Near St. Mary in the Marsh the Romney Marsh clays appear to overlap a broad outcrop of fine sand. A semblance of creek-ridges occurs on the eastern margin of this sand, but such deposits are poorly developed until south of a line of walling which follows the roadway from near Ivychurch to the road fork south-east of Yoakes Court,

---

[1] Remarks in brackets are by W. Whitaker, from specimens.

thence towards Honeychild Manor and finally follows the New Sewer for about 400 to 500 yd. The trace of the wall there becomes uncertain, but it may cross country to Slinches and follow the roadway to St. Mary's Bay (Fig. 14). The faint line of this wall appears to coincide with the change of deposits, but no continuation north of Ivychurch along the western margin of the clays has been detected.

At Brodnyx road-junction (072275) 1½ ft of fine khaki sand were sharply overlain by 2½ ft of greyish brown sandy clay with calcareous pellets, below 1½ ft of brown sandy soil with a few flint pebbles. In the New Sewer (071267), 550 yd S. of Slinches, 1½ ft of fine micaceous khaki sand were exposed below 1½ ft of khaki and grey sandy clay, overlain by 1 ft of brown loamy soil. The above two exposures showed the feather edge of the Romney Marsh clays. Clays have been dug on a small scale near the latter locality, presumably for brickmaking. A trial pit (086271) 1530 yd E. 19° S. of Brodnyx encountered firm brown mottled fissured clay, 8 ft ; on laminated soft grey fissured clay and sand, 6 ft ; resting on grey clayey sand, 1½ ft ; on fine grey sandy silt, 3 ft ; soft grey clayey sand, 2 ft ; ballast, 1 ft ; resting on grey sand, 1½ ft.

Hawthorn Corner—Westbroke Farm—New Romney.

Fine brown sand 4 to 5 ft thick was noted in a section (008245) 520 yd S. 22° W. of Coldharbour. A breach in the sea-wall is present 950 yd W. 25° N. of Hawthorn Corner ; on the landward (north) side a gully 150 yd wide has been scoured out and is now occupied partly by fresh-water alluvium. The repair wall, which extended in a bow 100 yd north of the roadway, has been levelled since 1956.

There is a section across one of the large creek-ridges and the pool area clays in the White Kemp Sewer (039240), 1200 yd N. 17° E. of the church ruin near Hawthorn Corner. The grey and brown 'pool' clays were exposed for 4½ to 5 ft and were commonly silty. The sand was laminated, brown and grey coloured and it contained seams of clay and drifted peat fragments ; it passed insensibly into the pool area clays. Peat occurred sporadically at about water level (p. 282), and Green and Askew (1958b, p. 21) noted that it was absent under the large ridges hereabouts. Among the sand, at or just above water-level, there were fragile ferruginous concretions with sideways extensions coinciding with certain laminae in the sands. Two reddened flint pebbles occurred in the sand.

Green and Askew (1958b, pp. 21–2) recorded sand older than peat and separated from it by a wedge of clays 0 to 2 ft thick. They stated that the older sand was apparently 4 ft or more thick directly underneath the peat at the bridge (030224) 650 yd W. 5° S. of Westbroke Farm. They recorded that this sand reached the surface locally.

Low mounds of sand rise 5 ft or more above the surrounding ground near New Romney (Fig. 14). On the south side of the port they appear to be banked against and over the shingle fulls and they become difficult to separate from the Blown Sand. The sands occupy a low ridge on the eastern side of the shingle south-west of Belgar and extend north of the shingle almost to Caldecot. Inland the sand mounds become lower, less widely distributed and less distinctive, and many are undercut by subsequent channelling. The seaward mounds, those at New Romney and the ridge near Belgar, have steep faces upon their eastern sides, where they appear to have been eroded by waves. The contained sand is fine to medium-grained, fawn in colour and contains shell fragments. The mounds are interpreted as estuary sand-banks and beach sands. At New Romney church (065248) brown medium-grained shelly sand with a few flint pebbles was temporarily exposed for 5 ft.

The sand adjacent to the mounds forms part of a tidal flat which extends to the shingle near Greatstone on Sea and inland to the roadway, perhaps marking a wall, running north–south through Caldecot. On the west of this roadway creek-ridges occur. A patch of clay lies among the creek-ridge formations immediately west of New Romney. At least part of this clay occupies a channel cut in the sand mounds and sand outcrop. The eastern end of the Rhee Wall emptied into this channel (p. 253).

Small sections in the sand flat of Romney Salts show it to be laminated and sporadically interbedded with clay seams. Clays rarely more than 3 to 4 ft thick occur near New Romney and among the shingle recurves.

Dymchurch—Burmarsh—Hythe.

In Dymchurch, 490 yd S. 32° W. of the church with tower, a trial pit proved 3 ft of stiff fissured clay overlying 9 to 12 ft of laminated soft grey clay and grey sand, resting on 2 to 5 ft of loose grey sand, on 6 in of clayey sand, on 8½ ft of loose grey silty sand.   J.G.O.S.

At the south end of Dymchurch a 6-ft section in the bank of Hoorne's Sewer (100291) showed grey to buff clay, becoming silty downwards. About 250 yd west of here a ditch side showed 4 ft of stiff clay, and a pit excavated to 12 ft at a sewage works nearby met running sand at 9 ft.

Between Dymchurch and Burmarsh the surface layer of the Marsh deposits, more than 3½ ft thick, consists of grey or bluish grey clay weathered to shades of pale brown. Towards the coast, between Dymchurch and the southern end of the Hythe shingle mass, the clay has been dug from a large area to expose at the ground surface silts and fine sands. The landward boundary of these silts and sands now runs for over a mile along the foot of a bank which repre-sents the limit of excavation. The bank is 4 to 5 ft high at its southern end but decreases to 2 or 3 ft at its northern end. In its northern part the excava-tion is limited on its seaward side by a corresponding bank which extends past the tips of shingle ridges that reach in from the coast. Two clay-pits in Dymchurch, respectively 200 yd N.W. and 800 yd S.W. of the church, have indented outlines similar to that of the main excavation and were dug to 4 or 5 ft, though without reaching sand. These excavations together provided some 900 000 cu yd of clay, probably for the construction of the Dymchurch Wall.

Elliott (1847, pl. 42b) gave a plan and cross-sections of the Dymchurch Wall in its original state in 1837, and showed also the position and cross-section of the new wall. From these diagrams the volume of the old wall can be calculated as approximately 350 000 cu yd, and that of the new wall as approximately 600 000 cu yd. These figures suggest that part of the excavation was made in the long period of years prior to 1839 during which the sea-wall was built up in piecemeal fashion. Elliott gave many particulars of the ragstone facing of the wall, but did not mention local excavations, although he stated (1847, p. 466) that Roman pottery was discovered over an area of several acres near Dymchurch when some alterations were made to the line of the wall.

From Burmarsh eastwards to the coast extends a broad though very slight rise of ground, on parts of which pale grey or brownish grey loamy silt or fine sand comes to within 2 or 3 ft of the surface ; these areas are shown as sand on the geological map. The sand is generally overlain by brown clay, which near the coast has been dug in the northern part of the excavation described above. About ¾ mile east of Donkey Street a narrow depression heads north-wards into the rise, and 350 yd farther east a depression cuts right across the rise. Since both are floored by alluvial clay they are probably natural features. The more easterly is now followed by the Willop Sewer.   B.C.W.

The deepest of several trial boreholes (117320) 1780 yd E. 3° S. of Burmarsh church proved the following downward sequence: soil, 1 ft; firm mottled clay, 3 ft; silty clay, 3½ ft; sand, 33½ ft; loose clayey grey silt with clay lumps and shells, 39 ft. A trial pit (120314) 2360 yd E. 20° S. of Burmarsh church proved 6 ft of brown sand with thin seams of clay, overlying, in downward order, laminated blue clay and sand, 1½ ft; loose grey sand, 4 ft; laminated blue clay and sand, 1 ft; loose grey sand, 3 ft 3 in; soft blue clay, 3 in; loose grey sand with a hard seam of decaying wood at the base (?peat horizon), 2 ft; and ballast, 2 ft.   J.G.O.S.

North of Burmarsh, in a low-lying part of the Marsh extending past Botolph's Bridge to Hythe, is bluish grey to brown clay, which in many places overlies silty clay or silt at a depth of about 3 ft. The banks of a ditch (097333) 700 yd N.W. of Abbott's Court show 5 ft of grey to buff clay with finely broken shell fragments. Buff silty clay exposed in the side of a main drainage channel (118338), 450 yd N. 35° W. of the inn at Botolph's Bridge, contained *Cardium* (*Cerastoderma*) *edule* Linné and *Scrobicularia plana* (da Costa).

East of Botolph's Bridge the clay reaches into hollows between shingle ridges that extend inland from the coast. A trial borehole (133332) in one of the broader of these hollows 1250 yd E. 10° S. of Botolph's Bridge proved top soil, 1 ft; on brown clay, 5 ft; on blue clay, 2 ft; on shingle (p. 289). A borehole (149344) at a laundry on the Dymchurch road just west of Hythe proved made ground, 6 ft; on light blue clay, 6 ft; on shingle and other deposits (p. 289).

From Hythe eastwards to Seabrook a low-lying tract behind the sea-wall consists of shingle covered by a layer of clay 1½ to 2 ft thick except in some patches, shown on the geological map, where the clay is more than 3½ ft thick. Most of these patches correspond with slight depressions of the ground surface. A tongue of clay coinciding with the easternmost mile or so of the Royal Military Canal is evidently the alluvium of the Seabrook stream, which formerly flowed into Hythe harbour, as is shown on Symonson's map of Kent of 1596.

In a brickyard (172347) 1100 yd E. 10° S. of Hythe parish church, on the south side of the Seabrook Road, Topley noted (in MS. dated 1864) 1½ to 4 ft of stiff loam, almost a clay in places, resting on shingle.   B.C.W.

STORM GRAVEL BEACH DEPOSITS

Birdskitchen—Greatstone on Sea—New Romney—Warren House.

Probably the oldest shingle within this district occurs as a low ridge (046220) 500 yd S. 25° E. of Birdskitchen. This shingle is exposed for 4 ft, but the original fulls cannot be seen. The ridge plunges below the level of the marsh deposits at the extreme southern boundary of this district, but shingle which is in alignment with it occurs 1½ miles distant to the south-west at Scotney Court, where sections reveal that the marshland peat dies out against the fulls of shingle (see also Green and Askew 1958b, p. 22). The shingle at Birdskitchen is, therefore, assumed also to be older than the peat. Northwards this shingle falls below the marsh deposits at the New Romney–Lydd roadway west-south-west of Belgar, where indefinite stumps of recurves were noted. Shingle can also be found buried among the sands near the roadway 1000 yd northwards of this place. Shingle of slightly younger age crops out 700 yd S.E. of Birdskitchen; again no definite fulls have been seen but two clear recurves are present. At its northern end, near the Isolation Hospital, this shingle is truncated by higher ridges of shingle, which are followed by the New Romney–Lydd roadway. At least ten major fulls form this high mass of shingle, which plunges below the marshland immediately east of the roadway to the south-west of Belgar. No recurves were seen. The fulls have been extensively dug for gravel.

Between this high ridge and Greatstone on Sea long jagged outcrops of shingle extend among the sand and clay of the old tidal flat. Two saw-tooth spits almost reach to Belgar itself, and a total of over 70 recurves have been mapped in this area, many containing two or more fulls. About 4 ft of shingle, bedded and with some intermingled sand, were noted 100 yd E. 20° S. of Northlade (067222) and 3 ft of unbedded shingle 400 yd S. 40° E. of this place. Gravel workings (077215) 1350 yd E. 32° S. of Northlade expose up to 7 ft of shingle above the water level.

At Greatstone on Sea the shingle is obscured by Blown Sand. Parts of northerly trending fulls, however, are discernible in the blow-outs. At Littlestone on Sea the coast is protected by a sea-wall with a mass of shingle upon its landward side. No line of fulls or recurve has been observed in this shingle, which has apparently been disturbed during the building of the resort; part of the outcrop may be artificial. In the northern part of the resort, for 400 yd southward of the (water) Tower, five short recurves are truncated by shingle lying immediately behind the sea-wall. These recurves are probably the southerly portions of spits which once extended northwards into Dymchurch Bay and to Hythe. One has a hooked inshore end which suggests that its shingle has been derived from the south, but the hook may be artificial. The slight curves of the other fulls show that they were derived from the north.

A finger of Blown Sand conceals parts of more recurved spits to the north of the (water) Tower. The lines of the fulls among the shingle mass at The Warren (Fig. 14) are indistinct, perhaps because the fulls are masked by relics of Blown Sand. Recurves are well displayed at the inshore end of the outcrop and show derivation from the north.

The roadway from New Romney to St. Mary's Bay follows a high ridge of shingle. For the most part traces of the individual fulls have been destroyed by quarrying and building operations, but they are well represented at the northern end of the ridge. A lower ridge branches off the main ridge near Warren House and follows the north-western edge of the port until it is lost north-north-west of the church. Sand commonly occurs among the shingle at the New Romney end of this ridge. Foundation digging (067251) 400 yd N. 30° E. of the church showed 4 ft of interbedded shingle and sand below 5 to 8 ft of medium-grained buff sand, current-bedded in places and with lines of comminuted shell fragments, overlain by 4 to 5 ft of made ground. This deposit is regarded as a passage from the shingle to the north into the sands occurring on the south-east of the port (p. 285). A low spit leads southwards from the main ridge, for a short distance only, at Warren Farm, where it consists of interbedded sand and shingle which are exposed for 3 ft. A pit (076259) 250 yd W. 23° S. of Warren House exposes 8 to 10 ft of bedded shingle and sand.   J.G.O.S.

Dymchurch—Burmarsh—Botolph's Bridge—Hythe—Seabrook.

Elliott (1847, p. 467) stated that shingle banks ran under and inland of the present Dymchurch Wall throughout its length, nearly at right angles to the line of coast. However, he gave no details of the positions of these banks. At the present day shingle ridges can be discerned at the surface from about 1 mile north-east of Dymchurch church northwards. All trend in a north-north-westerly direction, except for the most southerly, which trends about W. 10° N. They must represent the successive recurved ends of a former spit, sited off the present coast, that grew north-eastwards. These ridges rise only a foot or two from a flat surface of yellow sand which extends inland for about 300 yd, approximately as far as the tip of the longest ridge. A ditch crossing this ridge (123318) showed shingle to 4 ft below ground surface, with larger pebbles (3 to 4 inches across) below and smaller ones (1 to 1½ inches) above, forming a bank 40 yd wide. On each side of the bank the shingle fingered off into yellow sand, which suggested that the sand was of the same age as the shingle. The height reached by these ridges appeared to be lower by a few feet than that of the main Hythe shingle mass, and they may in consequence have been formed at an earlier date than the latter. The general resemblance of the sand to sands found beneath peat elsewhere in the marsh possibly indicates that the ridges are of a pre-peat date (p. 257).

The surface of the main part of the Hythe shingle mass formerly showed ridges which, according to Drew (1864, p. 18), were about 30 yd apart and 2 ft or so above the troughs. The whole tract south of the Dymchurch road has since been levelled, but traces of the ridges still show on aerial photographs; they follow the trends indicated by Drew on the Old Series 1-inch geological map, Sheet 4, radiating fanwise from the coast, by which they are truncated, their directions veering from north-north-westerly at the southern end of the spread to north-easterly near Hythe. The shingle near Hythe is composed mostly of small brown flint pebbles ½ to 1 inch in diameter, with some larger, grey, wave-battered flint pebbles 2 to 3 inches in diameter.

From the southern end of the mass ridges separated by clay-filled hollows project inland. The southernmost ridge extends for 600 yd and its continuation in a west-north-westerly direction for a further 600 yd, buried beneath clay, is indicated by shingle in the banks of the Canal Cutting, a main drainage channel, at (124327) 900 yd S.S.E. of Botolph's Bridge, and again in the bottom of a pond (121328) 750 yd S. of Botolph's Bridge. This buried ridge thus runs along the north side of the sand tract that extends inland to Burmarsh (p. 286). A parallel ridge some 200 yd north, and traceable as far west as the Canal Cutting, is covered by 1 to 2 ft of clay for most of its length and makes only a slight feature at the ground surface. In 1953 four trial boreholes were drilled at 140-yd intervals along a shingle ridge trending north-west and reaching nearly to the railway, south-west of the ballast pit ¼ mile E. of Botolph's Bridge. The southernmost borehole (133331) proved shingle, 14 ft; on clay, 3 ft; on shingle, 12½ ft; on silt seen for ½ ft. The other boreholes proved both the clay and the silt to expand northwards, and the shingle beneath the clay to die out, but that above it to be continuous and 11 to 17½ ft thick. A fifth borehole, in the hollow between this ridge and the ballast pit, proved, beneth 8 ft of clay (p. 287), 7 ft of shingle on 5 ft of sandy silt.

A laundry borehole (149344) to the west of Hythe proved the following deposits, in descending order: made ground and clay, 12 ft; "beach gravel", 15½ ft; "dark beach sand with shells", 7 ft; "rock boulders", 2½ ft; on "blue clay and stones", 11½ ft. The last mentioned was water-bearing and was probably the basal drift deposit, resting on Weald Clay.

A trial borehole (161346) for the foundations of a bridge over the Royal Military Canal, 350 yd S.S.W. of Hythe church, proved the following: made ground, 9 ft; shingle, 10 ft; compact sand, 10 ft; broken ragstone, 3 ft; blue clay, 7 ft; hard brown clay, 2½ ft. Ground level was 9 ft above O.D. The clays below 32 ft depth are probably part of the Weald Clay.

A distinct bank runs southward from the site of this bridge to the sea-shore; the ground surface drops sharply from a level surface of bare shingle on the west to a surface some feet lower on the east, over which the shingle is covered by 1½ to 2 ft or more of clay (p. 287). The bank, which more or less coincides in position with Stade Street, presumably formed the western wall of the old harbour of Hythe at its final stage of development.

North of the Royal Military Canal between Hythe and Seabrook are some low ridges in which shingle comes to the surface. Shallow wells in this tract obtain water from the shingle.   B.C.W.

Folkestone—Dover Harbour

The accumulation of shingle at Folkestone was described by Drew (1864, p. 24) and Wheeler (1902, pp. 207–8). Before the harbour was bought by the railway company in 1843, shingle was banked against a short pier on the south-west and extended across the harbour mouth to behind the east pier, so that vessels drawing 9 or 10 ft or more could not enter. In 1856 a west pier was extended and checked the drift for a time. In 17 years the beach advanced 120 ft seawards over a length of 500 ft. A further pier extension in a south-easterly direction caused an eddy which tended to cut out the shingle from the beach and drifted it towards the harbour entrance. Some accumulation took place however. A groyne constructed in the angle between the harbour wall and the beach checked the eddy and caused further deposition of the shingle. The pier was again lengthened in the early 20th century, and has checked the shingle drift.

In Folkestone made ground now masks the tiny estuary of the Pent Stream, which is carried in a culvert near Tontine Street.   J.G.O.S.

The sea-shore from ¼ mile north of Copt Point to a similar distance beyond the eastern end of Folkestone Warren is patchily covered by thin transient deposits of sand and shingle, the distribution and movement of which is controlled to some extent by groynes. The paucity of the beach deposits is due at least in part to the construction of Folkestone pier, which hinders the eastward progress of the shingle ; the consequent depletion of supplies to the east leaves the coastline there comparatively unprotected against wave erosion. Furthermore the restraining load on the toes of the Folkestone Warren landslips is lightened (p. 295).

Storm Gravel Beach Deposits, consisting of coarse flint gravel, extend almost continuously from 800 yd east of the eastern end of Folkestone Warren to Dover Harbour. In places they form a raised ridge or storm beach, enclosing areas at a lower level in which pools form.   G.B.

BLOWN SAND

Hawthorn Corner—Greatstone on Sea—New Romney.

Two ancient patches of Blown Sand near Hawthorn Corner and another at Birdskitchen, form marked rises in ground compared with the surrounding marshland ; rain-wash has destroyed the original outlines of the dunes, so that only a slight unevenness of the ground surface remains. The patches near Hawthorn Corner have formed in the lee of a sea-wall although the more southerly extends somewhat on to the seaward side of the wall. This sand is red to fawn in colour and medium to coarse in grain. Traces of peat were found at a depth of 7½ ft in one auger hole and at 9 ft in another, resting on dark fine-grained sand.

A small patch of Blown Sand on the southern boundary of this district is piled against the seaward side of the shingle fulls, 500 yd S. 32° E. of the Isolation Hospital. Another patch thereabouts occurs in the elbow between two shingle spits and traces were also noted upon the seaward side of the sand-ridge feature (p. 285).

Tiny patches are trapped by the shingle recurves near Belgar and towards Greatstone on Sea. At the latter place dunes of sand rise 15 to 20 ft above the high water mark and move at the present day. These dunes contain blow-outs in which the underlying shingle is seen.

Patches of Blown Sand occur amongst the shingle at Littlestone on Sea while to the north high dunes have formed on the landward side of the present sea-wall and among the shingle fulls and older sea-walls. The Warren contains only relics of Blown Sand, but northwards dunes occur in the angle formed by the New Romney shingle ridge and the present sea-wall.

In New Romney the ground is uneven, sandy and reminiscent of the older patches of Blown Sand near Hawthorn Corner. Some of the unevenness may be due to the presence of made ground, so that the extent of the Blown Sand is uncertain, particularly in view of the marine sand occurring in the town (p. 288).

Dymchurch—Botolph's Bridge—West Hythe.

North-east of Dymchurch village, yellow sand, probably blown from the beach, was noted in places on a broad ridge a few feet high, parallel to the coast.

Between West Hythe and Botolph's Bridge a group of old dunes rises some 12 ft above the surrounding flat clay land. The sand forms a main ridge trending south-west to north-east, across which a lesser ridge trends south to north. A sand-pit (121338) 600 yd S.W. of the bridge at West Hythe exposed in 1955 about 10 ft of fine-grained pale brownish grey sand. About 8 ft above the floor of the pit a dark earthy layer some 6 inches thick was full of the debris of human occupation, including shells of edible molluscs and fragments of bone, carbonized wood and pottery. At the time of this occupation layer the dunes had presumably reached much their present form. Drew (1864, p. 20) identified this patch of blown sand with land referred to in a charter of A.D. 833 as situated at Sandtun, and bounded on the south by the " river Limen ".   B.C.W.

Landslip

London Clay outcrop—Singleton Manor—Collier's Hill—Aldington.

Most of the slopes of London Clay seem to be so unstable that earthflows may occur if even a small excavation is made. The tumbled uneven ground which reveals their presence is commonly covered by a heavy reddish brown brickearth, probably formed by weathering of the slipped material. Earthflows also affect the Atherfield Clay and Weald Clay ¼ mile south-south-east of Singleton Manor, on the south side of Collier's Hill, 1¼ miles south-west of Mersham, at Little Stock, 1 mile south-east of this village, and at Handen and Upper Park Farm, Aldington. With the exception of those at Singleton Manor and Collier's Hill, the instability at these localities is caused by water issuing from springs at the base of the Hythe Beds and saturating the clays.

Ruckinge—Aldington Frith—Lympne—Hythe—Folkestone.

Steep valley sides in the deeply dissected Weald Clay between Ruckinge and Aldington Frith have also slipped. Apart from intervals of 170 yd at Seabrook and 350 yd at Sandgate, the escarpment from Aldington to Hythe and the coast from there to Folkestone Harbour are in a landslipped condition, a total distance of some eleven miles. From Aldington to north-east of Hythe the outcrops of Weald Clay and Atherfield Clay are affected, while the Hythe Beds form a nearly vertical slope above the slipped ground at the crest of the escarpment. Blocks of Hythe Beds, tilted by slipping were, however, noted ¼ mile south-west of Aldington church, at Fostums Land and near Honeypot.

Three large landslips on Aldington Knoll have left only a very small undisturbed remnant.   J.G.O.S.

In the terraced gardens below Port Lympne (102349), laid out in about 1930, some stone retaining walls had become tilted by 1955 and a stone staircase was cracked in places. The greatest amount of subsidence noted was 2 ft.

French House (112347), situated just below the cliff face of Hythe Beds at Lympne, is said locally to have subsided to that position as a result of land-slipping. This is possibly the slip described by Collinson (1728, p. 552) as having occurred about two years before his time of writing as a consequence of a very wet season. The brow of the hill, carrying the house, sank 40 or 50 ft in one night, whilst ground on the lower part of the hill, bordering Romney Marsh, was raised (Topley 1875, p. 316).

The whole slope below Lympne has a hummocky and irregular surface. At its foot is a steep bank 30 to 50 ft high, bordering Romney Marsh, above which are the ruins of Stutfall Castle. This Roman fortification must have been built in approximately its present position, on ground probably previously affected by slipping. The castle walls, disrupted by landslipping, are preserved in separate massive blocks, many of which tilt downhill. It appears that the fallen masonry was in part buried by slipped or downwashed clay (Smith and Elliott 1852, p. 20).

At the top of the landslipped slope south-east of Pedlinge Court sharp-crested ridges each about 50 to 100 yd long represent blocks of Hythe Beds detached from the escarpment and tilted backwards. Some blocks, as at ½ mile S.S.W. of Pedlinge Court, have sunk vertically 10 to 20 ft down the cliff face without drifting away from it. In the deep valley of the Brockhill Stream (150353) 1000 yd east of Pedlinge Court tumbled blocks of ragstone up to 4 ft across are seen in a matrix of sandy clay or light grey glauconitic clay.

Much of the town of Hythe is built on a landslipped slope, but according to the Borough Surveyor, in 1955, only one building, a house near the crest of the slope, has shown signs of slipping in living memory. Probably the slope has become stabilized by efficient drainage.

A landslip on the west side of the Seabrook valley north-west of Horn Street was possibly caused in part by the stream undercutting the slope of Sandgate Beds. The hummocky topography of the northern half of the slip is more subdued than that in the southern half, which appears to have moved more recently.

                                                                                    B.C.W.

The town of Sandgate is largely built on landslip; slight movement was reported in 1959. The slips now are probably restricted to the Sandgate Beds and Folkestone Beds, but the presence of two reefs of Hythe Beds at about low-tide level reveals that this formation has also slipped in the past. The ragstone in the reefs dips inland at angles which may be as high as 50°. It is probable that each reef is formed of Hythe Beds caught up in a single slip. Topley (1893, p. 43) noted on the foreshore, a band of clay which he thought was Atherfield Clay with Hythe Beds apparently dipping over and below it. This clay was probably involved in these slips. Blake's (1893, p. 468) opinion that the Hythe Beds on the foreshore were disturbed by faulting is untenable.

According to Topley (1893, p. 44) a slip occurred during 1827 from the church with spire to about the large house (Encombe House) 500 yd to the west (199353). This area was subsequently drained. In March, 1893, a slip 920 yd long and extending 233 yd back from the sea-wall, occurred west of the eastern boundary of Encombe House grounds. The shift in the ground, Topley stated, was not more than 10 ft vertically and a few feet horizontally. The maximum displace-ment of the sea-wall was only 18 inches. Many houses were damaged. The clay on the foreshore continued to move for a day or more after the main slip.

The cliff from Sandgate to Folkestone Harbour is double: an undercliff, more or less degraded, rises to about 80 ft O.D. at the western end and to about 30 ft O.D. at the eastern end ; at the top of this undercliff is an uneven surface less than 100 yd broad on which the Lower Sandgate Road has been built. The main cliff, composed of Sandgate Beds, Folkestone Beds and Drift, rises from the inland side of this ledge to about 150 ft O.D.

Embayments in this upper cliff, mostly about 100 yd wide, are scars left by landslipping. Although most of the uneven surface of the undercliff and ledge is now artificial, the whole is composed of slipped Sandgate Beds and Folkestone Beds. Recent slipping here has been confined to these two formations, but older slips, like those at Sandgate, have also disturbed the underlying Hythe Beds. Two rock-reefs composed probably of Hythe Beds, extend eastwards towards the harbour from the point (Mill Point) south of the central 'Toll' on the Lower Sandgate Road. The ragstone at Mill Point is only slightly disturbed, but westwards of the point ragstone dips inland at 30° or more. Descriptions of landslips here are contained in the Philosophical Transactions of the Royal Society for 1716 and 1786. The latter work described a rotational slip which also moved the offshore reefs.

Topley (1875, p. 316) considered that removal of stones from the shore for building purposes may have indirectly caused some of the slipping on this stretch of coast. However, some movement took place in 1957–8, despite the sea-front being protected by a promenade and groynes. Further drainage and coast-defence work has since been undertaken to check this slipping.

A further reef of rocks, which partially protrudes at low tide, commences about 450 yd N.E. of Folkestone Harbour Station. This reef, which is known locally as the 'Mole', follows a south-easterly direction for about 100 yd and then turns gradually to continue a little north of east until it apparently connects with the fallen blocks of Folkestone Beds off Copt Point. The reef probably marks a further landslip but it is not known if it is composed of Hythe Beds or Folkestone Beds.   J.G.O.S.

Danton Farm—Sugarloaf Hill—Folkestone Warren.

At intervals along the base of the Chalk escarpment between Danton Farm and Sugarloaf Hill the topography suggests that landslips have occurred, involving Chalk and Gault: scars can be traced around the backs of the slips, and at the toes the Gault has been bulged up, seemingly in a combination of the 'slump' and 'earthflow' forms of Sharpe (1938). These are distinct from occurrences such as the 'avalanche' which destroyed a house at the foot of the escarpment north of Danton Farm (p. 231).

Landslips have occurred in the Gault on the seaward slopes north of Copt Point, and these give way north-eastward to the major rotational slips or slumps of Folkestone Warren. The Warren is the name given to the area of landslipped undercliff that extends along the coast between Folkestone and Abbot's Cliff (Fig. 17 and Plate VIA). It has a length of 2 miles and a maximum width of some 400 yd, and it is traversed by the Folkestone–Dover railway, which was opened in 1844, and which passes through the Martello and Abbot's Cliff tunnels on the west and east respectively. On the north and north-west, The Warren is bounded by cliffs of Lower and Middle Chalk, which rise in part to over 500 ft above sea-level, while almost the whole of the slipped ground surface is below 200 ft. The Gault is not seen in the cliffs above The Warren, being concealed by slips and falls ; the Gault–Lower Chalk junction is exposed in the low cliff 850 yd N. of Copt Point, but at Abbot's Cliff it is below beach level, the dip in this area being about 1 in 60 and the strike W. 33° N. (Wood 1955, p. 411). The Sulphur Band protrudes through thin beach deposits on the foreshore 750 yd

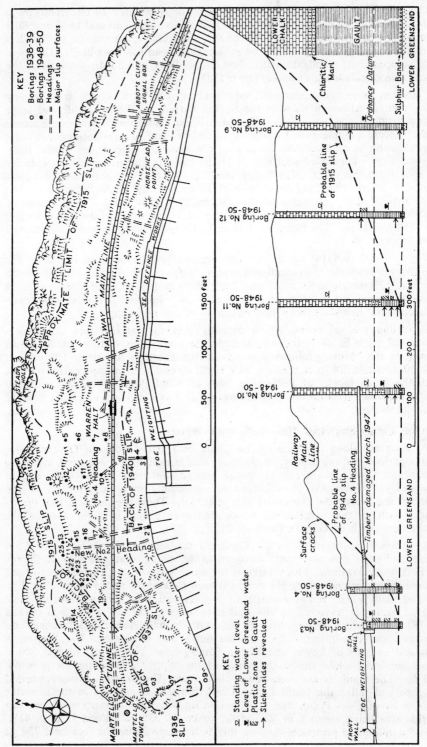

Fig. 17. *Sketch-map and section of Folkestone Warren, based on A. H. Toms, A. M. Muir Wood and N. E. V. Viner-Brady*

N.N.E. of Copt Point. Seawards of The Warren the foreshore is composed of a confused assemblage of Gault and Chalk, in which it is possible to trace numerous slip-planes; strata, both on the shore and in the low cliffs above the sea-wall, exhibit landward dips of up to 55°.

Landslips and falls of varying severity are known to have occurred in The Warren on at least nineteen occasions between 1765 and 1940. The occurrence of each slip was accompanied by upheaval of the foreshore into ridges, which were later eroded by the sea. A slip was also commonly accompanied by the appearance of a 'set' or crack, near the cliff-top, bounding an area of slight subsidence; this mass of loosened chalk later tended to collapse as a rock-fall. It was further remarked that slipping was initiated at times of low tide, when hydrostatic pressure on the foreshore was relieved.

Osman (1917, p. 61) recorded that in 1877 a slip surface cut off about 100 yd of the east end of the Martello Tunnel, and was followed two days later by a fall of chalk from the High Cliff between Capel Lodge and Abbotscliff House, which blocked the railway cutting to a depth of from 50 to 60 ft (see also McDakin 1894, p. 136). In 1915 a major landslip occurred, when almost the whole area of The Warren moved towards the sea; the railway line was displaced a maximum of about 55 yd near Warren Halt, and was buried by associated falls of chalk, with such damage that it was not reopened until 1919; Gault on the foreshore was thrown up into ridges, forming a chain of islands over ½ mile long and reaching up to 30 ft above water at low tide, according to a manuscript map (possibly drawn by Osman) in the Geological Survey library. In 1936 and 1937 further slips took place at the western end of The Warren, the later slip affecting about one-third of the area of The Warren, and moving part of the sea-wall 90 ft seaward.

An investigation into the nature and causes of the landslipping was instituted by the Southern Railway Company in 1939, interrupted by war, and resumed in 1948. The earlier work was concentrated at the western end of The Warren, and was described by Toms (1946). Five boreholes were sunk on the seaward side of the Martello Tunnel, on a line roughly at right angles to it at a place about 180 yd west of the east portal, and tests on samples of the strata encountered in these indicated the probable forms and positions of several slip surfaces; these sloped steeply seawards at the back, but levelled out at the base of the Gault. Toms (1946, p. 14) wrote: "It will be seen . . . that the inferred slip surfaces all pass through. the base of the Gault and that all slips have occurred by horizontal shearing of the Gault immediately above the Greensand. In each case the horizontal movement of the seaward part of the mass is accompanied by a corresponding rotational settlement of the driving mass at the back of the slips." Furthermore it was found that water in the Lower Greensand beneath The Warren was under artesian pressure: the effect of this was to apply hydrostatic uplift to the Gault and at the same time to lubricate the Lower Greensand–Gault junction. (It is noteworthy that fresh-water springs emerge on the shore near the base of the Gault.) Water emerging at the concealed Gault–Chalk junction, reinforced by direct rainfall, also tended to precipitate slipping by increasing the load at the back of the slips and by lubrication.

As a remedial measure parts of the undercliff were regraded, material from the back of the slips on the seaward side of the railway being removed and distributed on the seaward portions of the slips, so as to add to the stabilizing load. In addition, to prevent erosion of the stabilizing mass by the sea, the sea-wall and groynes were extended and large quantities of beach material were dumped on the foreshore to make good the natural deficiency. It was deemed impracticable to lower the Lower Greensand water-table, but headings, some originally driven after the 1877 slip, were used to drain water from

U

within the slipped mass. It was also impossible to check the collapse of chalk masses from the high cliffs, which increase the driving force at the back of the slips.

When research was resumed in 1948 (Wood 1955), attention was directed first to an area south-west of Warren Halt, where movements had been detected since 1940, and later to the ground on the landward side of the railway, west of the Halt. Twenty-five boreholes were sunk and samples from these were examined lithologically and palaeontologically, to detect stratigraphical discontinuities, and were subjected to comprehensive soil tests. In addition, the form and fluctuations of the ground-water tables were studied and the flow of water from headings was gauged.

It was confirmed that "the Folkestone Warren landslips penetrate to the base of the Gault and that failure is largely confined to a plastic sheet of Gault that immediately overlies the sulphur band" (Wood 1955, p. 424). At the toe the mode of upheaval of the foreshore changes with the depth of burial of the Sulphur Band: at the west end of the Warren, where depth is small, upheaval is by folding of the opposing strata; at the east end, where the thickness of Gault is greater, the forward-thrusting mass causes upheaval of the foreshore by driving in wedge fashion below pre-existing slip surfaces. From stability analyses it was concluded that perceptible advantage could be achieved by lowering the water-table at the back of the Warren and by toe-weighting on the foreshore, so that upheaval of the toe was caused to occur more remotely from the land and so that erosion at the toe was limited. The remedial measures developed and carried out as a result of these researches were described by Viner-Brady (1955).

East of The Warren rock-falls of chalk occur from the cliffs, due to undermining by the sea, leaving marked scars in the face of the cliffs and mounds of chalk debris on the shore. These are eroded and dispersed by the sea, giving rise to areas of scattered chalk boulders and flint nodules on the wave-cut platform.   G.B.

Chapter XI

ECONOMIC GEOLOGY

Agriculture and Soils

A GENERAL description of the agriculture of this district is contained in the account by Hall and Russell (1911), which includes many mechanical and chemical analyses of soils from within the area.

The predominantly heavy soils of the Weald Clay do not form good land for arable farming unless drainage is efficient. Much of the Weald Clay out-crop remains as pasture, some of indifferent quality. The Wealden strata also carry a high proportion of woodland.

Arable farming is pre-eminent upon the Lower Greensand. Fruit-growing is of secondary importance, due to the absence of the deep loams upon the Hythe Beds which are a favourite fruit-soil westward of this district.

Much of the intractable Gault soil is laid down to pasture, which is of excellent quality where drainage and farm-management are good. Many centres of arable farming utilize patches of drift, such as Coombe Deposits, but some large-scale arable farms on the Gault itself appear to be successful.

The Chalk outcrop, by virtue of its diverse drift deposits, carries a variety of soils. It is much wooded, but the majority of its woodland is little more than beech scrub. Arable exploitation is often intense, and its degree of success appears to depend on management rather than on the quality of the soil, which is generally adequate or potentially so. The drainage of the soils is excellent, perhaps excessively so in dry seasons, with the exception of those on the Clay-with-flints. The abundance of flints causes hard wear on implements. The barer chalk slopes are left to pasture.

The Thanet Beds give rise to soils which are light and loamy except on the clayey beds. Fruit farming and hop growing are primarily associated with the Thanet Beds outcrop, but it appears that this association is with the broad spreads of brickearth which are derived from the Thanet Beds, rather than with the beds themselves. Soils on the higher Eocene strata are of less merit ; the London Clay outcrop is almost covered by oak and chestnut forest.

Among the drift soils the deep, light, free-draining brickearth forms a fine soil for hops and fruit, particularly cherries. It is widely given over to these crops, except upon the exposed higher spreads on the Chalk. The Marine Alluvium, both the sands and the clays, form rich soils, rewarding for both arable and pasture purposes. These soils have been studied by Green and Askew (1955–1960).   J.G.O.S.

Brickmaking

Brickearth and clay strata appear to have been exploited at many places for brickmaking on a small scale. In Canterbury there are numerous old brickworks which drew material from the brickearth deposits. Degraded faces of a large brickpit lying in the City and west of the Stour are evidently of considerable antiquity.

Bricks are now made at only three localities. Wire-cut facing bricks, and ridge and roofing tiles are made at a works between Hinxhill and Wye, which draws supplies from the Upper Gault there and occasionally also from the Lower Gault near Brabourne.

At Folkestone Brickworks wire-cut bricks, used mostly for footing and foundation work, are made from the Gault. The clay is mixed with sand and breeze, and forced through $\frac{1}{8}$ to $\frac{3}{4}$-inch slots before passing successively between pairs of high-speed rolls set $\frac{1}{4}$ inch and $\frac{1}{16}$ inch apart. This product is compressed and extruded through a die in the form of a continuous column, which is cut to brick-size by taut wires; the bricks are dried in slow heat for six days and finally burnt and cooled for seven days.

Natural red hand-made sand-faced bricks and 'rustic'-finished multi-coloured bricks, as well as briquettes with similar finishes, are made from brickearth at Hawkinge. The brickearth is fed through a pug-mill, where very fine coke is added. The mixture is then moulded into bricks. These are dried for three to four days and then placed in intermittent down-draught kilns and burnt; a kiln takes about three days to reach the vitrification heat of 950 to 1050° C, and it is then left to 'soak' and cool for another seven days. The colouring of the rustic-finished bricks is controlled by the addition of ceramic colouring agents in the moulding process.   G.B., J.G.O.S.

BUILDING SAND

Building sand is quarried from the Folkestone Beds at Sandling Junction (p. 89), from the Woolwich Beds and Oldhaven Beds at Shelford (pp. 193, 197), from the Thanet Beds and Woolwich Beds at Trenleypark Wood (pp. 187, 193), and from the Woolwich Beds at White Wall (p. 192).

GRAVEL

Gravel for use in concrete aggregates is dredged from river gravel occurring below water level at Chartham and Fordwich and from Storm Gravel Beach Deposits at West Hythe and Greatstone on Sea. High level river gravel is dug at Chartham Hatch and Trenleypark Wood. According to the report of a Ministry of Housing and Local Government Advisory Committee (1954, p. 72), the average yield of the inland deposits is 10 000 cu yd per acre and that of the Storm Gravel Beach Deposits is 25 000 cu yd per acre. In this report, which considered an area of East Kent somewhat wider than that dealt with in this account, it was estimated that the present working areas would be exhausted in 25 years, and that thereafter gravel for north Kent would be imported by sea. As gravel becomes scarcer it may become economical to work some of the smaller patches shown on the geological maps. However, many of these patches are of poor quality and a large number have been worked out. Reserves which apparently have not been explored, are present in parts of the Great Stour valley between Canterbury and Milton Manor and south-west of Fordwich, in the valley between Petham and Chartham, and in the Little Stour valley, mainly north of Bridge.

BUILDING STONE

In this area building stone is not quarried in any great quantity. Occasional demands, mainly for ornamental purposes and for walling stone, are met from the Hythe Beds and from the stone doggers in the Folkestone Beds at Sandling Junction.   J.G.O.S.

CALCIUM PHOSPHATE

The phosphatic nodules near the base of the Gault (p. 57) were formerly worked at Cheriton as a source of calcium phosphate for use as a fertilizer. Topley (1875, p. 390) mentioned that the nodules occurred in two beds, separated by 2 ft of clay with scattered nodules: the lower bed was "the junction bed of the Gault and Lower Greensand", 12 to 15 inches thick; the upper bed was 3 to 6 inches thick. Trenches 6 ft wide were excavated and the scattered nodules were picked out by hand, while the nodule beds were dug, washed and sifted. The industry did not pay, however, for it was discontinued in 1876 (Jukes-Browne 1900, p. 83). The pits (206372), 1200 yd W. 42° S. of Castle Hill, have now been effaced for the erection of Harcourt School.

COAL

At Snowdown Colliery the Kent No. 1 (Beresford) Seam was reached in 1913, and it was exploited until 1924. Dines (1933, p. 40) recorded that north-west of the shafts the seam was split, while in the south-east dirt partings and water were troublesome. He remarked that the workings in the seam were limited, due to the proximity of the Coal Measures surface, but in 1945 he put the reserves of coal in this seam within the colliery take at about 11 million tons. Ritchie (1920, p. 189) stated that the coal from this seam was excellent for steam and gas purposes, but rather friable and productive of a large percentage of slack.

The Kent No. 3 Seam was known as the Snowdown Hard Seam at Snowdown Colliery; Dines (1933, p. 40) recorded that development in it consisted only of headings driven in the shaft pillar; a parting of hard sandstone thickened towards the south-east and rendered the seam unworkable.

The Kent No. 6 (Millyard) Seam has been mined at Snowdown Colliery since 1927, and the workings now extend $2\frac{1}{2}$ miles west-north-west, $1\frac{1}{4}$ miles north-east, 2 miles south-east and 1 mile south-west of the shafts. The seam yields prime coking coal with strong caking properties.[1] Within the worked area ash content varies from 3·8 to 7·9 per cent. Volatile matter calculated on the dry ash-free basis ranges from 26·8 per cent in the north-west to 20·8 per cent in the south-east. Calorific value is between 15 510 and 15 660 B.t.u./lb. in the dry ash-free state. Total sulphur is higher in the north-west than in the south-east, ranging from 2·42 to 1·01 per cent. Content of carbon in the dry mineral matter-free coal is between 89·5 and 91·1 per cent, of hydrogen between 5 and 4·5 per cent, of oxygen between 3·4 and 2 per cent, and of phosphorus between 0·016 and 0·006 per cent.

---

[1] Chemical and other data relating to the coal seams have been taken from reports and direct communications by officers of the Coal Survey, National Coal Board.

Some 438 gallons per minute of water from all sources are pumped from
Snowdown Colliery, in addition to 21·15 gallons per minute removed as water
vapour by ventilation (Plumptre 1959, pp. 162, 169). The number of men
employed at the colliery is 1589, of whom 1316 work underground. The
longwall system of mining is employed throughout, the coal on some faces
being broken by means of coal-ploughs. In the five-year period up to
December, 1964, gross annual output averaged 598 600 tons of coal, of which
476 600 tons per annum were saleable. The large graded coal is used by
general industry ; the small coal is used in the manufacture of electricity,
cement and paper.

Where the Chislet Colliery workings in the Kent No. 7 (Chislet No. 5) Seam
extend into the area of the Canterbury (289) Sheet the ash content of this
seam is sometimes as high as 10 per cent. Volatile matter on the dry ash-free
basis is between 28 and 30 per cent ; calorific value on the same basis ranges
from 15 480 to 15 650 B.t.u./lb. Total sulphur content is less than 0·9 per
cent ; content of carbon on the dry mineral matter-free basis is about 89 per
cent, of hydrogen about 5 per cent, of oxygen (plus organic sulphur) about
4·5 per cent, and of phosphorus about 0·05 per cent. The coal has strong
caking properties ; due to its friable nature it gives rise to a high proportion
of fines. It is used mainly in the production of cement, gas and electricity.

G.B.

LIME, CHALK AND WHITING

Chalk is quarried for agricultural lime and whiting in Canterbury, near
Bramling and near Pean, and on a smaller scale at Crundale and East
Brabourne. The old methods of working, with their legacy of numerous
geological sections, have now been largely superseded by the use of tractors
and disc-harrows to break up the chalk surface, the loose chalk being
collected by scrapers.

ROADSTONE

The ragstones of the Hythe Beds are exploited for road-making, both as
hardcore and as tar-macadam. Workings are located at Willesborough,
Mersham, Clap Hill, Aldington and Otterpool. A further quarry near Great
Chart ceased production between 1955 and 1960.   J.G.O.S.

WATER SUPPLY

Overground sources are generally unused in the district and water-supply
requirements are met from ground-water supplies. These are available
from a variety of aquifers, though yields of any size are obtainable only
from the Chalk and the sandy formations of the Lower Greensand. Details
of some of the wells in the area are given in publications by Whitaker (1908),
Buchan and others (1940), and Margaret Davies and others (1964).

The following table shows the annual abstraction (million gallons) from
the most important aquifers in the area under review. The information is
based on annual returns for 1962 made under Section 6 of the Water Act
1945. There is additionally some non-returnable abstraction which is estimated at 5 per cent of the returned abstraction.

Aquifer	Public Supply	Others	Total
Total: Drift	—	5	5
Chalk (undivided)	2390	49	2439
Upper Chalk	1695	20	1715
Middle Chalk	1182	—	1182
Lower Chalk	406	—	406
Total: Chalk	5673	69	5742
Lower Greensand (undivided)	54	—	54
Folkestone Beds	38	16	54
Sandgate Beds	20	—	20
Hythe Beds	23	—	23
Total: Lower Greensand	135	16	151
Grand Totals:	5808	90	5898

A few wells have been sunk into Hastings Beds but yields are generally small: e.g. 600 gal/hr from a 7-in bore at Ashford (022412); overflow of 60 gal/hr from a 6-in bore near Ruckinge (032313). The water is soft, and commonly contains iron and manganese salts.

Apart from the Atherfield Clay, each formation of the Lower Greensand contributes supplies that are of local importance. Yields from the Hythe Beds are variable and range upwards from 1000 gal/hr from a 10-in bore at Folkestone (216354) to 30 000 gal/hr from interconnected shafts and bores at Henwood Pumping Station, Ashford (021429), while a minimum of 13 240 gal/day issues from the Town Springs at Hythe (163351), and is used for public supply purposes. The less permeable, loamy Sandgate Beds yield poor supplies from wells as shown by the range from 400 to 1500 gal/day from a number of shallow shafts near Ashford. Two spring sources near Hythe (187356, 180350) have yields of up to 25 000 and 75 000 gal/day respectively. The water from the latter is hard (total hardness 331 p.p.m. as calcium carbonate), and might therefore be derived from the Hythe Beds.

The most important aquifer in the Lower Greensand is the Folkestone Beds, a sandy unit overlain and underlain by relatively impermeable horizons. Overflows are recorded from 3-in bores at Wye (055469) and Brook (061442), the former yielding 480 gal/hr. Elsewhere, pumping yields range from 3000 gal/hr from a 12-in bore reducing to 6-in at Folkestone (232368) and from a 30-in bore reducing to 10-in at Postling (148377) to 13 000 gal/hr (maximum 328 800 gal/day) from a 21-in bore reducing to 18-in at Hinxhill P.S. (055431). Where wells cut more than one aquifer, yields are much improved; two connected shafts at the Shearway Well, Folkestone (207375), yield 28 000 gal/hr, though $\frac{3}{4}$ million gallons per day (m.gal/day) have been produced under favourable conditions.

Ground waters from individual Lower Greensand formations show much variation in their chemical characteristics; there is some evidence that changes take place both along the outcrop and under cover. Lithological changes may be important in determining these characteristics. There is not necessarily any difference between ground water from Folkestone and

Hythe beds. Total hardness is high (210 to 434 parts per million (p.p.m.) as calcium carbonate) with non-carbonate hardness varying between 70 and 184 p.p.m., and carbonate hardness between 87 and 260 p.p.m. The degree of mineralization of these waters, expressed as total solids, ranges from 250 to 636 p.p.m. The Sandgate Beds water has a higher chloride content (62 to 73 p.p.m.) than the largely sandy formations (30 to 40 p.p.m.).

The Chalk is the foremost aquifer in the district, both in thickness and extent of the resultant outcrop, and in quantitative abstraction. Upper Chalk occupies by far the greater part of the outcrop, consisting of chalk with nodular horizons and numerous layers of flint. The latter are fewer in the Middle Chalk and marly bands more frequent, though the hard nodular, basal, Melbourn Rock is present. The Lower Chalk is largely soft and marly. Inasmuch as the permeability of the Chalk is more dependent on flow along fractures and fissures rather than seepage through the mass of the rock, yield is related to the number of fissures cut during well-sinking operations. Such systems are better developed in the harder rocks or those with well-marked bedding planes and flint courses.

As might be expected, yields tend to be superior from wells penetrating the upper two divisions of the Chalk. A maximum yield of 3 M. gal/day from a 14-ft diameter shaft with headings at Wingham P.S., Adisham (243553), is the highest recorded yield from the Upper Chalk of the district. Other Upper Chalk yields include 40 000 gal/hr from a 24-in bore at Ospringe P.S. (002602), and 10 000 gal/hr from a now disused 10-in bore at Canterbury (148573). Two 10-in bores at Elverland Nurseries, Ospringe (978583), yielded 1500 gal/hr or less on test, whereas 8-in bores at Wingham, Canterbury (157567), and Bekesbourne (179562) each show yields in excess of 3700 gal/hr. The largest yield from Middle Chalk is one of 105 000 gal/hr from a 10-ft diameter shaft with headings at Drellingore P.S., Alkham (243412). A 15-ft diameter shaft with a 15-in bore from the base at Lyoak Farm P.S., Lydden (253446), gave only a small yield, while a 12-in bore at the same site yielded negligible quantities. Wells penetrating both Upper and Middle Chalk divisions give improved yields. The site at Barham P.S., Kingston (199509), is capable of yielding more than 100 000 gal/hr from four shafts, one of which has yielded 75 000 gal/hr on test. Similarly at Newnham P.S., Norton (975597), each of three wells can yield in excess of 50 000 gal/hr.

Lower Chalk wells produce relatively poor yields, and in the south of the district where they are the main source of supply, large works are necessary to supply adequate quantities. Thus a 6-in bore at Monks Horton, Stanford (141407), yields 50 gal/hr, and a 6-ft diameter shaft with heading at Ottinge P.S., Elham (172425), yielded only 2300 gal/hr. The latter site is now connected to an adjacent shaft and bore, and yields 20 000 gal/hr. The adit known as 'Terlingham Tunnel' at Folkestone (211381) has an average yield of 150 000 gal/day, but may exceed 500 000 gal/day for short periods in a wet winter and in a dry summer drop to as little as 40 000 gal/day. Yields from combined Middle and Lower Chalk units show a marked improvement. The connected shafts at Lower Standen P.S., Capel-le-Ferne (241404), are capable of yielding 2 M. gal/day in winter and half that amount in summer, but are affected by pumping at Drellingore P.S.

($\frac{1}{2}$ mile away); the effective summer yield is 400 000 gal/day at a rate of 44 000 gal/hr.

Wells that penetrate all three divisions of the Chalk are few in number but some give exceptionally good yields. A 24-in bore at Thanington P.S., Canterbury (138567), initially yielded 100 000 gal/hr. In contrast a 27-in bore at Denton P.S, Barham (224481), formerly yielded only 12 100 gal/hr for a drawdown of 112 ft, but this yield was improved after acid treatment to 20 000 gal/hr for a drawdown of 39 ft. Preliminary test pumping proved good supplies from the three abstraction sites associated with the Belmont Scheme. At Belmont Park, Ospringe (980572), a 40-in bore reducing to 12-in gave a maximum yield of 149 600 gal/hr, and at Selling P.S. (035575) three bores of less than 27-in diameter together yielded in excess of 50 000 gal/hr. Conditions at Throwley P.S (995559) are not as favourable and an initial yield of 18 500 gal/hr from the main well was improved by surging, acid treatment, subsidiary borings, and shot-firing at depth to 86 620 gal/hr.

Chalk ground-waters in the district differ from those of the Lower Greensand. They are invariably hard, with total hardness varying from 240 to 320 p.p.m., the bulk of this being represented by carbonate hardness (213 to 260 p.p.m.). Figures for total hardness from the three divisions of the Chalk overlap each other and only in the non-carbonate content is there any obvious distinction. Upper and Lower Chalk ground-waters appear to have a non-carbonate hardness in excess of 35 p.p.m., whereas this figure is the maximum for Middle Chalk water from the district. The total dissolved solids content ranges between 290 and 360 p.p.m., and there is some evidence that values for Middle Chalk may be greater than those for Lower and Upper Chalk. The chloride content of Upper and Middle Chalk waters varies from 18 to 23 p.p.m., that for Lower Chalk from 23 to 29 p.p.m.

Few wells have been sunk into the Lower London Tertiaries and those mainly into Thanet Beds. Details of yields are not known but must have been small. A few cases of overflowing shallow bores are recorded. G.P.J.

REFERENCES

AGER, D.V. 1954. The Genus *Gibbirhynchia* in the British Domerian. *Proc. Geol. Assoc.*, **65**, 25–51.

ALLEN, P. 1954. Geology and Geography of the London–North Sea Uplands in Wealden Times. *Geol. Mag.*, **91**, 498–508.

—— 1955. Age of the Wealden in North-Western Europe. *Geol. Mag.*, **92**, 265–81.

—— 1959. The Wealden environment : Anglo-Paris basin. *Phil. Trans. Roy. Soc.* (B), **242**, 283–346.

ANDERSON, F. W. 1956. In *Sum. Prog. Geol. Surv.* for 1955, 53.

ANDREAE, S. J. F. 1958. A Dutchman looks at Romney Marsh. *Tijdschr. ned. aardrijksk. Genoot*, **75**, No. 3, 230–8.

ARBER, E. A. N. 1909. On the Fossil Plants of the Waldershare and Fredville Series of the Kent Coalfield. *Quart. J. Geol. Soc.*, **65**, 21–39.

—— 1912. A note on some Fossil Plants from the Kent Coalfield. *Geol. Mag.*, (5), **9**, 97–9.

—— 1914a. On the Fossil Flora of the Kent Coalfield. *Quart. J. Geol. Soc.*, **70**, 54–81.

—— 1914b. The Geology of the Kent Coalfield. *Trans. Inst. Min. Engrs.*, **47** for 1913–1914, 677–714.

—— 1915. The Thickness of the Lower Carboniferous Rocks proved in the Trapham Boring, Kent. *Colliery Guard.*, **110**, 738.

ARKELL, W. J. 1933. *The Jurassic System in Great Britain.* Oxford.

BAKER, H. A. 1920. Structural Features of the East Kent Coalfield. *Iron Coal Tr. Rev.*, **101**, 785–8.

BANNISTER, M. A. 1932. The distinction of pyrite from marcasite in nodular growths. *Miner Mag.*, **33**, No. 138, 179–87.

BARROIS, C. 1876. Recherches sur le Terrain Crétacé Supérieur de l'Angleterre et de l'Irlande. *Mém. Soc. géol. Nord*, **1**.

BATHER, F. A. 1911. Upper Cretaceous Terebelloids from England. *Geol. Mag.* (5), **8**, 481–8, 549–56.

BEDWELL, F. A. 1874. The Isle of Thanet. The Ammonite Zone, the Depth of the Chalk in Section, and the Continuity of its Flint Floorings. *Geol. Mag.* (2), **1**, 16–22.

BELL, T. 1858–1862. Monograph of the fossil Malacostracous Crustacea of Great Britain. *Palaeont. Soc.*

BLACK, M. 1953. The Constitution of the Chalk. *Proc. Geol. Soc.*, No. 1499, lxxxi–ii, lxxxv–vi.

BLAKE, J. F. 1893. The Landslip at Sandgate. *Nature, Lond.*, **47**, 467–9.

BOLTON, H. 1915. The Fauna and Stratigraphy of the Kent Coalfield. *Trans. Inst. Min. Engrs.*, **49**, 643–702.

—— 1916. Discussion on The Fauna and Stratigraphy of the Kent Coalfield. *Trans. Inst. Min. Engrs*, **50**, 294–319.

BOSWELL, P. G. H. 1917. The Stratigraphy and Petrology of the Lower Eocene Deposits of the North-Eastern Part of the London Basin. *Quart. J. Geol. Soc.*, **71** for 1915, 536–91.

BOWEN, R. 1961. Oxygen isotope paleotemperature measurements on Cretaceous Belemnoidea from Europe, India, and Japan. *J. Paleont.*, **35**, 1077–84.

BRAJNIKOV, B. and FURON, R. 1934. Sur l'Argile à silex. *C. R. Soc. géol. Fr.*, 202–3.

BRAJNIKOV, B. and URBAIN, P. 1934. Sur le spectre de rayons X de la phyllite de l'argile à silex de Vernon (Eure). *C. R. Soc. géol. Fr.*, 203-5.

BREISTROFFER, M. 1947. Sur les zones d'ammonites de l'Albien de France et d'Angleterre. *Trav. Lab. géol. Grenoble*, **26**, 17--104.

BRITISH STANDARDS INSTITUTION. 1961. Methods of Testing Soils for Civil Engineering Purposes. *British Standard No. 1377 : 1961.* London.

BROMEHEAD, C. E. N. 1922. Excursion to Sole Street, Holly Hill and East Malling. *Proc. Geol. Assoc.*, **33**, 324-7.

BROWN, E. O. FORSTER. 1923. Underground Waters in the Kent Coal-field and their Incidence in Mining Development. *Proc. Inst. Civ. Engrs.*, **215**, pt. i, 27-114.

BUCHAN, S., ROBBIE, J. A., HOLMES, S. C. A., EARP, J. R., BUNT, E. F. and MORRIS, L. S. O. 1940. Water Supply of South-East England from Underground Sources. *Geol. Surv. Wartime Pamphlet* No. 10, Pt. VI.

BUCKMAN, S. S. 1910. *Yorkshire Type Ammonites*, **1**, pt 2, xiii–xvi, descriptions 9–22, 12 plates. London.

BURR, M. 1909. The South-Eastern Coalfield, its Discovery and Development. *Science Progress*, **3**, 379–409.

—— 1913. Ten Deep Borings in East Kent. *Colliery Guard.*, **106**, No. 2754, 731-4.

BURROWS, A. J. 1885. Romney Marsh, Past and Present. *Trans. Surv. Instn*, **17** for 1884–5, 335–76.

CARRUTHERS, W. 1866. On some fossil coniferous fruits. *Geol. Mag.*, **3**, 534–46.

—— 1869. On some undescribed coniferous fruits from the Secondary rocks of Britain. *Geol. Mag.*, **6**, 1–7.

—— 1871. On two undescribed coniferous fruits from the Secondary rocks of Britain. *Geol. Mag.*, **8**, 540–4.

CASEY, R. 1936. Recent additions to the Albian ammonoid faunas of Folkestone. *Geol. Mag.*, **73**, 444–8.

—— 1939. The Upper Part of the Lower Greensand around Folkestone. *Proc. Geol. Assoc.*, **50**, 362–78.

—— 1949a. In *South-Eastern Naturalist and Antiquary*, **54**, xxi–xxii.

—— 1949b. The ammonite genus *Uhligella* in the English Albian. *Geol. Mag.*, **86**, 333–45.

—— 1950. The junction of the Gault and Lower Greensand in East Sussex and at Folkestone, Kent. *Proc. Geol. Assoc.*, **61**, 268–98.

—— 1954. *Falciferella*, a new genus of Gault ammonites, with a review of the family Aconeceratidae in the British Cretaceous. *Proc. Geol. Assoc.*, **65**, 262–77.

—— 1955. The Neomiodontidae, a new family of the Arcticacea (Pelecypoda). *Proc. Malac. Soc.*, **31**, 208–222.

—— 1959. In *Sum. Prog. Geol. Surv.* for 1958, 48.

—— 1960a. *Hengestites*, a new genus of Gault ammonites. *Palaeontology*, **2**, 200–9.

—— 1960b. A Lower Cretaceous gastropod with fossilized intestines. *Palaeontology*, **2**, 270–6.

—— 1960–5. A monograph of the Ammonoidea of the Lower Greensand. Pts. 1–6. *Palaeont. Soc.*

—— 1961. The stratigraphical palaeontology of the Lower Greensand. *Palaeontology*, **3**, 487–621.

—— 1963. The Dawn of the Cretaceous Period in Britain. *Bull. S.-E. Un. Sci. Socs.*, No. 117.

CHAPMAN, F. 1891–8. The Foraminifera of the Gault of Folkestone. Pts. I–X. *J. Roy. Micr. Soc.* for 1891–8.

CHAPMAN, F. 1898. The probable depths of the Gault sea as indicated by its rhizopodal fauna. *Natural Science,* **13,** No. 81, 305–12.

—— and SHERBORN, C. D. 1893. On the Ostracoda of the Gault at Folkestone. *Geol. Mag.* (3), **10,** 345–9.

COLEMAN, ALICE. 1952. Some Aspects of the Development of the Lower Stour, Kent. *Proc. Geol. Assoc.,* **63,** 63–86.

—— 1954. The Relief and Drainage Evolution of the Blean. *Proc. Geol. Assoc.,* **65,** 52–63.

COLLINS, I. and GILL, D. M. C. 1923. Notes on the Geology of Boughton-under-Blean. *Proc. Geol. Assoc.,* **34,** 300–13.

COLLINSON, P. 1728. An uncommon Sinking of the Ground in Kent. *Phil. Trans. Roy. Soc.,* **35,** 551–2.

CROOKALL, R. 1933. Contributions to the Geology of the Kent Coalfield. II. The Fossil Flora of the Kent Coalfield. *Sum. Prog. Geol. Surv.* for 1932, pt. II, 44–70.

DARWIN, C. 1851–5. A monograph of the fossil Cirripedes of Great Britain (Lepadidae, Balanidae, Verrucidae). *Palaeont. Soc.*

DAVIDSON, T. 1852. A monograph of the British fossil Brachiopoda. Pt. II. No. 1, 1–54, pls. i–v. *Palaeont. Soc.*

—— 1855. A monograph of the British fossil Brachiopoda. Pt. II. No. 2, 55–117, pls. vi–xii. *Palaeont. Soc.*

DAVIES, MARGARET C. and others. 1964. Records of Wells in the Area of New Series One-Inch (Geological) Canterbury (289) Sheet. *Water Supply Papers, Geol. Surv. G.B. Well Cat. Ser.*

DAVIS, A. G. and ELLIOTT, G. F. 1958. The Palaeogeography of the London Clay Sea. *Proc. Geol. Assoc.,* **68** for 1957–8, 255–77.

DAWKINS, W. B. 1900. On the South-Eastern Coalfield. *Rep. Brit. Assoc.* for 1899, 734–8.

—— 1905. In *Royal Commission on Coal Supplies:* Final Report, Pt. X, 26–41.

—— 1907. The Discovery of the South-Eastern Coalfield. *J. Soc. Arts,* **55,** No. 2833, 450–60.

—— 1913. The South-Eastern Coalfield, the Associated Rocks, and the Buried Plateau. *Trans. Inst. Min. Engrs.,* **44,** Pt. 2, 350–78.

DE RANCE, C. E. 1868. On the Albian or Gault of Folkestone. *Geol. Mag.,* **5,** 163–71.

DESOR, E. 1842. *Monographies d'Echinodermes vivants et fossiles. Des Galérites.* Neuchatel.

DEWEY, H. 1926. The River Gravels of the South of England, their relationship to Palaeolithic Man and to the Glacial Period. *C.R. XIII Congr. géol. international,* 1922, 1429–46.

——, BROMEHEAD, C. E. N., CHATWIN, C. P. and DINES, H. G. 1924. The Geology of the Country around Dartford. *Mem. Geol. Surv.*

—— and SMITH, R. A. 1925. Flints from the Sturry gravels, Kent. *Archaeologia,* **74,** 117–36.

——, WOOLDRIDGE, S. W., CORNES, H. W. and BROWN, E. E. S. 1925. The Geology of the Canterbury District and Report of Excursion to Canterbury. *Proc. Geol. Assoc.,* **36,** 257–90.

DINES, H. G. 1933. Contributions to the Geology of the Kent Coalfield. I. The Sequence and Structure of the Kent Coalfield. *Sum. Prog. Geol. Surv.* for 1932, pt. II, 15–43.

—— 1945. Report of the Geological Survey in *Kent Coalfield Regional Survey Report, Ministry of Fuel and Power,* 7–25.

——, HOLMES, S. C. A. and ROBBIE, J. A. 1954. Geology of the Country around Chatham. *Mem. Geol. Surv.*

DIRECTORY OF BRITISH FOSSILIFEROUS LOCALITIES. 1954. *Palaeont. Soc.*

DIX, EMILY and TRUEMAN, A. E. 1931. Some Non-marine Lamellibranchs from the Upper Part of the Coal Measures. *Quart. J. Geol. Soc.,* **87,** 180–211, pl. 17.

DIXON, T. 1850. *The geology and fossils of the Tertiary and Cretaceous formations of Sussex.* London.

DOWKER, G. 1866. On the Junction of the Chalk with the Tertiary Beds in East Kent. *Geol. Mag.,* **3,** 210–3.

—— 1897. On Romney Marsh. *Proc. Geol. Assoc.,* **15** for 1897–8, 211–23.

DREW, F. 1864. The Geology of the Country between Folkestone and Rye, including the whole of Romney Marsh. *Mem. Geol. Surv.*

DUGDALE, W. 1662. *The History of Imbanking and Draining of divers Fens and Marshes both in Foreign Parts and in the Kingdom, and of the Improvements thereby.* 2nd edit., 1772, by C. N. Cole.

EDELMAN, C. H. 1933. Petrologische Provincies in het Nederlandsche Kwartair. *Meded. geol. Inst. Univ. Amst.,* **43,** 1–104.

—— 1950. *Soils of the Netherlands.* Amsterdam.

—— and DOEGLAS, D. J. 1933. Bijdrage tot de Petrologie van het Nederlandsche Tertiair. *Verh. geol.-mijnb. Genoot. Ned. Kolon., Geol. Ser.,* **10,** 1–38.

EDEN, R. A., STEVENSON, I. P. and EDWARDS, W. 1957. Geology of the Country around Sheffield. *Mem. Geol. Surv.*

EDMUNDS, F. H. 1948. The Wealden District. 2nd edit. *British Regional Geology, Geol. Surv.*

—— 1953. In *Sum. Prog. Geol. Surv.* for 1951, 24.

ELLIOTT, J. Jnr. 1847. Account of the Dymchurch Wall which forms the Sea Defences of Romney Marsh. *Proc. Inst. Civ. Engrs.,* **6,** 466–78.

—— 1852. *Ancient State of the Romney Marshes.* With Roach Smith, *Report on Excavations made on the site of the Roman Castrum at Lymne.* London. Privately printed.

ETHERIDGE, R. 1867. On the stratigraphical position of *Acanthopholis horridus* (Huxley). *Geol. Mag.,* **4,** 67–9.

—— 1900. On the Relation between the Dover and Franco-Belgian Coal Basins. *Rep. Brit. Assoc.* for 1899, 730–4.

FAGG, C. C. 1954. The Coombes and Embayments of the Chalk Escarpments. *Proc. Croydon Nat. Hist. and Sci. Soc.,* **12,** 117–31.

FALCONER, H. 1868. Notes on a Collection of Fossil Bones discovered in a Section of Gravel in excavating the Folkestone Battery. *Palaeontological Memoirs and Notes,* **2,** 564–9.

FITTON, W. H. 1836. Observations on some of the Strata between the Chalk and the Oxford Oolite, in the South-East of England. *Trans. Geol. Soc.* (2), **4,** 103–388.

FOORD, A. H. 1891. Catalogue of the fossil Cephalopoda in the British Museum (Natural History). Pt. II. Nautiloidea. *Brit. Mus. (Nat. Hist.).*

FURLEY, R. 1874. *A History of the Weald of Kent,* **2,** Pt. 1. Ashford and London.

GARDNER, J. S. 1873. Notes on the genus *Rostellaria* (or *Aphorrhais?*) of the Gault. *Geol. Mag.,* **10,** 161–3.

—— 1875. On the Cretaceous Aporrhaidae. *Geol. Mag.,* **12,** 392–400.

—— 1876. Cretaceous Gasteropoda. *Geol. Mag.,* **13,** 75–8, 105–14, 160–3.

—— 1877a. Notes on Cretaceous Gasteropoda. *Geol. Mag.* (2), **4,** 556–7.

—— 1877b. On the British Cretaceous Patellidae and other families of patelloid Gasteropoda. *Quart. J. Geol. Soc.,* **33,** 192–206.

—— 1878. On the Cretaceous Dentaliidae. *Quart. J. Geol. Soc.,* **34,** 56–65.

—— 1880. Cretaceous Gasteropoda. *Geol. Mag.* (2), **7,** 49–54.

GARDNER, J. S. 1883. On the Lower Eocene Section between Reculvers and Herne Bay, and on some Modifications in the Classification of the Lower London Tertiaries. *Quart. J. Geol. Soc.,* **39,** 197–210.

GARDNER, MARGARET I. 1888. The Greensand at the base of the Thanet Sand. *Quart. J. Geol. Soc.,* **44,** 755–60.

GILBERT, C. J. 1930. Earth Movements during the Closing Stages of the Neolithic Depression. *Quart. J. Geol. Soc.,* **86,** 94–5.

—— 1933. The Evolution of Romney Marsh. *Archaeol. cantiana,* **45,** 246–72.

GODWIN, H. 1961. Radiocarbon dating and Quaternary history in Britain. *Proc. Roy. Soc.* (B), **153,** 287–320.

—— and WILLIS, E. H. 1960. Cambridge University Natural Radiocarbon Measurements II. *Amer. J. Sci. Radiocarbon Suppl.,* **2,** 62–72.

—— —— 1961. Cambridge University Natural Radiocarbon Measurements III. *Amer. J. Sci. Radiocarbon Suppl.,* **3,** 60–76.

GODWIN-AUSTEN, R. 1856. On the Possible Extension of the Coal Measures beneath the South-Eastern Part of England. *Quart. J. Geol. Soc.,* **12,** 38–73.

GREEN, R. D. and ASKEW, P. W. 1955. In *Rep. Soil Surv. Gt. Brit,* No. 7, 11–4.

—— 1956. In *Ibid.,* No. 8, 16–7.

—— 1958a. In *Ibid.,* No. 9, 27–30.

—— 1958b. In *Ibid.,* No. 10, 21–5.

—— 1959. In *Ibid.,* No. 11, 22–7.

—— 1960. In *Ibid.,* No. 12, 35.

GREGORY, J. W. 1895. On a Collection of fossils from the Lower Greensand of Great Chart in Kent. *Geol. Mag.,* **32,** 97–103, 187–9.

GROVES, A. W. 1931. The Unroofing of the Dartmoor Granite and the Distribution of its Detritus in the Sediments of Southern England. *Quart. J. Geol. Soc.,* **87,** 62–96.

GUILCHER, A. 1958. *Coastal and Submarine Morphology.* Translated by B. W. Sparks and R. H. W. Kneese. London.

GULLIVER, F. P. 1897. Dungeness Foreland. *Geogr. J.,* **9,** 536–46.

GUNTHER, A. 1864. Description of a new fossil fish from the Lower Chalk. *Geol. Mag.,* **1,** 114–8.

HALL, A. D. and RUSSELL, E. J. 1911. *Report on the Agriculture and Soils of Kent, Surrey and Sussex.* Board of Agriculture and Fisheries.

HALLIMOND, A. F. 1925. Iron Ores: Bedded ores of England and Wales. Petrography and Chemistry. *Spec. Rep. Min. Res. Gt. Brit.,* **29, Mem. Geol. Surv.**

HAYNES, J. 1955. Pelagic foraminifera in the Thanet beds, and the use of Thanetian as a stage name. *Micropaleontology,* **1,** No. 2, 189.

—— 1956–8. Certain smaller British Paleocene Foraminifera, Pts. I–V. *Contr. Cushman Fdn.,* **7,** Pt. 3, 79–101 ; **8,** Pt. 2, 45–53 ; **9,** Pt. 1, 4–16 ; **9,** Pt. 3, 58–77 ; **9,** Pt. 4, 83–92.

—— and EL-NAGGAR, Z. R. M. 1964. Reworked Upper Cretaceous and Danian planktonic foraminifera in the type Thanetian. *Micropaleontology,* **10,** No. 3, 354–6.

HÉBERT, E. 1874. Comparaison de la Craie des côtes d'Angleterre avec celle de France. *Bull. Soc. géol. Fr.* (3), **2,** 416–28.

HESTER, S. W. 1965. Stratigraphy and Palaeogeography of the Woolwich and Reading Beds. *Bull. Geol. Surv. Gt. Brit.,* No. 23, 117–37.

HILL, W. 1886. On the Beds between the Upper and Lower Chalk of Dover. *Quart. J. Geol. Soc.,* **42,** 232–48.

HINDE, G. J. 1883. Catalogue of the fossil sponges in the geological department of the British Museum (Natural History). *Brit. Mus. (Nat. Hist.)*

HOLLOWAY, W. 1854. On Romney Marsh. *J. Brit. Arch. Assoc.,* **9** for 1853–4, 376–85.

HOLMES, T. R. 1907. *Ancient Britain and the Invasion of Julius Caesar.* Oxford.

HOMAN, W. M. 1938. The Marshes between Hythe and Pett. *Sussex Arch. Coll.,* **79,** 199–223.

HUDSON, R. G. S. and MITCHELL, G. H. 1937. The Carboniferous Geology of the Skipton Anticline. *Sum. Prog. Geol. Surv.* for 1935, pt. II, 1–45.

HUGHES, T. McK. 1866. Note on the Junction of the Thanet Sand and the Chalk, and of the Sandgate Beds and Kentish Rag. *Quart. J. Geol. Soc.,* **22,** 402–4.

HULL, E. and WHITAKER, W. 1861. The Geology of parts of Oxfordshire and Berkshire. *Mem. Geol. Surv.*

HUME, W. F. 1914. Professor Walthur's Erosion in the Desert considered. *Geol. Mag.,* **51,** 18–22.

HUTCHINSON, J. N. (In preparation). Surveys of coastal landslides: Kent. *Natn. Bldg. Stud., Spec. Rep., Building Res. Stn.*

HUXLEY, T. H. 1867. On *Acanthopholis horridus,* a new reptile from the Chalk-Marl. *Geol. Mag.,* **4,** 65–7.

INESON, J. 1962. In *Sum. Prog. Geol. Surv.* for 1961, 62–4.

—— and GRAY, D. A. 1963. Electrical Investigations of Borehole Fluids. *J. Hydrology,* **1,** 204–18.

JEFFERIES, R. P. S. 1962. The Palaeoecology of the *Actinocamax plenus* Sub-zone (Lowest Turonian) in the Anglo-Paris Basin. *Palaeontology,* **4,** 609–47.

—— 1963. The Stratigraphy of the *Actinocamax plenus* Subzone (Turonian) in the Anglo-Paris Basin. *Proc. Geol. Assoc.,* **74,** 1–33.

JONES, T. R. 1850. Monograph of the Entomostraca of the Cretaceous formations of England. *Palaeont. Soc.*

—— and HINDE, G. J. 1890. A supplementary monograph of the Cretaceous Entomostraca of England and Ireland. *Palaeont. Soc.*

JUKES-BROWNE, A. J. 1900. The Cretaceous Rocks of Britain, **1,** The Gault and Upper Greensand of England. *Mem. Geol. Surv.*

—— 1903. Op. cit., **2,** The Lower and Middle Chalk of England. *Mem. Geol. Surv.*

—— 1904. Op. cit., **3,** The Upper Chalk of England. *Mem. Geol. Surv.*

KERNEY, M. P. 1963. Late-glacial deposits on the Chalk of South-East England. *Phil. Trans. Roy. Soc.* (B), **246,** 203–54.

——, BROWN, E. H. and CHANDLER, T. J. 1964. The late-glacial and post-glacial history of the Chalk escarpment near Brook, Kent. *Phil. Trans. Roy. Soc.* (B), **248,** 135–204.

KHAN, M. H. 1950a. On some new foraminifera from the Lower Gault of Southern England. *J. Roy. Micr. Soc.,* **70,** 268–79.

—— 1950b. Note on the depth and temperature of the Gault sea as indicated by Foraminifera. *Geol. Mag.,* **87,** 175–80.

—— 1952. Zonal analysis of the Lower Gault of Kent based on Foraminifera. *Contr. Cushman Fdn.,* **3,** 71–80.

KIDSTON, R. 1919. List of the Fossil Plants from the Coal Measures of the Borings at Bere Farm, Elham, Folkestone and Lydden Valley, Kent. *Sum. Prog. Geol. Surv.* for 1918, Appendix II, 46–9.

KIRKALDY, J. F. 1937. The Overstep of the Sandgate Beds in the Eastern Weald. *Quart. J. Geol. Soc.,* **93,** 94–126.

—— 1950. Solution of the Chalk in the Mimms Valley, Herts. *Proc. Geol. Assoc.,* **61,** 219–24.

KITCHIN, F. L. 1934. In *Sum. Prog. Geol. Surv.* for 1933, pt. I, 77–85.

LAMPLUGH, G. W. 1917. The Underground Range of the Lower Cretaceous Rocks in East Kent. *Sum. Prog. Geol. Surv.* for 1916, 45–52.

—— and KITCHIN, F. L. 1911. On the Mesozoic Rocks in some of the Coal Explorations in Kent. *Mem. Geol. Surv.*

—— —— and PRINGLE, J. 1923. The Concealed Mesozoic Rocks in Kent. *Mem. Geol. Surv.*

——, WEDD, C. B. and PRINGLE, J. 1920. Iron Ores (*contd.*)—Bedded Ores of the Lias, Oolites and Later Formations in England. *Spec. Rep. Min. Res. Gt. Brit.*, **12**, *Mem. Geol. Surv.*

LEES, G. M. and COX, P. T. 1937. The Geological Basis of the Present Search for Oil in Great Britain by the D'Arcy Exploration Company, Limited. *Quart. J. Geol. Soc.*, **93**, 156–94.

LEWIN, T. 1862. *The Invasion of Britain by Julius Caesar.* 2nd edit. London.

—— 1866. On the Position of the Portus Lemanis of the Romans. *Archaeologia*, **40**, 2, 361–74.

LEWIS, W. V. 1932. The Formation of Dungeness Foreland. *Geogr. J.*, **80**, 309–24.

—— and BALCHIN, W. G. V. 1940. Past Sea-Levels at Dungeness. *Geogr. J.*, **96**, 258–85.

LOVEDAY, J. 1962. Plateau Deposits of the Southern Chiltern Hills. *Proc. Geol. Assoc.*, **73**, 83–102.

LUCAS, J. 1908. The Hydrogeology of the Dover Basin, Dover Harbour and the Channel Tunnel. *Trans. Surv. Instn.*, **40**, 455–82.

LYDEKKER, R. 1888–90. Catalogue of the fossil Reptilia and Amphibia in the British Museum. Pts. I–IV. *Brit. Mus. (Nat. Hist.).*

MACKIE, S. J. 1851. On a Deposit at Folkestone containing Bones of Mammalia. *Quart. J. Geol. Soc.*, **7**, 257–62.

—— 1860. Geology of Folkestone. *Geologist*, **3**, 41–5, 81–90, 121–31, 201–7, 281–4, 321–7, 353–7, 393–6.

—— 1863. On a new species of Hybodus from the Lower Chalk. *Geologist*, **6**, 241–6.

—— 1865. *In* English, J. *A Handbook of Folkestone for Visitors.* 4th edit.

MCDAKIN, J. G. 1888. On the occurrence of Manganese in some gravel beds under Bigberry Wood, nr. Canterbury. *Trans. E. Kent Nat. Hist. Soc.*, Nos. 1–4, 1885–1899, 134.

—— 1893. Landslip near Folkestone! *Trans. Assoc. Nat. Hist. Soc. S.-E. England. S. East. Nat.*, **1**, pt. 3, 97–100.

—— 1894. Coast erosion, and landslips in the neighbourhood of Dover. *Trans. Assoc. Nat. Hist. Soc. S.-E. England. S. East. Nat.*, **1**, pt. 4, 132–6.

MANTELL, G. 1822. *The Fossils of the South Downs; or Illustrations of the Geology of Sussex.* London.

—— 1833. *The Geology of the South-East of England.* London.

MILNER, H. B. and BULL, A. J. 1925. Excursion to Eastbourne–Hastings. *Proc. Geol. Assoc.*, **36**, 317–20.

MINISTRY OF HOUSING AND LOCAL GOVERNMENT. 1954. *Report of the Advisory Committee on Sand and Gravel.* Pt. 13. Kent and East Sussex.

MURCHISON, R. I. 1851. On the Distribution of the Flint Drift of the South-East of England, on the Flanks of the Weald, and over the Surface of the South and North Downs. *Quart. J. Geol. Soc.*, **7**, 349–98.

NEWTON, E. T. 1878. Description of a new fish from the Lower Chalk of Dover. *Quart. J. Geol. Soc.*, **34**, 439–46.

OLLIER, C. D. and THOMASSON, A. J. 1957. Asymmetrical Valleys of the Chiltern Hills. *Geogr. J.*, **123**, 71–80.

OSMAN, C. W. 1917. The Landslips of Folkestone Warren and Thickness of the Lower Chalk and Gault near Dover. *Proc. Geol. Assoc.*, **28**, 59–82.

PARKINSON, J. 1811. *The Organic Remains of a Former World*. **3.** London.
—— 1819. Remarks on the fossils collected by Mr. Phillips near Dover and Folkestone. *Trans. Geol. Soc.* (i), **5,** Pt. 1, 52–9.

PENNING, W. H. and JUKES-BROWNE, A. J. 1881. Geology of the Neighbourhood of Cambridge. *Mem. Geol. Surv.*

PETTITT, N. E. 1954. A monograph on the Rhynchonellidae of the British Chalk. Pt. II, 27–52, pl. iii. *Palaeont. Soc.*

PHILLIPS, W. 1818. Remarks on the Chalk Cliffs in the neighbourhood of Dover, and on the Blue Marle covering the Green Sand, near Folkestone. *Trans. Geol. Soc.* (1), **5,** Pt. 1, 16–46.

PITCHER, W. S., SHEARMAN, D. J. and PUGH, D. C. 1954. The Loess of Pegwell Bay, Kent, and its Associated Frost Soils. *Geol. Mag.,* **91,** 308–14.

PLUMPTRE, J. H. 1959. Underground Waters of the Kent Coalfield. *Trans. Inst. Min. Engrs.,* **119,** 155–69.

POPIEL-BARCZYK, E. 1958. Jezowce rodzaju *Conulus* z Turonu okolic Krakowa, Miechowa i Wolbromia. *Prace Muz. Ziemi,* **2,** 41–79, pls. i–v.

POTIER, A. and DE LAPPARENT, A. 1875. *Rapports sur les Explorations Géologiques (Chemin de Fer Sous-Marin)*. Paris.

PRESTWICH, J. 1850. On the Structure of the Strata between the London Clay and the Chalk in the London and Hampshire Tertiary Systems. Pt. I. *Quart. J. Geol. Soc.,* **6,** 252–81.
—— 1852. Op. cit., Pt. III. The Thanet Sands. *Quart. J. Geol. Soc.,* **8,** 235–64.
—— 1854a. Op. cit., Pt. II. The Woolwich and Reading Series. *Quart. J. Geol. Soc.,* **10,** 75–172.
—— 1854b. On Some Swallow Holes in the Chalk Hills near Canterbury. *Quart. J. Geol. Soc.,* **10,** 222–4.
—— 1854c. On the Thickness of the London Clay ; etc. *Quart. J. Geol. Soc.,* **10,** 401–19.
—— 1858. On the Age of some Sands and Iron-Sandstones on the North Downs. *Quart. J. Geol. Soc.,* **14,** 322–35.

PRICE, F. G. H. 1874a. On the Gault of Folkestone. *Quart. J. Geol. Soc.,* **30,** 342–68.
—— 1874b. On the Lower Greensand and Gault of Folkestone. *Proc. Geol. Assoc.,* **4,** 135–50.
—— 1875. On the probable depth of the Gault sea. *Proc. Geol. Assoc.,* **4,** 269–78.
—— 1876. Excursion to Sandgate and Folkestone. *Proc. Geol. Assoc.,* **4,** 554–6.
—— 1877. On the Beds between the Gault and Upper Chalk near Folkestone. *Quart. J. Geol. Soc.,* **33,** 431–48.
—— 1879. *The Gault*. London.
—— 1894. Excursion to Hythe, Sandgate, and Folkestone. *Proc. Geol. Assoc.,* **13,** 142–51.

PRINGLE, J. 1917. On Deep Borings for Coal and Ironstone at Bere Farm, Elham and Folkestone, Kent. *Sum. Prog. Geol. Surv.* for 1916, 34–40.
—— 1928. The Adisham Boring, Kent. *Sum. Prog. Geol. Surv.* for 1927, pt. II, 78–80.

REDMAN, J. B. 1854. On the Alluvial Formations, and the Local Changes of the South Coast of England. *Proc. Inst. Civ. Engrs.,* **11** for 1851–2, 162–226.

REID, C. 1887. On the Origin of Dry Chalk Valleys and of Coombe Rock. *Quart. J. Geol. Soc.,* **43,** 364-73.
—— 1890. The Pliocene Deposits of Britain. *Mem. Geol. Surv.*
—— 1899. The Geology of the Country around Dorchester. *Mem. Geol. Surv.*

REID, J. 1891. A short account of some Bones and Teeth found in the Valley Drift of the River Stour, near Canterbury. *Trans. Assoc. Nat. Hist. Soc. S.-E. England. S. East Nat.,* **1,** pt. 2, 51-3.

REYNOLDS, D. H. B. 1948. The Movement of Water in the Middle and Lower Chalk of the River Dour Catchment. *J. Inst. Civ. Engrs.,* **29** for 1947–8, 73–108.

REYNOLDS, S. H. 1921. The Lithological Succession of the Avonian at Clifton. *Quart. J. Geol. Soc.,* **77,** 213–43, pls. 8–14.

—— 1926. Progress on the Study of the Lower Carboniferous (Avonian) Rocks of England and Wales. *Rep. Brit. Assoc.,* 65–101.

—— and VAUGHAN, A. 1911. Faunal and Lithological Sequence in the Carboniferous Limestone Series (Avonian) of Burrington Combe (Somerset). *Quart. J. Geol. Soc.,* **67,** 342–92, pls. 28–31.

RITCHIE, A. E. 1920. *The Kent Coalfield, Its Evolution and Development.* London.

ROBERTSON, T. 1932. The Geology of the South Wales Coalfield. Part V. The Country around Merthyr Tydfil. 2nd edit. *Mem. Geol. Surv.*

ROWE, A. 1900. The Zones of the White Chalk of the English coast. 1.–Kent and Sussex. *Proc. Geol. Assoc.,* **16,** 289–368, pls. viii–x.

—— 1908. The Zones of the White Chalk of the English coast. 5.–The Isle of Wight. *Proc. Geol. Assoc.,* **20,** 209–352, pls. viii–xxiii.

RUSSELL, R. J. 1944. Lower Mississippi Valley Loess. *Bull. Geol. Soc. Amer.,* **55,** 1–40.

SABINE, P. A., YOUNG, B. R. and DANGERFIELD, J. 1963. Brickearth and Clay-with-flints from Kent. *Clay Minerals Bull.,* **5,** No. 29, 248–53.

SAHNI, M. R. 1929. A monograph of the Terebratulidae of the British Chalk. i–vi, 1–62, pls. 1–10. *Palaeont. Soc.*

SCHLUETER, C. 1872–6. Die Cephalopoda der oberen deutschen Kreide. *Palaeontographica,* **21,** 1–120, pls. I–XXXV.

SCROPE, G. P. 1866. The Terraces of the Chalk Downs. *Geol. Mag.,* **3,** 293–6.

SHARPE, C. F. S. 1938. *Landslides and Related Phenomena.* New York.

SHARPE, D. 1857. Description of the fossil remains of Mollusca found in the Chalk of England, Pt. III, 37–68, pls. xvii–xxvii. *Palaeont. Soc.*

SHEPHARD-THORN, E. R., SMART, J. G. O., BISSON, G. and EDMONDS, E. A. 1966. Geology of the Country around Tenterden. *Mem. Geol. Surv.*

SHERLOCK, R. L. 1912. On the Glacial Origin of the Clay-with-flints of Buckinghamshire and on a Former Course of the Thames. *Quart. J. Geol. Soc.,* **68,** 199–209.

SIMMS, F. W. 1843. Account of a section of the strata between the Chalk and the Wealden Clay in the vicinity of Hythe, Kent. *Proc. Geol. Soc.,* **4,** 206–8.

—— 1860. *Practical Tunnelling; . . . as exemplified by the particulars of Bletchingley and Saltwood Tunnels.* 2nd edit. London.

SLADEN, W. P. and SPENCER, W. K. 1891–1908. A monograph on the British fossil Echinodermata from the Cretaceous formations. **2.** The Asteroidea and Ophiuroidea. Pt. I, 1–28, pls. i–viii, by W. P. Sladen (1891). Pt. IV, 91–132, pls. xxvii–xxix, by W. K. Spencer (1907). *Palaeont. Soc.*

SMART, J. G. O., SABINE, P. A. and BULLERWELL, W. 1964. The Geological Survey Exploratory Borehole at Canvey Island, Essex. *Bull. Geol. Surv. Gt. Brit.,* No. 21, 1–36.

SMITH, C. R. and ELLIOTT, J. 1852. *Report on Excavations made on the Site of the Roman Castrum at Lymne, in Kent, in 1850.* London.

SMITH, R. A. 1918. Prehistoric and Anglo-Saxon remains discovered by Capt. L. Moysey at Howletts, near Bridge, Kent. *Proc. Soc. Antiq.* (2), **30** for 1917–18, 102–13.

—— 1926. A Guide to Antiquities of the Stone Age in the Department of British and Mediaeval Antiquities. 3rd edit. *British Museum.*

SMITH, R. A. 1933. Implements from the High Level Gravel near Canterbury. *Proc. Prehist. Soc. E. Anglia*, **7**, 165–70.

SNELL, F. C. 1938. *The Intermittent (or Nailbourne) Streams of East Kent*. Canterbury.

SOWERBY, J. and SOWERBY, J. DE C. 1812–46. *The Mineral Conchology of Great Britain*. 7 vols. pls. 1–383 (1812–22) by J. Sowerby ; pls. 384–648 (1823–46) by J. de C. Sowerby. London.

SOWERBY, J. DE C. 1836. See FITTON, W. H.

SPATH, L. F. 1923a. Excursion to Folkestone, with notes on the zones of the Gault. *Proc. Geol. Assoc.*, **34**, 70–6.

—— 1923b. On the ammonite horizons of the Gault and contiguous deposits. Appendix II. *Sum. Prog. Geol. Surv.* for 1922, 139–49.

—— 1923–43. A monograph of the Ammonoidea of the Gault. Pts. 1–16. *Palaeont. Soc.*

—— 1924. On a new ammonite (*Engonoceras iris* sp. n.) from the Gault of Folkestone. *Ann. Mag. Nat. Hist.* (9), **14**, 504–8.

—— 1937. The Canadian ammonite genus *Gastroplites* in the English Gault. *Ann. Mag. Nat. Hist.* (10), **19**, 257–60.

STAMP, L. D. 1921. On the Beds at the Base of the Ypresian (London Clay) in the Anglo-Franco-Belgian Basin. *Proc. Geol. Assoc.*, **32**, 57–108.

STEERS, J. A. 1946. *The Coastline of England and Wales*. Cambridge.

STUBBLEFIELD, C. J. 1933. Contributions to the Geology of the Kent Coalfield. III. Notes on the Fauna of the Coal Measures of Kent. *Sum. Prog. Geol. Surv.* for 1932, pt. II, 71–7.

—— 1953. In *Sum., Prog. Geol. Surv.* for 1952, 41–4.

—— and TROTTER, F. M. 1957. Divisions of the Coal Measures on Geological Survey Maps of England and Wales. *Bull. Geol. Surv. Gt. Brit.*, No. 13, 1–5.

—— and TRUEMAN, A. E. 1946. The Faunal Sequence in the Kent Coalfield. *Geol. Mag.*, **83**, 266–79.

SWINNERTON, H. H. 1955. A monograph of British Lower Cretaceous Belemnites, Pt. V. *Palaeont. Soc.*

TAITT, A. H. and KENT, P. E. 1958. Deep boreholes at Portsdown (Hants) and Henfield (Sussex). *Tech. Pub. B.P. Co. Ltd*. London.

THOMASSON, A. J. 1961. Some Aspects of the Drift Deposits and Geomorphology of South-East Hertfordshire. *Proc. Geol. Assoc.*, **72**, 287–302.

TOMS, A. H. 1946. Folkestone Warren Landslips: Research carried out in 1939 by the Southern Railway Company. Railway Paper No. 19. *Proc. Inst. Civ. Engrs.*

TOPLEY, W. 1875. The Geology of the Weald. *Mem. Geol. Surv.*

—— 1893. The Landslip at Sandgate. *Proc. Geol. Assoc.*, **13**, 40–7.

TRIMMER, J. 1841. On the Locality and Geological Position of *Cucullaea decussata*. *Proc. Geol. Soc.*, **3**, 456–7.

TRUEMAN, A. E. 1933. A Suggested Correlation of the Coal Measures of England and Wales. *Proc. S. Wales Inst. Engrs.*, **49**, 63–94.

—— 1946. Stratigraphical Problems in the Coal Measures of Europe and North America (Presidential Address). *Quart. J. Geol. Soc.*, **102**, xlix–xcii.

—— and WEIR, J. 1946. A Monograph of British Carboniferous Non-marine Lamellibranchia. Pt. I. *Palaeont. Soc.*

VAUGHAN, A. 1905. The Palaeontological Sequence in the Carboniferous Limestone of the Bristol Area. *Quart. J. Geol. Soc.*, **61**, 181–305, pls. 22–9.

VINER-BRADY, N. E. V. 1955. Folkestone Warren Landslips: Remedial Measures, 1948–1954. Railway Paper No. 57, *Proc. Inst. Civ. Engrs.*

WARD, G. 1931a. Saxon Lydd. *Archaeol. cantiana*, **43**, 29–37.

—— 1931b. Sand Tunes Boc. *Archaeol. cantiana*, **43**, 39–47.

WARD, G. 1933a. The River Limen at Ruckinge. *Archaeol. cantiana*, **45**, 129–32.

—— 1933b. The Saxon Charters of Burmarsh. *Archaeol. cantiana*, **45**, 133–41.

—— 1936. The Wilmington Charter of A.D. 700. *Archaeol. cantiana*. **48**, 11–28.

WHEELER, W. H. 1902. *The Sea Coast*. London.

WHITAKER, W. 1866. On the Lower London Tertiaries of Kent. *Quart. J. Geol. Soc.*, **22**, 404–35.

—— 1872. The Geology of the London Basin. **4**, Pt. I. *Mem. Geol. Surv.*

—— 1889. The Geology of London. **1**. *Mem. Geol. Surv.*

—— 1908. The Water Supply of Kent. *Mem. Geol. Surv.*

—— and DAVIES, G. MACD. 1920. The Section at Worms Heath, Surrey. *Quart. J. Geol. Soc.*, **75**, 7–31.

WHITE, H. J. O. 1924. The Geology of the Country near Brighton and Worthing. *Mem. Geol. Surv.*

—— 1928a. The Geology of the Country near Ramsgate and Dover. *Mem. Geol. Surv.*

—— 1928b. The Geology of the Country near Hastings and Dungeness. *Mem. Geol. Surv.*

WHITTARD, W. F. and SMITH, S. 1943. Geology of a recent Borehole at Filton, Glos. *Proc. Bristol Nat. Soc.* (4), **9**, 434–50.

WITHERS, T. H. 1935. Catalogue of fossil Cirripedia in the Department of Geology. **2**. Cretaceous. *Brit. Mus.* (*Nat. Hist.*).

WOOD, A. M. Muir. 1955. Folkestone Warren Landslips: Investigations, 1948–50. Railway Paper No. 56, *Proc. Inst. Civ. Engrs.*

WOOD, P. D. 1961. Strip lynchets reconsidered. *Geogr. J.*, **127**, 449–59.

WOODS, H. 1897. The Mollusca of the Chalk Rock: Part II. *Quart. J. Geol. Soc.*, **53**, 377–404.

—— 1899–1913. A monograph of the Cretaceous Lamellibranchia of England. 2 vols. *Palaeont. Soc.*

—— 1924–31. A monograph of the fossil Macrurous Crustacea of England. Pts. I–VII. *Palaeont. Soc.*

WOODWARD, A. S. 1889–1901. Catalogue of the fossil fishes in the British Museum (Natural History). Pts. I–IV. *Brit. Mus.* (*Nat. Hist.*).

—— 1902–1912. The fossil fishes of the English Chalk. *Palaeont. Soc.*

WOODWARD, H. 1872. On a new species of *Rostellaria* from the Grey Chalk, Folkestone. *Geol. Mag.*, **9**, 97–9.

WOOLDRIDGE, S. W. 1926. The Structural Evolution of the London Basin. *Proc. Geol. Assoc.*, **37**, 162–96.

—— 1927. The Pliocene History of the London Basin. *Proc. Geol. Assoc.*, **38**, 49–132.

—— and KIRKALDY, J. F. 1936. River Profiles and Denudation-Chronology in Southern England. *Geol. Mag.*, **73**, 1–16.

—— and LINTON, D. L. 1955. *Structure, Surface and Drainage in South-East England*. London.

WORRALL, G. A. 1954. The Lower Greensand in East Kent. *Proc. Geol. Assoc.*, **65** for 1954–5, 185–202.

—— 1957. The Mineralogy of Some Lower Greensand Borehole Samples in Kent. *Proc. Geol. Assoc.*, **67** for 1956, 138–41.

WORSSAM, B. C. 1963. Geology of the Country around Maidstone. *Mem. Geol. Surv.*

WRIGHT, C. W. and WRIGHT, E. V. 1951. A survey of the fossil Cephalopoda of the Chalk of Great Britain. *Palaeont. Soc.*

WRIGHT, T. 1864–1881. Monograph of the British fossil Echinodermata from the Cretaceous formations. **1**. The Echinoidea. Pts. I–IX. *Palaeont. Soc.*

WRIGLEY, A. 1949. The Thanet Sands. *South-Eastern Naturalist and Antiquary*, **54**, 41–6.

Appendix

LIST OF GEOLOGICAL SURVEY PHOTOGRAPHS

Taken by Mr. J. Rhodes or Mr. J. M. Pulsford

Copies of these photographs are deposited for reference in the library of the Geological Survey and Museum, South Kensington, London, S.W.7. Black and white prints and lantern slides may be supplied at a fixed tariff, and in addition colour prints and transparencies are available for many of the photographs with numbers higher than 9409. All the numbers belong to Series A.

ONE-INCH SHEET 289

4001–2 River gravels and brickearth. Yeoman's Gravel-pit, on the north side of Sturry Road, Canterbury.

8798 River gravels of the 4th Terrace of the Great Stour River. Gravel-pit (111573) near Chartham Hatch.

9142 River gravels of the 3rd Terrace of the Great Stour River. Gravel-pit (185591) at Stodmarsh Road, near Fordwich.

9143 Brickearth and gravel resting on Upper Chalk at Wingham Well (231566).

8797 Solution pipes in Upper Chalk. Railway cutting (103557) at Chartham siding.

9588 Head-filled solution pipe in Upper Chalk. Railway cutting (257484) 250 yd N. 18° W. of Shepherdswell Station.

9590 View across Swingfield Minnis (234442).

9140 A deep coombe (the Devil's Kneadingtrough) (076452) in the Chalk escarpment near Wye. (Plate VIB).

8795–6 Woolwich Beds and Oldhaven Beds. Sand-pit (065576) at Winterbourne.

8799–801 Woolwich Beds overlain by brickearth. Sand-pit (089575) 350 yd W. 12° S. of White Wall.

9139, 9141 Thanet Beds, Woolwich Beds, gravel and brickearth. Trenleypark Wood Sand-pit (192592). (A 9139–Plate VB).

9145 Worm borings in Thanet Beds. Cutting (161560) on disused Canterbury–Folkestone railway.

8794 View towards the Blean from The Mount, Shottenden Hill.

8802 View from Golden Hill, Harbledown, showing valleys cut in Eocene strata.

8803 View of the outlier of Eocene beds near Selling, from Old Wives Lees.

9662–4 Upper Chalk, *Micraster coranguinum* and *Uintacrinus socialis* zones. Denne's Limeworks Pit (147567), Canterbury.

315

9587   Upper Chalk, *Micraster coranguinum* Zone, showing Whitaker's Three-inch Flint Band.   Railway cutting (255488) 800 yd N. 18° W. of Shepherdswell Station.

9586   Upper Chalk, *Micraster coranguinum* Zone.   Chalk-pit (207490) 350 yd S. of the inn at Derringstone.

9589   Upper Chalk, at or near the junction of the *Holaster planus* and *Micraster cortestudinarium* zones.   Chalk-pit (214460) in Denton Wood.

9144   Effect of frost-action on Upper Chalk.   Chalk-pit (127559) at Cockering Road, near Canterbury.

9215–6   Middle Chalk–Upper Chalk junction.   Chalk-pit (028486) in Eastwell Park.

9585   Middle Chalk, *Terebratulina lata* Zone.   Chalk-pit (168457) 600 yd N. 37° W. of Upper Parkgate Farm.

9676   View of valleys in the Middle Chalk at Elham.

9217   Lower Chalk–Middle Chalk junction.   Chalk-pit (034482) 200 yd S.E. of Boughton Aluph church.

9661   View of Chalk escarpment near Brook.

9665   Dry valleys (141398) in the Chalk escarpment 500 yd N. of The Pent, Postling, showing an 'elbow of capture'.

8804–6   Current-bedded Folkestone Beds.   Sand-pit (005457) ½ mile W. of Lenacre Hall.

8807   Hythe Beds.   Quarry (983408) ¼ mile S.S.E. of Great Chart.

8808,   Steeply-dipping and faulted Hythe Beds.   Goldwell Quarry (973426),
9218–9   ½ mile N.W. of Great Chart.

9220   Hythe Beds.   Quarry (981411) ½ mile S. of Great Chart.   (Plate IIIA).

9677   Headgear and winding-engine house at Snowdown Colliery (246514).

ONE-INCH SHEETS 305 AND 306

5388   Coombe Deposits filling a depression in Chalk.   Cutting (235376) to the west of Dover Hill.

9591   Sand in Clay-with-flints filling a solution pipe in Middle Chalk.   Cliff (244379) above Folkestone Warren, 1650 yd N. 15° E. of Copt Point.

9674   Sand in Clay-with-flints filling solution pipes in Middle Chalk.   Chalk-pit (238377) on west side of Dover Hill.

9224   View of Gibbins' Brook valley (116385), ¾ mile E.N.E. of Sellindge Lees.

9675   View of New Romney (062243).

9949   View of Romney Marsh from Lympne.

9225–6   Views of Romney Marsh south of Lympne, showing the ruins of Stutfall Castle (118342).

9227   Debris of the Saxon settlement of Sandtun (121338).

9228   Ruins of Stutfall Castle (118342), south of Lympne.

9229–30   Landslipped slope (145346) of Weald Clay, Atherfield Clay and Hythe Beds bordering Romney Marsh 1 mile W. of Hythe.

9138   Landslip topography (067361) near Aldington.

9135–6 Damage caused by landslipping of Lower Greensand. West Cliff (224355), Folkestone.

9592, Views of Folkestone Warren from the west. (A 10045–Plate VIA).
9673, 10045

9583 Steeply-dipping chalk involved in landslip. Horsehead Point (256381), Folkestone Warren.

9668 Steeply-dipping chalk involved in landslip. Foreshore (263383) 1·8 miles N.E. of Copt Point.

9584 Lower Chalk overlain by Middle Chalk. Chalk-pit (225380) 250 yd E.N.E. of Sugarloaf Hill.

9666 Lower Chalk–Middle Chalk contact at Shakespeare Cliff (307398). (Plate VA).

9667 Lower Chalk–Middle Chalk contact displaced by a normal fault. Shakespeare Cliff (307398).

3986, Shakespeare Cliff, 1 mile S.W. of Dover harbour, viewed from the
10050 east-north-east.

5389 Chalk cliffs between Folkestone and Dover, viewed from Copt Point.

5390 Coombes in the Chalk escarpment south of Etchinghill.

5391–2, Summerhouse Hill (167377), a Chalk Hill separated from the main
9231, 9750 escarpment by a fault.

9672 Gault and Chloritic Marl overlain by loam and soil. Cliff (241371) 50 yd N.E. of the Roman Villa near Copt Point.

5393–4, Lower Greensand overlain by Gault at Copt Point (about 243364),
9924 Folkestone.

5395 The *Douvilleiceras mammillatum* nodule bed at Copt Point (about 243364), Folkestone.

9137 Current-bedded Folkestone Beds overlain by Gault and river gravels. Granary Court Sand-pit (090401), near Brabourne Lees. (Plate IIIB).

9232–3 Folkestone Beds overlain by Gault. Sand-pit (119392) on Swan Lane, 450 yd N. 15° E. of Hope Farm, Stanford.

9221 Folkestone Beds: current-bedded sands with chert layers and ' doggers ' of sandy limestone. Sandling Junction Sand-pit (147370).

5396–7, Folkestone Beds. Sand-pit (180369) 600 yd S.S.W. of Newington
  Sand-pit (147370).

5396–7, Folkestone Beds. Sand-pit (180369) 600 yd S.S.W. of Newington
9223 church.

9234 Hythe Beds. Otterpool Quarry (112366), 1 mile W.N.W. of New Inn Green.

9669 Hythe Beds. Quarry (063369) 300 yd N. 10° W. of cross-roads at Aldington Corner.

9670–1 Hythe Beds. Quarry (046411) 1050 yd E. 20° N. of Sevington church.

9410–1 Phosphatic nodule bed in uppermost Hythe Beds. Foreshore at Mill Point (221351), Folkestone.

INDEX

Abbot's Cliff, 163, 220; Chalk at, 119, 131, 134–5
—— —— Borehole, 31, 42, 43, 60, 65, 79, 101
—— —— Tunnel, 13, 129, 293
Abbotscliff Trial Pit, fossils from Chalk in, 129
Abbott's Court, 287
Acheulian implements, 271, 274, 275
Acrise, 141, 236
Actinocamax plenus Marls, see *plenus* Marls
—— —— Subzone, 114, 117, 121, 164
Adisham, 159, 161; Borehole, 10, 20, 25, 29, 31, 33, 35–7, 39, 46, 49, 61, 66, 68, 84, 101, 116
AGER, D. V., 33
Agriculture, 297
Aker's Steps, 121, 122; Chalk at, 119, 131–2, 136–7, 147
Albian, 54; Lower, 50; Upper, 58
Aldington, 12, 14, 51, 74, 291; quarries at, 52, 75, 300
—— Corner, quarry at, 12, 74
—— Frith, 291
—— Knoll, 64, 75, 280, 292
Alkham, 134, 142, 150, 229; nailbourne, 3, 278; valley, 3, 4, 133
ALLEN, H. A., 128
ALLEN, P., 46, 48
Allerød interstadial, 230, 232
Alluvium, 200, 201, 247, **278–80**
——, Marine, 249–63, 282–7; soils and agriculture on, 297
Aluminium (in Clay-with-flints), 204
Amage Farm, 129
Amaltheus margaritatus Zone, 33
Ammonite Beds (Chalk), 125
Ammonites rhotomagensis and *A. varians* Zone, 117
Ammonoidea, from Chalk, 170, 173; from Gault, 106–11
Amphibole, 206, 207, 216
Anahoplites daviesi Subzone, 58; *A. intermedius* Subzone, 58, 99
Anatase, 205–7
Andalusite, 205–7
ANDERSON, F. W., iii, 46, 48, 64–6, 203, 214
ANDREAE, S. J. F., 250
Ankerite, 22
Annelida, from Chalk, 167; from Gault, 104

" Annulatus " Zone, 34
Anthozoa, from Chalk, 165; from Gault, 105
Anthraconaia lenisulcata Zone, 23, 25; *A. modiolaris* Zone, 23–5; *A. prolifera* Zone, 23
Anthraconauta phillipsii Zone, 23, 24, 28; *A. tenuis* Zone, 9, 23, 24, 28, 30
Anthracosia similis and *Anthraconaia pulchra* Zone, 23–6, 28
Appledore, 254, 260; Station, 250, 255, 259, 281, 282
Aptian, 49, 50
ARBER, E. A. N., 9, 19, 21, 23
ARKELL, W. J., 34–6, 40, 41, 43
Arpinge, 122, 127
Arrhaphoceras substuderi Subzone, 58
Ashdown Beds, 6, 45, 46
Ashenfield Farm, 5, 13
Ashford, 1, 3, 4, 10, 49, 51–4, 66, 71, 233, 243, 245, 265, 268; boreholes at, 44, 45, 59, 62, 301
Asholt Wood, 249
Ashton Group, 23
ASKEW, P. W., 250, 253–61, 280, 282, 285, 287, 297
Asteroidea, from Chalk, 166, 171
Atherfield Clay, 12, 46, 49, **50–1, 66–8,** 71, 73, 81; landslip of, 291, 292; thickness of, 6, 51; unconformity at base of, 46, 49, 62
Augite, 206
Aycliff (near Dover), 278
—— Borehole, see Dover No. 1 Borehole
Aylesham, 161, 220; Halt, 161, 220

Backhouse Wood, 12
Badlesmere, 202; Court, 151
Bagham, 143, 208, 225, 226, 234
Bajocian, 32, 36
BAKER, H. A., 17
Baker's Gap (Folkestone), 95
BALCHIN, W. G. V., 249, 260
Baldwin's Innings, 261
Bank Farm, 74
BANNISTER, M. A., 120f.n.
Barfreston Borehole, 24
Bargrove, 99, 281
Barham, 4, 10, 267; Downs, 160, 175, 246; Pumping Station, wells at, 228, 302

318

Y

Dd. 505182 K8

Printed in England by Commercial Colour Press for Her Majesty's Stationery Office.